Lecture Notes in Networks and Systems

Volume 210

Series Editor

Janusz Kacprzyk, Systems Research Institute, Polish Academy of Sciences,
Warsaw, Poland

Advisory Editors

Fernando Gomide, Department of Computer Engineering and Automation—DCA,
School of Electrical and Computer Engineering—FEEC, University of Campinas—
UNICAMP, São Paulo, Brazil

Okyay Kaynak, Department of Electrical and Electronic Engineering,
Bogazici University, Istanbul, Turkey

Derong Liu, Department of Electrical and Computer Engineering, University
of Illinois at Chicago, Chicago, USA; Institute of Automation, Chinese Academy
of Sciences, Beijing, China

Witold Pedrycz, Department of Electrical and Computer Engineering
University of Alberta, Alberta, Canada; Systems Research Institute
Polish Academy of Sciences, Warsaw, Poland

Marios M. Polycarpou, Department of Electrical and Computer Engineering,
KIOS Research Center for Intelligent Systems and Networks, University of Cyprus,
Nicosia, Cyprus

Imre J. Rudas, Óbuda University, Budapest, Hungary

Jun Wang, Department of Computer Science, City University of Hong Kong,
Kowloon, Hong Kong

The series "Lecture Notes in Networks and Systems" publishes the latest developments in Networks and Systems—quickly, informally and with high quality. Original research reported in proceedings and post-proceedings represents the core of LNNS.

Volumes published in LNNS embrace all aspects and subfields of, as well as new challenges in, Networks and Systems.

The series contains proceedings and edited volumes in systems and networks, spanning the areas of Cyber-Physical Systems, Autonomous Systems, Sensor Networks, Control Systems, Energy Systems, Automotive Systems, Biological Systems, Vehicular Networking and Connected Vehicles, Aerospace Systems, Automation, Manufacturing, Smart Grids, Nonlinear Systems, Power Systems, Robotics, Social Systems, Economic Systems and other. Of particular value to both the contributors and the readership are the short publication timeframe and the world-wide distribution and exposure which enable both a wide and rapid dissemination of research output.

The series covers the theory, applications, and perspectives on the state of the art and future developments relevant to systems and networks, decision making, control, complex processes and related areas, as embedded in the fields of interdisciplinary and applied sciences, engineering, computer science, physics, economics, social, and life sciences, as well as the paradigms and methodologies behind them.

Indexed by SCOPUS, INSPEC, WTI Frankfurt eG, zbMATH, SCImago.

All books published in the series are submitted for consideration in Web of Science.

More information about this series at http://www.springer.com/series/15179

Sanjoy Kumar Saha · Paul S. Pang ·
Debnath Bhattacharyya
Editors

Smart Technologies in Data Science and Communication

Proceedings of SMART-DSC 2021

 Springer

Editors
Sanjoy Kumar Saha
Department of Computer Science
and Engineering
Jadavpur University
Kolkata, West Bengal, India

Paul S. Pang
School of Science, Engineering
and Information Technology
Federation University
Ballarat, VIC, Australia

Debnath Bhattacharyya
Department of Computer Science
and Engineering
K L Deemed to be University
Vaddeswaram, Andhra Pradesh, India

ISSN 2367-3370   ISSN 2367-3389 (electronic)
Lecture Notes in Networks and Systems
ISBN 978-981-16-1775-1   ISBN 978-981-16-1773-7 (eBook)
https://doi.org/10.1007/978-981-16-1773-7

This Springer imprint is published by the registered company Springer Nature Singapore Pte Ltd.
The registered company address is: 152 Beach Road, #21-01/04 Gateway East, Singapore 189721,
Singapore

Conference Committee Members

Organizing Committee

General Chairs

Yu-Chen Hu, Providence University, Taiwan
Debnath Bhattacharyya, Koneru Lakshmaiah Education Foundation, Vaddeswaram,
Guntur, Andhra Pradesh, India

Advisory Board

Paul S. Pang, Unitec Institute of Technology, New Zealand
Andrzej Goscinski, Deakin University, Australia
Osvaldo Gervasi, Perugia University, Perugia, Italy
Jason Levy, University of Hawaii, Hawaii, USA
Tai-hoon Kim, BJTU, China
Sabah Mohammed, Lakehead University, Ontario, Canada
Tarek Sobh, University of Bridgeport, USA
Jinan Fiaidhi, Lakehead University, Ontario, Canada
Y. Byun, Jeju National University, South Korea
Amiya Bhaumick, LUC KL, Malaysia
Divya, LUC KL, Malaysia
Dipti Prasad Mukherjee, ISI Kolkata, India
Sanjoy Kumar Saha, Jadavpur University, Kolkata, India
Sekhar Verma, IIIT Allahabad, India
C. V. Jawhar, IIIT Hyderabad, India
Pabitra Mitra, IIT Kharagpur, India
Joydeep Chandra, IIT Patna, India
Koneru Satyanarayana, KLEF, Guntur, India

K. Siva Kanchana Latha, KLEF, Guntur, India
Koneru Lakshman Havish, KLEF, Guntur, India
Koneru Raja Hareen, KLEF, Guntur, India
S. S. Mantha, KLEF, Guntur, India
L. S. S. Reddy, KLEF, Guntur, India
Y. V. S. S. S. V. Prasada Rao, KLEF, Guntur, India

Editorial Board

Debnath Bhattacharyya, K L Deemed to be University, India
Paul S. Pang, Federation University, Australia
Sanjoy Kumar Saha, Jadavpur University, India

Program Chairs

Tai-hoon Kim, BJTU, China
Hari Kiran Vege, K L Deemed to be University, Guntur, India

Management Co-chairs

Qin Xin, University of Faroe Island, Faroe Island
V. Srikanth, K L Deemed to be University, Guntur, India

Publicity Committee

G. Yvette, Philippines
K. Amarendra, K L Deemed to be University, Guntur, India
V. Naresh, K L Deemed to be University, Guntur, India
Deburup Banerjee, K L Deemed to be University, Guntur, India
K. V. Prasad, K L Deemed to be University, Guntur, India
N. Raghvendra Sai, K L Deemed to be University, Guntur, India
Sougatamoy Biswas, K L Deemed to be University, Guntur, India

Finance Committee

N. Srinivasu, K L Deemed to be University, Guntur, India
P. V. R. D. Prasada Rao, K L Deemed to be University, Guntur, India
M. Srinivas, K L Deemed to be University, Guntur, India
B. Vijay Kumar, K L Deemed to be University, Guntur, India
P. Haran Babu, K L Deemed to be University, Guntur, India
Rahul Mahadeo Sahane, K L Deemed to be University, Guntur, India

Local Arrangements Committee

P. Raja Rajeswari, K L Deemed to be University, Guntur, India
Venkata Naresh Mandhala, K L Deemed to be University, Guntur, India
Shahana Bano, K L Deemed to be University, Guntur, India
Chayan Paul, K L Deemed to be University, Guntur, India
C. Karthikeyan, K L Deemed to be University, Guntur, India

Technical Programme Committee

Sanjoy Kumar Saha, Professor, Jadavpur University, Kolkata
Hans Werner, Associate Professor, University of Munich, Munich, Germany
Goutam Saha, Scientist, CDAC, Kolkata, India
Samir Kumar Bandyopadhyay, Professor, University of Calcutta, India
Ronnie D. Caytiles, Associate Professor, Hannam University, Republic of Korea
Y. Byun, Professor, Jeju National University, Jeju Island, Republic of Korea
Alhad Kuwadekar, Professor, University of South Wales, UK
Debasri Chakraborty, Assistant Professor, BIET, Suri, West Bengal, India
Poulami Das, Assistant Professor, Heritage Institute of Technology, Kolkata, India
Indra Kanta Maitra, Associate Professor, St. Xavier University, Kolkata, India
Divya Midhun Chakravarty, Professor, LUC, KL, Malaysia
F. C. Morabito, Professor, Mediterranea University of Reggio Calabria, Reggio Calabria RC, Italy
Hamed Kasmaei, Islamic Azad University, Tehran, Iran
Nancy A. Bonner, University of Mary Hardin-Baylor, Belton, TX 76513, USA
Alfonsas Misevicius, Professor, Kaunas University of Technology, Lithuania
Ratul Bhattacharjee, AxiomSL, Singapore
Lunjin Lu, Professor, Computer Science and Engineering, Oakland University, Rochester, MI 48309-4401, USA
Ajay Deshpande, CTO, Rakya Technologies, Pune, India
Debapriya Hazra, Jeju National University, Jeju Island, South Korea

Preface

Knowledge in engineering sciences is about sharing our ideas of research with others. In engineering, it is exhibited in many ways in that conference is the best way to propose your idea of research and its future scope and it is the best way to add energy to build a strong and innovative future. So, we are here to give a small support from our side to confer your ideas by an "International Conference on Smart Technologies in Data Science And Communication (Smart-DSC 2021)", related to Electrical, Electronics, Information Technology and Computer Science. It is not confined to a specific topic or region, and you can exhibit your ideas in similar or mixed or related technologies bloomed from anywhere around the world because "an idea can change the future, and its implementation can build it". KLEF Deemed to be University is a great platform to make your idea(s) penetrated into the world. We give as best as we can in every related aspect. Our environment leads you to a path on your idea, our people will lead your confidence, and finally, we give our best to make yours. Our intention is to make intelligence in engineering fly higher and higher. That is why we are dropping our completeness into the event. You can trust us on your confidentiality. Our review process is double-blinded through EasyChair.

At last, we pay the highest regard to the Koneru Lakshmaiah Education Foundation, K L Deemed to be University, from Guntur and Hyderabad for extending support for the financial management of 4th Smart-DSC 2021.

	Best Wishes from:
Vaddeswaram, India	Debnath Bhattacharyya
Ballarat, Australia	Paul S. Pang
Kolkata, India	Sanjoy Kumar Saha

Acknowledgements

The editors wish to extend heartfelt acknowledgement to all contributing authors, esteemed reviewers for their timely response, members of the various organizing committee, and production staff whose diligent work put shape to the 4th Smart-DSC 2021 proceedings. We especially thank our dedicated reviewers for their volunteering efforts to check the manuscript thoroughly to maintain the technical quality and for useful suggestions.

We thank all the following invited speakers who extended their support by sharing knowledge in their area of expertise.

Dr. Fatiha Merazka, Full Professor, University of Science and Technology Houari Boumediene, Algeria.

Dr. Sally ELGhamrawy, Head of Communications and Computer Engineering Department, MISR Higher Institute for Engineering and Technology, Egypt

Dr. Hosna Salmani, Healthcare IT, Iran University of Medical Sciences (IUMS), Tehran, Iran

Dr. Saptarshi Das, Pennsylvania State University, University Park, Pennsylvania, USA.

Dr. Susmit Shannigrahi, Tennessee Technological University, Nashville Metropolitan Area, USA.

Dr. Zhihan Lv, Qingdao University, Qingdao, Shandong, China.

Dr. Debashis Ganguly, Software and Hardware Modeling Engineer, Apple, Pittsburgh, Pennsylvania, USA.

Dr. Bahman Javadi, Western Sydney University, Greater Sydney Area, Australia.

Dr. Hamed Daei Kasmaei, Faculty of Science and Engineering, IAUCTB, Tehran, Iran.

Debnath Bhattacharyya
Paul S. Pang
Sanjoy Kumar Saha

Contents

Editors and Contributors

About the Editors

Dr. Sanjoy Kumar Saha currently associated as Professor with the Department of Computer Science and Engineering, Jadavpur University, Kolkata, India. He did is B.E. and M.E. from Jadavpur University and completed his Ph.D. from IIEST Shibpur, West Bengal, India.

His Research interests include Image, Video and Audio Data Processing, Physiological Sensor Signal Processing and Data Analytics. He published more than hundred articles in various International Journals and Conferences of repute. He has guided eleven Ph.D. Students. He holds four US patents. He is a member of IEEE Computer Society, Indian Unit for Pattern Recognition and Artificial Intelligence, ACM. He has served TCS innovation Lab, Kolkata, India as advisor for the signal processing group.

Dr. Paul S. Pang is an Associate Professor of cyber security at the School of Engineering, Information Technology and Physical Sciences, Federation University Australia. Before joining Federation University, he was a Professor of Data Analytics and Director of Center Computational Intelligence for Cybersecurity at the Unitec Institute of Technology, New Zealand. He acted as a Principle Investigator for over 13 research grant projects, totalling more than NZD $3.5 million in funding by the Ministry of Business, Employment and Innovation, NZ (MBIE), the Ministry for Primary Industries, NZ (MPI), the Health Research Council, NZ (HRC), the National Institute of Information and Communications Technology, Japan (NICT), Telecom NZ, Mitsubishi Electric Japan, LuojiaDeyi Technology China, and Lucent & Bell Lab USA.

Dr. Debnath Bhattacharyya received Ph.D. (Tech., CSE) from University of Calcutta, Kolkata, India. Currently, Dr Bhattacharyya associated with Koneru Lakshmaiah Education Foundation, K L Deemed to be University, Vaddeswaram, Guntur, Andhra Pradesh, India as Professor from May 2021 and Dean R&D, VIIT from the year 2015–May 2020. His research areas include Image Processing, Pattern recognition, Bio-Informatics, Computational Biology, Evolutionary Computing and Security. He published 200+ research papers in various reputed International Journals and Conferences. He published six textbooks for Computer Science as well. He is the member of IEEE, ACM, ACM SIGKDD, IAENG, and IACSIT.

Contributors

Kombathula Abhishek Department of Computer Science and Engineering, Vignan's Institute of Information Technology, Visakhapatnam, India

K. Amarendra Department of Computer Science and Engineering, Koneru Lakshmaiah Education Foundation, Guntur, Andhra Pradesh, India

T. Archana Acharya Department of Management Studies, Vignan's Institute of Information Technology (A), Duvvada, Visakhapatnam, India

K. Asish Vardhan Department of Computer Science and Engineering, Dr. Lankapalli Bullayya College of Engineering, Visakhapatnam, AP, India

M. Babu Rao Department of Computer Science & Engineering, Gudlavalleru Engineering College, Vijayawada, India

Debnath Bhattacharyya Department of Computer Science and Engineering, Koneru Lakshmaiah Education Foundation (K L Deemed to be University), Vaddeswaram, Guntur, Andhra Pradesh, India

Pathipati Bhavya Department of Electronics and Communication Engineering, Koneru Lakshmaiah Education Foundation, Vaddeswaram, AP, India

Pavan Nageswar Reddy Bodavarapu Department of CSE, Koneru Lakshmaiah Education Foundation, Guntur, India

Lakshmi Ramani Burra Department of Computer Science and Engineering, PVP Siddhartha Institute of Technology, Kanuru, Vijayawada, India

Midhun Chakkravarthy Department of Computer Science and Multimedia, Lincoln University College, Kuala Lumpur, Malaysia

Illapu Tarun Chand Department of Computer Science and Engineering, Vignan's Institute of Information Technology, Visakhapatnam, India

Chilukuri Sai Revanth Chowdary Department of Computer Science and Engineering, Vignan's Institute of Information Technology, Visakhapatnam, India

Srikanth Dasari Department of CSE, Dr. Lankapalli Bullayya College of Engineering, Visakhapatnam, Andhra Pradesh, India

Matta Bharathi Devi Department of CSE, KoneruLakshmaiah Education Foundation, Guntur, Andhra Pradesh, India

G. Dheeraj Chaitanya Department of Computer Science and Engineering, Koneru Lakshmaiah Education Foundation, Guntur, Andhra Pradesh, India

V. Dheeraj Varma Department of Computer Science and Engineering, Koneru Lakshmaiah Education Foundation, Vaddeswaram, AP, India

V. Dhiraj Department of Computer Science and Engineering, Koneru Lakshmaiah Education Foundation, Vaddeswaram, AP, India

D. Dinesh Kumar Department of Computer Science and Engineering, Koneru Lakshmaiah Education Foundation, Guntur, Andhra Pradesh, India

B. Dinesh Reddy Department of Computer Science and Engineering, Vignan's Institute of Information Technology (A), Vishakhapatnam, Andhra Pradesh, India

Bhanu Prakash Doppala Vignan's Institute of Information Technology (A), Visakhapatnam, India

P. Febin Koshy Department of Computer Science and Engineering, Koneru Lakshmaiah Education Foundation, Guntur, Andhra Pradesh, India

Vithya Ganesan Department of Computer Science and Engineering, Department of Electronics and Communication Engineering, Koneru Lakshmaiah Education Foundation, Vaddeswaram, AP, India

Rajendra Kumar Ganiya Department of Computer Science and Engineering, Vignan's Institute of Information Technology (A), Vishakhapatnam, Andhra Pradesh, India

Anil B. Gavade The Department of Electronics and Communication Engineering, KLS Gogte Institute of Technology, Belagavi, Karnataka, India

Shridhar Ghagane Department of Urology and Radiology, JN Medical College, KLE Academy of Higher Education and Research (Deemed-To-Be-University), Belagavi, Karnataka, India

Deepak Gupta Department of Computer Science and Engineering, Engineering College Ajmer, Ajmer, India

Shikha Gupta Department of Computer Science and Engineering, Engineering College Ajmer, Ajmer, India

G. Himanshu Department of Computer Science and Engineering, Koneru Lakshmaiah Education Foundation, Guntur, Andhra Pradesh, India

S. Hrushikesava Raju Department of Computer Science and Engineering, Koneru Lakshmaiah Education Foundation, Vaddeswaram, Guntur, India

Eali Stephen Neal Joshua Department of Computer Science and Multimedia, Lincoln University College, Kuala Lumpur, Malaysia

B. Karthik Department of Mechanical Engineering, VR Siddhartha Engineering College, Vijayawada, India

Gudipati Bharadwaja Sri Karthik Department of Computer Science and Engineering, Koneru Lakshmaiah Education Foundation, Vaddeswaram, AP, India

Hye-jin Kim Kookmin University, Seoul, Republic of Korea

Tai-hoon Kim School of Economics and Management, Beijing Jiaotong University, Chungwon-daero, Beijing, China

Srilalitha Kopparapu Department of Computer Science and Engineering, Koneru Lakshmaiah Education Foundation, Vaddeswaram, Andhra Pradesh, India

Muddada Murali Krishna Department of CSE, Dr. Lankapalli Bullayya College of Engineering, Visakhapatnam, Andhra Pradesh, India

Panja Hemanth Kumar Department of Computer Science and Engineering, Vignan's Institute of Information Technology, Visakhapatnam, India

V. Laxmi Narasamma   Department of Computer Science and Engineering, Koneru Lakshmaiah Education Foundation, Vaddeswaram, Guntur, Andhra Pradesh, India

S. M. M. Naidu   Department of E&Tc, International Institute of Information Technology (I2IT), Hinjawadi, Pune, India

Mohan Mahanty   Department of Computer Science and Engineering, Vignan's Institute of Information Technology, Visakhapatnam, India;
Department of Computer Science and Multimedia, Lincoln University College, Kuala Lumpur, Malaysia

Venkata Naresh Mandhala   Department of Computer Science and Engineering, Koneru Lakshmaiah Education Foundation, Vaddeswaram, Guntur, Andhra Pradesh, India

Harshali Mane   Department of ECE, KLEF, Guntur, India;
Department of E&Tc, International Institute of Information Technology (I2IT), Hinjawadi, Pune, India

Sk. Meeravali   Department of Computer Science and Engineering, Malla Reddy University, Hyderabad, Telangana, India

Divya Midhunchakkaravarthy   Department of Computer Science and Multimedia, Lincoln University College, Kuala Lumpur, Malaysia

Manisha Miriyala   Department of Computer Science and Engineering, Koneru Lakshmaiah Education Foundation, Vaddeswaram, Andhra Pradesh, India

Pragnyaban Mishra   Department of CSE, Koneru Lakshmaiah Education Foundation, Guntur, India

H. Naga Chandrika   Research Scholar, Dr. A.P.J. Abdul Kalam University, Indore, Madhya Pradesh, India

S. NagaMallik Raj   Vignan's Institute of Information Technology (A), Visakhapatnam, India;
Department of Computer Science and Multimedia, Lincoln University College, Kuala Lumpur, Malaysia

Rajendra B. Nerli   Department of Urology and Radiology, JN Medical College, KLE Academy of Higher Education and Research (Deemed-To-Be-University), Belagavi, Karnataka, India

Aleemullakhan Pathan   Department of Computer Science and Engineering, Vignan's Institute of Information Technology, Visakhapatnam, India

A. V. S. Pavan Kumar   Department of Computer Science and Engineering, GIET University, Gunupur, India

Sowjanya Pentakota   Research Scholar, Vishakhapatnam, Andhra Pradesh, India

Ch. M. L. Prasanna Department of Computer Science and Engineering, Department of Electronics and Communication Engineering, Koneru Lakshmaiah Education Foundation, Vaddeswaram, AP, India

Nikitha Rajanedi Department of Computer Science and Engineering, Koneru Lakshmaiah Education Foundation, Vaddeswaram, Andhra Pradesh, India

G. Rajendra Kumar Department of Computer Science & Engineering, Vignan's Institute of Information Technology (A), Visakhapatnam, Andhra Pradesh, India

V. Rajesh Chowdhary Department of E&Tc, International Institute of Information Technology (I2IT), Hinjawadi, Pune, India

V. Rajesh Department of Electronics and Communication Engineering, Koneru Lakshmaiah Education Foundation, Vaddeswaram, AP, India

Puppala Ramya Department of Computer Science & Engineering, Faculty of Engineering and Technology, Annamalai University, Chidambaram, Tamil Nadu, India

V. Ramya Department of Computer Science & Engineering, Faculty of Engineering and Technology, Annamalai University, Chidambaram, Tamil Nadu, India

Matcha Venu Gopala Rao Department of ECE, KLEF, Guntur, India

Nakka Thirupathi Rao Department of Computer Science and Engineering, Vignan's Institute of Information Technology, Visakhapatnam, India

Rakesh Rathi Department of Computer Science and Engineering, Engineering College Ajmer, Ajmer, India

Y. Sai Sadhvi Department of Computer Science and Engineering, Department of Electronics and Communication Engineering, Koneru Lakshmaiah Education Foundation, Vaddeswaram, AP, India

Udimudi Satish Varma Department of Computer Science and Engineering, Koneru Lakshmaiah Education Foundation, Vaddeswaram, AP, India

K. V. Satyanarayana Department of Computer Science and Engineering, Raghu Engineering College, Visakhapatnam, AP, India

Mummana Satyanarayana Department of Computer Science and Engineering, KLEF, K L Deemed to be University, Guntur, India

P. Seetha Rama Krishna Department of Computer Science and Engineering, Koneru Lakshmaiah Education Foundation, Vaddeswaram, Guntur, India

B. Sekhar Babu Department of Computer Science and Engineering, Koneru Lakshmaiah Education Foundation, Vaddeswaram, AP, India

Neetu Sharma Department of Computer Science and Engineering, Engineering College Ajmer, Ajmer, India

Ch. Sharmila Department of Computer Science and Engineering, Department of Electronics and Communication Engineering, Koneru Lakshmaiah Education Foundation, Vaddeswaram, AP, India

Sachin S. Shinde Koneru Lakshimaiah Education Foundation, KLEF, Vaddeswram, Guntur, Andhra Pradesh, India

M. Sreedevi Department of Computer Science and Engineering, Koneru Lakshmaiah Education Foundation, Vaddeswaram, Guntur, Andhra Pradesh, India

P. S. V. S. Sridhar Department of Computer Science and Engineering, Koneru Lakshmaiah Education Foundation, Vaddeswaram, Andhra Pradesh, India

M. Srilatha Department of CSE, Koneru Lakshmaiah Education Foundation, Guntur, India;
VR Siddhartha Engineering College, Vijayawada, India

P. V. V. S. Srinivas Department of CSE, Koneru Lakshmaiah Education Foundation, Guntur, India

P. Srinivasa Rao Department of Computer Science and Systems Engineering, AU College of Engineering, Andhra University, Visakhapatnam, AP, India

N. Srinivasu Department of CSE, Koneru Lakshmaiah Education Foundation, Guntur, India

Eali Stephen Neal Joshua Vignan's Institute of Information Technology (A), Visakhapatnam, India

G. Subbarao Department of Computer Science and Engineering, Koneru Lakshmaiah Education Foundation, Vaddeswaram, Guntur, India

D. Sushma Department of Computer Science and Engineering, Vignan's Institute of Information Technology (A), Visakhapatnam, AP, India

Manjeti Sushma Department of Computer Science and Engineering, Anil Neerukonda Institute of Technology & Sciences, Visakhapatnam, India

Kalam Swathi Department of Computer Science and Engineering, Vignan's Institute of Information Technology, Visakhapatnam, India

Madhuri Thimmapuram Assistant Professor, Department of Computer Science and Engineering, Vardhaman College of Engineering (A), Shamshabad, Hyderabad, Telangana State, India

N. Thirupathi Rao Department of Computer Science and Engineering, Vignan's Institute of Information Technology (A), Visakhapatnam, Andhra Pradesh, India

Bhimavarapu Usharani Department of Computer Science and Engineering, Koneru Lakshmaiah Education Foundation, Vaddeswaram, Andhra Pradesh, India

Bandi Vamsi Department of Computer Science and Multimedia, Lincoln University College, Kuala Lumpur, Malaysia;

Department of Computer Science and Engineering, Vignan's Institute of Information Technology, Visakhapatnam, India

Jayavani Vankara Department of CSE, Dr. Lankapalli Bullayya College of Engineering, Visakhapatnam, Andhra Pradesh, India

P. Veda Upasan Department of Computer Science and Systems Engineering, College of Engineering (A), Andhra University, Visakhapatnam, India

G. Vijay Kumar Department of Computer Science and Engineering, Koneru Lakshmaiah Education Foundation, Vaddeswaram, Guntur, Andhra Pradesh, India

Sai Sameera Voleti Department of Computer Science and Engineering, Koneru Lakshmaiah Education Foundation, Vaddeswaram, Andhra Pradesh, India

A Detailed Study on Optimal Traffic Flow in Tandem Communication Networks

N. Thirupathi Rao, K. Asish Vardhan, P. Srinivasa Rao,
Debnath Bhattacharyya, and Hye-jin Kim

Abstract The utilization of communication networks is increasing in a rapid manner day to day. As the network models are growing a lot, the people using those networks are also increasing a lot. As the number of users is increasing, the provision of Internet and other network facilities also had to be increased. The increase of facilities will cost a lot, as the cost of this network equipment will be more. Hence, before going for the actual installation and utilization of these facilities, the models are based on the real-time situation scenarios and those network models are studied in detail for the better understanding and for better proper utilization and installation of the network facilities. Several models had been developed in the literature for various scenarios. In the current model, a particular situation of the network model was taken by choosing that the model was in forked model condition. Forked model condition means the two nodes are connected in parallel to each other, and the other third node was connected in series with the other two-node parallel combination. The flow of the network will first cross the two nodes which were connected in parallel to each other and then the flow will be carried to the third node in serial communication. The distribution to be considered here for the data flow was the Poisson process with binomial bulk arrivals. In that scenario, the performance of the network model is

N. Thirupathi Rao (✉)
Department of Computer Science and Engineering, Vignan's Institute of Information Technology, Visakhapatnam, AP, India

K. Asish Vardhan
Department of Computer Science and Engineering, Dr. Lankapalli Bullayya College of Engineering, Visakhapatnam, AP, India

P. Srinivasa Rao
Department of Computer Science and Systems Engineering, AU College of Engineering, Andhra University, Visakhapatnam, AP, India

D. Bhattacharyya
Department of Computer Science and Engineering, K L Deemed to be University, KLEF, Guntur, AP, India

H. Kim
Kookmin University, 77 Jeongneung-RO, Seongbuk-gu, Seoul 02707, Republic of Korea
e-mail: hyejinaa@daum.net

© The Author(s), under exclusive license to Springer Nature Singapore Pte Ltd. 2021   1
S. K. Saha et al. (eds.), *Smart Technologies in Data Science and Communication*,
Lecture Notes in Networks and Systems 210,
https://doi.org/10.1007/978-981-16-1773-7_1

calculated, and the sensitivity of the model also calculated. The results are discussed in detail in the results section.

Keywords Communication networks · Forked model · Parallel nodes · Serial node · Bulk arrivals · DBA · Throughput of the node · Optimal analysis

1 Introduction

Communication networks are playing a vital role in the development of communication in terms of data, transportation and other means of source between the transmitter and the receivers [1, 2]. All these developments of both wired networks and wireless networks are playing a vital role in the development of the country and standards of the people living. As the technology is to be going to each person in the country, the development of the public in the country can be increased and the living standards of the public also increase [3–5]. The queuing models are the best way of source to analyze the performance of these network models before going to be established in a bigger way. Once these networks are established, these cannot be moved or replaced. So, one should have a clear idea and picture about the placement of a network at a particular place, and also the cost of this network establishment is a bigger issue. Hence, the models are to be considered and studied first before going for the actual establishment of these networks.

In some other situations, arising at places like computer communications, data voice transmission, neuro-psychological problems, transportation systems, etc., both parallel and series queuing systems are connected in a single network [7, 8]. For example, in a communication system the request arrives at the first two buffers which are in parallel form for different type of arrivals of packets [9, 10]. After getting the process at either of the first two nodes, the packets will join the third node which is connected in tandem to the first two nodes. This type of queuing model network systems is known as parallel and series queuing systems or forked model queuing models [6, 7]. In these network models, the arrivals occur in batches of random size for both nodes in the initial stage.

Hence, in the current model these sorts of systems were considered and modeled to analyze the load dependent parallel and series queuing model with bulk arrivals. Here also it is assumed that the arrivals for both streams formulate compound Poisson processes with different parameters. It is further assumed that the number of service completions in each service station follow Poisson process with parameters β_1, β_2 and β_3 respectively. The load-dependent service rate implies that the service rate of the node depends on the number of packets in the queue connected to it. In order to analyze the performance of this considered model, several performance metrics are considered like the throughput of the nodes, delay at each node, mean number of packets at each node, and sensitivity of the network model. For all these network parameters, the results are calculated and the values are tabulated and graphical representation also presented for the same values, and the results are displayed in the results section in detail for the better understanding of the models considered.

2 Literature Review

From the previous works so far done in the same area was collected and presented in the current section as follows,

Suhasini et al. [9] had discussed the development and analysis of the two-node tandem queuing model with state- and time-dependent arrivals. The arrival of the packets is being characterized by using non-homogeneous Poisson process. The results are displayed in the results section for the better understanding of the considered model.

Thirupathi Rao et al. [10] had considered the two-node tandem communication networks with the arrivals directly connected with the next nodes in the considered model of the network. Using various difference differential equations, the models are developed and the performance of such network models is analyzed by using various packet arrival modes; the results are displayed in the results section in detail with graphical presentations also.

Satyanarayana et al. [11] had discussed the optimal traffic flow in two-node tandem communication networks. The two nodes in the current scenarios are connected in serial to each other, and the arrival of packets to each node will be separated. The first node will have a separate arrival of input packets, and the second node also will have an arrival of packets separately. The optimal values to be operated for the current considered model are calculated, and the results are displayed in results section with detail graphical models.

From the existing works studied and reviewed, it is observed that the arrivals are different forms in different articles when compared to each paper [1, 3, 4]. Some authors had considered the arrival of packets follow homogeneous Poisson process, and some others follow non-homogeneous Poisson process [9–11]. Most of the authors considered the models as two nodes considered in serial connected to other and for three nodes also, they are connected in serial. Only one author had considered the forked model of communication network with non-homogeneous arrivals. But they had not done on the basis of the binomial bulk arrivals with dynamic bandwidth allocation strategies. Hence, this point had given us a scope of considering the current work.

From the various articles and various models being considered by various authors, a detailed analysis of the models discussed so far is given, and the current model considered for the current work is also given at last for the better understanding of the current problem.

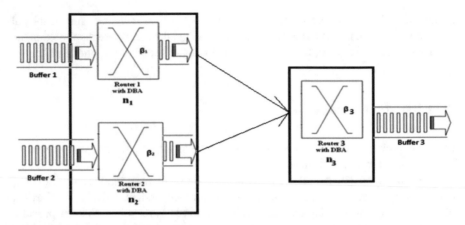

Fig. 1 Forked communication network model

3　Forked Communication Network Model

In the current model, three nodes are considered for the model. Among the three nodes, two nodes are connected in parallel to each other and the other third connected in serial to the combination of these two-node parallel combinations. The arrivals considered here are the compound Poisson binomial bulk arrivals with the utilization of dynamic bandwidth allocation. The bandwidth will be calculated and will be allocated based on the demand that is considered as the dynamic bandwidth allocation. The three nodes are represented n1, n2 and n3. Each node is having buffer with names as buffer 1, buffer 2 and buffer 3 for storage of packets in the queue such that all the packets to be processed in a line.

The entrance of packets at nodes is: α_1, α_2 and α_3.

The transmission of packets at nodes is: β_1, β_2 and β_3.

The model considered for the current work was as follows (Fig. 1).

4　Methodology and Working of the Current Model

Let α_1, α_2 and α_3 are the number of input packets being entering into the network. Then the probability that there will be α_1 arrivals in first node, α_2 arrivals in second node and α_3 arrivals at third node. The difference differential equations for the model are,

$$P'_{n_1,n_2,n_3}(t) = -(\alpha_1 + \alpha_2 + \alpha_1\beta_1 + \alpha_2\beta_2 + \alpha_3\beta_3)P_{n_1,n_2,n_3}(t)$$

$$+ \alpha_1 \left[\sum_{k=1}^{n_1} \frac{{}^mC_k p^k (1-p)^{m-k}}{1-(1-p)^m} . k P_{n_1-k,n_2,n_3}(t) \right]$$

The joint probability generating function of $P_{n_1,n_2,n_3}(t)$ is

$$P(z_1, z_2, z_3; t) = \sum_{n_1=0}^{\infty} \sum_{n_2=0}^{\infty} \sum_{n_3=0}^{\infty} z_1^{n_1} z_2^{n_2} z_3^{n_3} \, P_{n_1,n_2,n_3}(t)$$

Therefore, the joint probability generating function of the current model is as follows,

$$P(z_1, z_2, z_3; t) = \exp\Bigg\{ \alpha_1 \sum_{k=1}^{\infty} \sum_{r=1}^{k} \sum_{i=0}^{r} (-1)^i \frac{{}^m C_k p^k (1-p)^{m-k}}{1-(1-p)^m} . k \binom{k}{r}\binom{r}{i}$$

$$\left[(z_1 - 1) + \left(\frac{\beta_1(z_3 - 1)}{\beta_3 - \beta_1} \right) \right]^{r-1} \left(\frac{\beta_1(z_3 - 1)}{\beta_3 - \beta_1} \right)^i$$

$$\left[\frac{1 - \exp[-(i\beta_3 + (r-i)\beta_1)t]}{i\beta_3 + (r-i)\beta_1} \right]$$

$$+ \alpha_2 \sum_{l=1}^{\infty} \sum_{s=1}^{l} \sum_{j=0}^{s} (-1)^j C_l \binom{l}{s}\binom{s}{j}$$

$$\left[(z_2 - 1) + \left(\frac{\beta_2(z_3 - 1)}{\beta_3 - \beta_2} \right) \right]^{s-j}$$

$$\left(\frac{\beta_2(z_3 - 1)}{\beta_3 - \beta_2} \right)^j \left[\frac{1 - \exp[-(i\beta_3 + (r-i)\beta_1)t]}{j\beta_3 + (s-j)\beta_2} \right] \Bigg\};$$

$$\alpha_1, \alpha_2 < \min\{\beta_1, \beta_2, \beta_3\}$$

The performance metrics of the model are calculated for each node, and they were given as follows. The performance metrics of the first node are as follows,

$$P(z_1, t) = \exp\left[\alpha_1 \sum_{k=1}^{\infty} \sum_{r=1}^{k} \frac{{}^m C_k p^k (1-p)^{m-k}}{1-(1-p)^m} \binom{k}{r}(z_1 - 1)^r \left(\frac{1 - e^{-r\beta_1 t}}{r\beta_1} \right) \right]$$

$$L_1(t) = \frac{\alpha_1 \left(1 - e^{-\beta_1 t}\right)}{\beta_1} \sum_{k=1}^{\infty} k \frac{{}^m C_k p^k (1-p)^{m-k}}{1-(1-p)^m} = \frac{\alpha_1 \left(1 - e^{-\mu_1 t}\right)}{\beta_1} E(X_1)$$

$$U_1(t) = 1 - P_{0,\ldots.}(t).$$

$$Thp_1(t) = \mu_1 U_1(t)$$

$$W_1(t) = \frac{L_1(t)}{Thp_1(t)}$$

The performance metrics of the second node are as follows,

$$P(z_2, t) = \exp\left[\lambda_2 \sum_{l=1}^{\infty} \sum_{s=1}^{l} \frac{{}^m C_k p^k (1-p)^{m-k}}{1 - (1-p)^m} \binom{l}{s} (z_2 - 1)^s \left(\frac{1 - e^{-s\mu_2 t}}{s\mu_2}\right)\right]$$

$$L_2(t) = \frac{\lambda_2 \left(1 - e^{-\mu_2 t}\right)}{\mu_2} \sum_{l=1}^{\infty} l \frac{{}^m C_l pl (1-p)^{m-l}}{1 - (1-p)^m} = \frac{\lambda_2 \left(1 - e^{-\mu_2 t}\right)}{\mu_2} E(X_2)$$

$$U_2(t) = 1 - P_{.,0,.}(t)$$

$$Thp_2(t) = \mu_2 U_2(t)$$

$$W_2(t) = \frac{L_2(t)}{Thp_2(t)}$$

The performance metrics of the third node are as follows,

$$P(z_3; t) = \exp\left\{\lambda_1 \sum_{k=1}^{\infty} \sum_{r=1}^{k} \sum_{i=0}^{r} (-1)^i \frac{{}^m C_k p^k (1-p)^{m-k}}{1 - (1-p)^m} \binom{k}{r}\binom{r}{i} \left(\frac{\mu_1(z_3 - 1)}{\mu_3 - \mu_1}\right)^r\right.$$

$$\left[\frac{1 - \exp[-(i\mu_3 + (r-i)\mu_1)t]}{i\mu_3 + (r-i)\mu_1}\right]$$

$$+ \lambda_2 \sum_{l=1}^{\infty} \sum_{s=1}^{l} \sum_{j=0}^{s} (-1)^j \frac{{}^m C_l p^l (1-p)^{m-l}}{1 - (1-p)^m} \binom{l}{s}\binom{s}{j}$$

$$\left.\left(\frac{\mu_2(z_3 - 1)}{\mu_3 - \mu_2}\right)^s \left[\frac{1 - \exp[-(j\mu_3 + (s-j)\mu_2)t]}{j\mu_3 + (s-j)\mu_2}\right]\right\}$$

$$L_3(t) = \frac{\lambda_1 \left(1 - e^{-\mu_1 t}\right)}{\mu_3 - \mu_1} \sum_{k=1}^{\infty} k \frac{{}^m C_k p^k (1-p)^{m-k}}{1 - (1-p)^m}$$

$$+ \frac{\lambda_2 \left(1 - e^{-\mu_2 t}\right)}{\mu_3 - \mu_2} \sum_{l=1}^{\infty} l \frac{{}^m C_l p^l (1-p)^{m-l}}{1 - (1-p)^m}$$

$$= \frac{\lambda_1 \left(1 - e^{-\mu_1 t}\right)}{\mu_3 - \mu_1} E(X_1) + \frac{\lambda_2 \left(1 - e^{-\mu_2 t}\right)}{\mu_3 - \mu_2} E(X_2)$$

$$U_3(t) = 1 - P_{.,.,0}(t)$$

$$Thp_3(t) = \mu_3 U_3(t)$$

$$W_3(t) = \frac{L_3(t)}{Thp_3(t)}$$

In order to verify the working of the current model, several numerical values are considered to give as input and try to identify the performance of the model. By using the MatchCad software and the MathType software, the analysis of the model was done and the equations are being completed. Several values are considered for calculating the performance metrics like the throughput of the nodes, mean delay at nodes, utilization of nodes, mean number of packets, etc. The equation with the letter L1, L2, L3 will give us the mean number of packets at the node sin the network. U1, U2, U3 will give the utilization of the nodes in the network, Thp1, Tho2, Thp3 will give the throughput of the nodes and W1, W2 and W3 will give the waiting period at each node of the network model considered.

$$t = 0.2, \ 0.5, \ 0.8, \ 1.2, \ 2.5; \quad m = 1, \ 2, \ 3, \ 4, \ 5;$$
$$k = 10, \ 15, \ 20, \ 25, \ 30;$$
$$\alpha_1 = 1.5, \ 2, 2.5, 3, \ 3.5; \quad \alpha_2 = 1.0, 1.5, \ 2.0, \ 2.5, \ 3.0;$$
$$\beta_1 = 3, \ 4, \ 5, \ 6, \ 7;$$
$$\beta_2 = 8, \ 9, \ 11, \ 12, \ 13; \quad \text{and} \quad \beta_3 = 14, \ 16, \ 18, \ 20, \ 22.$$

The input parameters that considered are as above and the performance of the model are as follows (Table 1),

The performance was explained in graphical representation as follows (Fig. 2),

From the above set of values and the graphical representations, it is observed that as the time of arrivals changes the utilization of each node increases. Also, as time increases, the mean number of packets at each node also increases. Also, as the arrival number of nodes increases, the utilization of all the three nodes and the mean number of packets at each node increases (Table 2; Fig. 3).

From the above numerical analysis and the graphical analysis, it is clear that the impact of these input variables and their values has some impact on the performance or the behavior of these variables or the performance metrics. As the arrival time of the packet's changes, the throughput of the nodes also increases time to time. Similarly, the other input parameters values increase, the waiting time at each node decreases gradually based on the time and other important factors. Similarly, as the time and other input parameter values changes, the delays and the throughput at each decrease for delays and increases for throughputs at each nodes of the network.

Table 1 Mean number of packets and utilization of the model for three nodes

t	m	k	α_1	α_2	β_1	β_2	β_3	$L_1(t)$	$U_1(t)$	$L_2(t)$	$U_2(t)$	$L_3(t)$	$U_3(t)$	$L(t)$
0.2	5	25	2	2.5	4	10	20	4.1300	0.3294	3.2425	0.3825	4.2750	0.5245	11.6475
0.5	5	25	2	2.5	4	10	20	6.4850	0.6186	3.7247	0.5428	5.3460	0.7495	15.5557
0.8	5	25	2	2.5	4	10	20	7.1943	0.7415	3.7487	0.5533	5.5473	0.7892	16.4903
1.2	5	25	2	2.5	4	10	20	7.4383	0.7887	3.7500	0.5538	5.6095	0.8013	16.7978
2.5	5	25	2	2.5	4	10	20	7.4997	0.8009	3.7500	0.5539	5.6249	0.8043	16.8746
0.5	**1**	25	2	2.5	4	10	20	5.6203	0.5986	3.2281	0.5136	4.6332	0.7091	13.4816
0.5	**2**	25	2	2.5	4	10	20	5.8365	0.6066	3.3523	0.5231	4.8114	0.7213	14.0001
0.5	**3**	25	2	2.5	4	10	20	6.0527	0.6119	3.4764	0.5307	4.9896	0.7319	14.5187
0.5	**4**	25	2	2.5	4	10	20	6.2688	0.6157	3.6006	0.5371	5.1678	0.7412	15.0372
0.5	**5**	25	2	2.5	4	10	20	6.4850	0.6186	3.7247	0.5428	5.3460	0.7495	15.5557
0.5	5	**10**	2	2.5	4	10	20	3.2425	0.5967	1.8624	0.4758	2.6730	0.6366	7.77780
0.5	5	**15**	2	2.5	4	10	20	4.3233	0.6085	2.4832	0.5051	3.5640	0.6900	10.3705
0.5	5	**20**	2	2.5	4	10	20	5.4042	0.6148	3.1039	0.5263	4.4550	0.7249	12.9631
0.5	5	**25**	2	2.5	4	10	20	6.4850	0.6186	3.7247	0.5428	5.3460	0.7495	15.5557
0.5	5	**30**	2	2.5	4	10	20	7.5658	0.6212	4.3455	0.5561	6.2370	0.7677	18.1483
0.5	5	25	**1.5**	2.5	4	10	20	4.8637	0.5147	3.7247	0.5428	4.9407	0.7051	13.5291
0.5	5	25	**2.0**	2.5	4	10	20	6.4850	0.6186	3.7247	0.5428	5.3460	0.7495	15.5557
0.5	5	25	**2.5**	2.5	4	10	20	8.1062	0.7003	3.7247	0.5428	5.7513	0.7872	17.5823
0.5	5	25	**3.0**	2.5	4	10	20	9.7275	0.7645	3.7247	0.5428	6.1566	0.8192	19.6088
0.5	5	25	**3.5**	2.5	4	10	20	11.348	0.8149	3.7247	0.5428	6.5619	0.8464	21.6354
0.5	5	25	2	**1.0**	4	10	20	6.4850	0.6186	1.4899	0.2688	3.1111	0.6114	11.0360

(continued)

Table 1 (continued)

t	m	k	α_1	α_2	β_1	β_2	β_3	$L_1(t)$	$U_1(t)$	$L_2(t)$	$U_2(t)$	$L_3(t)$	$U_3(t)$	$L(t)$
0.5	5	25	2	**1.5**	4	10	20	6.4850	0.6186	2.2348	0.3747	3.8561	0.6643	12.5759
0.5	5	25	2	**2.0**	4	10	20	6.4850	0.6186	2.9798	0.4653	4.6010	0.7100	14.0658
0.5	5	25	2	**2.5**	4	10	20	6.4850	0.6186	3.7247	0.5428	5.3460	0.7495	15.5557
0.5	5	25	2	**3.0**	4	10	20	6.4850	0.6186	4.4697	0.6090	6.0909	0.7836	17.0456
0.5	5	25	2	2.5	**3**	10	20	7.7687	0.6274	3.7247	0.5428	5.0957	0.7429	16.5891
0.5	5	25	2	2.5	**4**	10	20	6.4850	0.6186	3.7247	0.5428	5.3460	0.7495	15.5557
0.5	5	25	2	2.5	**5**	10	20	5.5075	0.6031	3.7247	0.5428	5.5606	0.7507	14.7928
0.5	5	25	2	2.5	**6**	10	20	4.7511	0.5810	3.7247	0.5428	5.7609	0.7540	14.2367
0.5	5	25	2	2.5	**7**	10	20	4.1563	0.5540	3.7247	0.5428	5.9627	0.7547	13.8438
0.5	5	25	2	2.5	4	**8**	20	6.4850	0.6186	4.6016	0.6051	4.6890	0.7604	15.7756
0.5	5	25	2	2.5	4	**9**	20	6.4850	0.6186	4.1204	0.5737	4.9925	0.7550	15.5978
0.5	5	25	2	2.5	4	**11**	20	6.4850	0.6186	3.3952	0.5133	5.7709	0.7440	15.6510
0.5	5	25	2	2.5	4	**12**	20	6.4850	0.6186	3.1173	0.4857	6.2971	0.7387	15.8994
0.5	5	25	2	2.5	4	**13**	20	6.4850	0.6186	2.8803	0.4602	6.9703	0.7337	16.3356
0.5	5	25	2	2.5	4	10	**14**	6.4850	0.6186	3.7247	0.5428	11.905	0.8393	22.1155
0.5	5	25	2	2.5	4	10	**16**	6.4850	0.6186	3.7247	0.5428	8.3695	0.7864	18.5793
0.5	5	25	2	2.5	4	10	**18**	6.4850	0.6186	3.7247	0.5428	6.5088	0.7676	16.7185
0.5	5	25	2	2.5	4	10	**20**	6.4850	0.6186	3.7247	0.5428	5.3460	0.7495	15.5557
0.5	5	25	2	2.5	4	10	**22**	6.4850	0.6186	3.7247	0.5428	4.5451	0.7322	14.7548

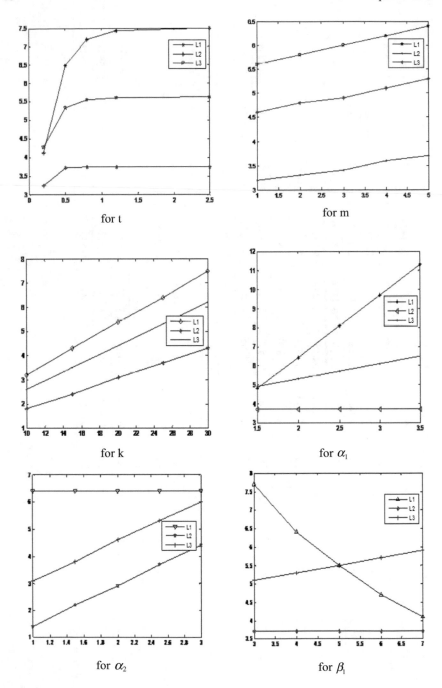

Fig. 2 Graphical representation of the utilization and mean number of packets at each node

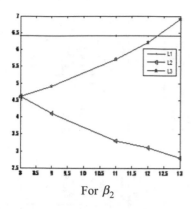

For β_2

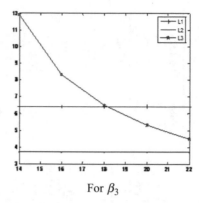

For β_3

Fig. 2 (continued)

5 Conclusion

In the current article, a new model of communication network had considered that was the forked communication network model which was different from the earlier model of communication networks. The input arrivals are following the compound Poisson binomial bulk arrivals. The input at first will go the first and second nodes and then after processing from those two nodes, the flow of packets will be flown as input to the other third node in the network model. Based on the loads at various nodes, the bandwidth will be allocated dynamically. The results are calculated for the above considered three nodes, and the results are tabulated and also the performance was given in graphical format also. This method of models is helpful in working of the LAN/WAN and MAN.

Table 2 Values of throughput and average delay for different values of the parameters

t	m	k	α_1	α_2	β_1	β_2	β_3	$Thp_1(t)$	$W_1(t)$	$Thp_2(t)$	$W_2(t)$	$Thp_3(t)$	$W_3(t)$
0.2	5	25	2	2.5	4	10	20	1.3176	3.1345	3.8246	0.8478	10.4897	0.4075
0.5	5	25	2	2.5	4	10	20	2.4746	2.6207	5.4276	0.6863	14.9896	0.3566
0.8	5	25	2	2.5	4	10	20	2.9658	2.4257	5.5330	0.6775	15.7834	0.3515
1.2	5	25	2	2.5	4	10	20	3.1549	2.3577	5.5385	0.6771	16.0261	0.3500
2.5	5	25	2	2.5	4	10	20	3.2036	2.3410	5.5386	0.6771	16.0862	0.3497
0.5	**1**	25	2	2.5	4	10	20	2.3944	2.3473	5.1365	0.6285	14.1817	0.3267
0.5	**2**	25	2	2.5	4	10	20	2.4262	2.4056	5.2313	0.6408	14.4257	0.3335
0.5	**3**	25	2	2.5	4	10	20	2.4475	2.4730	5.3072	0.6550	14.6381	0.3409
0.5	**4**	25	2	2.5	4	10	20	2.4629	2.5453	5.3714	0.6703	14.8246	0.3486
0.5	**5**	25	2	2.5	4	10	20	2.4746	2.6207	5.4276	0.6863	14.9896	0.3566
0.5	5	**10**	2	2.5	4	10	20	2.3869	1.3585	4.7582	0.3914	12.7311	0.2100
0.5	5	**15**	2	2.5	4	10	20	2.4339	1.7763	5.0506	0.4917	13.7994	0.2583
0.5	5	**20**	2	2.5	4	10	20	2.4592	2.1975	5.2629	0.5898	14.4981	0.3073
0.5	5	**25**	2	2.5	4	10	20	2.4746	2.6207	5.4276	0.6863	14.9896	0.3566
0.5	5	**30**	2	2.5	4	10	20	2.4846	3.0451	5.5609	0.7814	15.3540	0.4062
0.5	5	25	**1.5**	2.5	4	10	20	2.0588	2.3624	5.4276	0.6863	14.1018	0.3504
0.5	5	25	**2.0**	2.5	4	10	20	2.4746	2.6207	5.4276	0.6863	14.9896	0.3566
0.5	5	25	**2.5**	2.5	4	10	20	2.8013	2.8938	5.4276	0.6863	15.7438	0.3653
0.5	5	25	**3.0**	2.5	4	10	20	3.0580	3.1810	5.4276	0.6863	16.3844	0.3758
0.5	5	25	**3.5**	2.5	4	10	20	3.2597	3.4815	5.4276	0.6863	16.9287	0.3876
0.5	5	25	2	**1.0**	4	10	20	2.4746	2.6207	2.6876	0.5544	12.2280	0.2544

(continued)

Table 2 (continued)

t	m	k	α_1	α_2	β_1	β_2	β_3	$Thp_1(t)$	$W_1(t)$	$Thp_2(t)$	$W_2(t)$	$Thp_3(t)$	$W_3(t)$
0.5	5	25	2	**1.5**	4	10	20	2.4746	2.6207	3.7470	0.5964	13.2861	0.2902
0.5	5	25	2	**2.0**	4	10	20	2.4746	2.6207	4.6529	0.6404	14.2000	0.3240
0.5	5	25	2	**2.5**	4	10	20	2.4746	2.6207	5.4276	0.6863	14.9896	0.3566
0.5	5	25	2	**3.0**	4	10	20	2.4746	2.6207	6.0900	0.7339	15.6717	0.3887
0.5	5	25	2	2.5	**3**	10	20	1.8823	4.1271	5.4276	0.6863	14.8584	0.3429
0.5	5	25	2	2.5	**4**	10	20	2.4746	2.6207	5.4276	0.6863	14.9896	0.3566
0.5	5	25	2	2.5	**5**	10	20	3.0156	1.8264	5.4276	0.6863	15.0141	0.3704
0.5	5	25	2	2.5	**6**	10	20	3.4859	1.3630	5.4276	0.6863	15.0241	0.3846
0.5	5	25	2	2.5	**7**	10	20	3.8778	1.0718	5.4276	0.6863	15.0350	0.3998
0.5	5	25	2	2.5	4	**8**	20	2.4746	2.6207	4.8407	0.9506	15.2085	0.3083
0.5	5	25	2	2.5	4	**9**	20	2.4746	2.6207	5.1630	0.7981	15.1007	0.3306
0.5	5	25	2	2.5	4	**11**	20	2.4746	2.6207	5.6460	0.6013	14.8798	0.3878
0.5	5	25	2	2.5	4	**12**	20	2.4746	2.6207	5.8283	0.5348	14.7739	0.4262
0.5	5	25	2	2.5	4	**13**	20	2.4746	2.6207	5.9826	0.4814	14.6745	0.4750
0.5	5	25	2	2.5	4	10	**14**	2.4746	2.6207	5.4276	0.6863	11.7499	1.0133
0.5	5	25	2	2.5	4	10	**16**	2.4746	2.6207	5.4276	0.6863	12.5826	0.6652
0.5	5	25	2	2.5	4	10	**18**	2.4746	2.6207	5.4276	0.6863	13.8163	0.4711
0.5	5	25	2	2.5	4	10	**20**	2.4746	2.6207	5.4276	0.6863	14.9896	0.3566
0.5	5	25	2	2.5	4	10	**22**	2.4746	2.6207	5.4276	0.6863	16.1074	0.2822

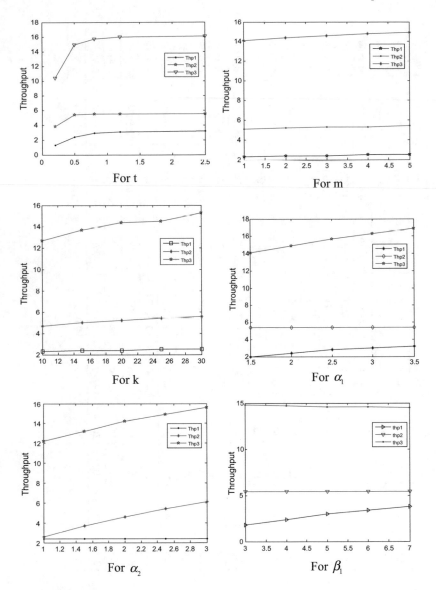

Fig. 3 Graphical representation of the throughput and mean delays at each node of the model

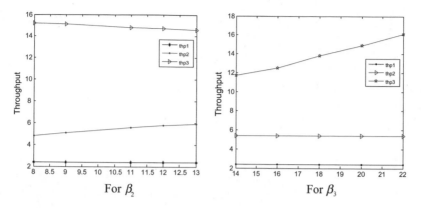

For β_2 For β_3

Fig. 3 (continued)

References

1. K. Srinivasa Rao, M.V. Ramasundari, P. Srinivasa Rao, P.S. Sureshvarma, Three node communication network model with modified phase type transmission under DBA having NHP arrivals. Int. J. Comput. Eng. **4**(1), 17–29 (2014)
2. M.V. Ramasundari, K. Srinivasa Rao, P. Srinivasa Rao, P.S. Sureshvarma, Three node communication network model with dynamic bandwidth allocation and non homogeneous Poisson arrivals. Int. J. Comput. Appl. **31**(1), 19–27 (2011)
3. K. Srinivasa Rao, P.S. Varma, P.V.G.D.P. Reddy, A communication network with random phases of transmission. Indian J. Math. Math. Sci. **3**(1), 39–46 (2007)
4. K.K. Leung, Load dependent service queues with application to congestion control in broadband networks. Perform. Eval. **509**(1–4), 27–40 (2002)
5. K. Srinivasa Rao, M. Govinda Raoand, P.Chandra Sekhar, Studies on interdependent tandem queuing models with modified phase type service and state dependent service rates. Int. J. Comput. Appl. **3**(4), 319–330 (2013)
6. K. Sriram, Methodologies for bandwidth allocation, transmission scheduling and congestion avoidance in broadband ATM networks.Comput. Netw. ISDN Syst. **26**, 43–59 (1997)
7. L. Kleinrock, R.R. Muntz, E. Rodemich, The processor-sharing queuing model for time-shared systems with bulk arrivals. Int. J. Netw. **1**(1), 1–13 (2006)
8. J. Durga Aparajitha, K. Srinivasa Rao, Two Node Tandem Queuing model with direct arrivals to both the service stations having stateand time dependent phase type service.Glob. J. Pure Appl. Math. **13**(11), 7999–8023 (2017)
9. A.V.S. Suhasini, K. Srinivasa Rao, P.R.S. Reddy, Transient analysis of Tandem Queuing model with non-homogenous poisson bulk arrivals having state dependent service rates. Int. J. Adv. Comput. Math. Sci. **3**(3), 272–289 (2018)
10. N. Thirupathi Rao, K. Srinivasa Rao, P. Srinivasa Rao, Optimal traffic flow in tandem communication networks. Int. J. Control Autom. **11**(12), 35–46 (2018)
11. K.V. Satyanarayana, K. Sudha, N. Thirupathi Rao, D. Bhattacharyya, *Analysis of Queuing Model Based Cloud Data Centers*. (LNNS, Springer, Berlin, 2020), vol. 105, pp. 293–309

Detection of Brain Stroke Based on the Family History Using Machine Learning Techniques

Bandi Vamsi⊙, **Debnath Bhattacharyya**⊙,
and **Divya Midhunchakkaravarthy**⊙

Abstract The death of cells occurs when the blood flow is reduced among the brain resulting in a weak condition causing a medical emergency usually defined as a stroke. Stroke is the major common condition which is nowadays resulting in the maximum number of deaths among the world in the present-day situations. Stroke is involved with various uncertain constituents also these are identified by examination of pretentious specific people. In analyzing and dividing the type of infections caused by these strokes, there have been various testings done to reduce the risk factors. Depending on these predictions, understanding the symptoms causing the stroke can be implemented by four machine learning algorithms by considering the person's physical and analysis of the medicals. Of all the strokes, ischemic stroke was the deadliest which is causing major damage to adults worldwide. The major part lies in separate projections in identifying the treatment decisions that are to be followed during the acute phase of the stroke. In identifying the treatment methodologies, there is definite model algorithm for cerebral ischemia after subarachnoid hemorrhage (SAH) which is predicted earlier. In such scenerio this is improvised by developing a temporal unsupervised feature engineering approach. From this we can conclude the improved precision over standard features with 80% trained data, and 20% testing data.

Keywords Ischemic stroke · Hemorrhagic stroke · NIHSS · Machine learning · Brain stroke

B. Vamsi (✉) · D. Midhunchakkaravarthy
Department of Computer Science and Multimedia, Lincoln University College, 47301 Kuala Lumpur, Malaysia

D. Midhunchakkaravarthy
e-mail: divya@lincoln.edu.my

D. Bhattacharyya
Department of Computer Science and Engineering, K L Deemed to be University, KLEF, Guntur 522502, India
e-mail: debnathb@kluniversity.in

1   Introduction

1.1   *Stroke*

The term stroke is defined as the flow of blood vessels among the nerves with its weaker flow which leads to causing the cell to die. Strokes are categorized into two types, namely ischemic stroke and hemorrhage stroke. These strokes encounter while there is a weak blood flow among the nerves resulting in blockage of blood to the nerves, and this is referred to as an ischemic stroke. On the other hand, if in this situation if the nerve causes bleeding, then it is termed as hemorrhage stroke [1, 18]. Whereas there also arises another stroke namely transient ischemic attack. The ischemic stroke is further classified into two types, namely embolic and thrombotic stroke. The blockage of blood flow may happen anywhere in the body and travel toward the brain resulting in clotting which is referred to as an embolic stroke. Whereas, when such type of clot arises in the brain, the flow of blood reduces among the nerves, thereby diminishing the blood flow among the arteries, and it is termed as thrombotic stroke [2]. Also, hemorrhagic stroke is broadly classified into two types, namely intracerebral hemorrhage and subarachnoid hemorrhage [3].

The stroke itself is a phrase often used in the common terminology for disfunctioning of certain nerves in the brain leading to the irregularity of the blood flow [4]. Strokes can be defined under various circumstances based upon the variant possibilities. Likewise, all over the world, strokes play an important vital congenital retaliation. Normally, the brain in the human body consists of a large number of neurons and glia usually around 100 billion to trillions those which are wrapped inside the tissues of 3 lb in which the memories and moments are set aside in the form of a network [5]. The brain has control over every minor thing in the human body including breathing and moments that are done by the human body. Globally, the number of people who die by these strokes is 5 times numerous than centuries which are developing for the past three decades which can be estimated to be twice that at 2025 [6].

The different types of strokes are common among the patients, out of which 15 to 20% of them can be foretold that it is due to hemorrhagic stroke, while on the other hand, the correctness is high to that of the ischemic stroke [7]. Hemorrhagic stroke can be categorized into two types, and they are subarachnoid hemorrhage and intracerebral hemorrhage strokes. The strokes which cause clotting in the brain commonly called mini-stroke are the result of transient ischemic stroke (TIA) [8]. This TIA is a short-term clotting stroke that is relevant to the rest of the strokes. Under these types of strokes, the symptoms will be recovered within 20 h and last for a limited period. This TIA is a temporary disturbance that does not damage the brain and its tissues over longer periods. TIA is a cautious sign in advance stating that the different types of strokes may occur in future. Irrelevant to the type of stroke, it might be a deadly disease [9]. In common, the affected area among the patience can be bought out in a slow manner evolving the symptoms that cause the type of stroke which can be identified from its root and identified either ischemic stroke or hemorrhagic stroke depending upon the root word [10].

Strokes are life-threatening factors that damage the entire human system. These strokes are balanced based upon the modulations of these various strokes that appear in the non-identical variations [11]. The developing technologies can evaluate and comprehend these various strokes and their components of threats. Among various algorithms, the machine learning algorithm can ameliorate people who are being affected by these strokes through advanced perception and ministrations [12]. Various types of machine learning algorithms are developed in determining the contrasting strokes that may attack the people, and some which are previously affected can be brought to the notice by the patients' objective data and analytical reports. A collection of these datasets of various types of strokes from the medical experts has been stored under a database which is authorized by them and in which these datasets are used under machine learning algorithms in detecting the type of stroke that has affected the human body [13]. These are developed based upon several categorizations. The outcomes of the machine learning algorithm are quite satisfactory and are implemented in the day-to-day activities of the patients who are affected by the strokes [14].

2   Related Work

According to the theories demonstrated by Govindarajan for detecting the type of stroke affected to the human body, he used an artificial neural network (ANN), support vector machine (SVM), decision tree, logistic regression, and ensemble methods (bagging and boosting). He collected the data from a super-specialty hospital from people who are affected by these different types of strokes who fall under the age of 30–90 years. Nearly 507 patients with various stroke disorders were identified in Sugam multi-specialty hospital located in India. They used the algorithm of novel stemmer in processing the data that they have gained and to maintain stability among the datasets. The results of these datasets witness that most of the people around 81.52% of them were attacked with ischemic stroke, whereas on the other hand, only 8.48% were attacked with hemorrhagic strokes. In terms of the accuracy and categorization of these strokes, an artificial neural network algorithm with a stochastic gradient descent learning algorithm plays a major role [15].

Stroke prediction can be found out based on a model of support vector machine, the theory proposed by Jeena and kumar [16]. The data that is used in this model has been gathered from the database of the international stroke trail. This data has been categorized into datasets in which it consists of each 12 attributes. To explain this model, 350 specimens are taken into a survey. These 350 specimens are utilized in the training part, and on the other hand, 50 specimens are utilized for testing purposes. There are different essential practices, namely polynomial, quadratic, radical basis functions, and also, linear functions are also applicable. The highest accuracy of this model is obtained to be 91%. By demonstrating a linear kernel which provides stability measure F1-score with 91.7%.

Another model like artificial neural network (ANN) in predicting the stroke has been proposed by Singh and Choudhary [17]. In this model, these datasets are divided into three sets and comprise 212 strokes (all three) and 52, 69, 79 non-strokes correspondingly. The resulting datasets comprise 357 attributes, 1842 entities, 212 phenomena of stroke factors. At this stage of characteristic collection, the C4.5 decision tree algorithm will be taken into consideration and principle component analysis (PCA) for extent depletion. In this ANN application, a backpropagation learning mechanism will be used. In this model, the exactness is 95, 95.2, and 97.7% for each dataset accordingly.

3 System Architecture and Methodology

3.1 Data Sources

Almost all the information of the people aging around 18–65 years was gathered through the National Institute of Health Stroke Scale (NHISS). The implemented dataset comprises required data that includes 991 male and 259 female datasets comprising overall 1250 separate patient records. The class of the stroke is been categorized into different types of strokes. Of which, these strokes are individually divided into 531 ischemic strokes, 421 intra hemorrhagic stroke (IHS), and 298 sub-hemorrhagic strokes [18].

3.2 Feature Selection

The results of the observation and predictions for the input data, in general, referred to a job in machine learning modeling studies by selecting the characteristics of the inputs [19]. The important aim of this feature is to attain the slightest feasible set of input factors that perform with the highest accuracy and shows predictive presentation. Also, this can be made easier in understanding and learning by decreasing the outcome of diminishing the overall layout and also in removing unused contents to distinct on instructive variables. This develops an exceptionally randomized trees algorithm to evaluate the selection job among the variables within the complete dataset. Upon each iteration, in the extremely randomized trees algorithm, we can choose different splitting of tree nodes both at the feature and cut-point. Based on this randomization of splitting the nodes, the concatenation of the attribute list can vary. It is suggested to replicate the extra forest attribute selection to obtain a vigorous selected attribute list. The attributes principal can be deliberated by the extra trees

Table 1 Features of stroke based on family history

Previous stroke	Stroke history	Description
0—No stroke	0—No stroke	Stroke combinations with respect to no stroke
	1—Diabetes	
	2—Heart disease	
	3—Stroke	
	4—Hypertension	
1—Ischemic Stroke	0—No stroke	Stroke combinations with respect to ischemic stroke
	1—Diabetes	
	2—Heart disease	
	3—Stroke	
	4—Hypertension	
2—Intracerebral Hemorrhagic (ICH) stroke	0—No stroke	Stroke combinations with respect to ICH stroke
	1—Diabetes	
	2—Heart disease	
	3—Stroke	
	4—Hypertension	
3—Subarachnoid Hemorrhagic (SAH) stroke	0—No stroke	Stroke combinations with respect to SAH stroke
	1—Diabetes	
	2—Heart disease	
	3—Stroke	
	4—Hypertension	

algorithm for each iteration while checking at the hold-out round [20]. The summary feature selection of is available under Table 1.

3.3 Data Preprocessing

The term itself states that it plays a key role to elaborate the characteristics of the data in terms of its usage. The main characteristic components that define the quality of the data include accuracy, interoperability, and reliability. Preprocessing of the data requires several phases which include data cleaning, data integration, data reduction, and data transformation. Eliminating the resonate in the data, finding the missed terms, and sorting out diametric are the things in which data cleaning involves. The process of combining different forms into single compatible data can be done by data integration. The major drawback commonly observed in preprocessing is its redundancy that majorly occurs in data integration, and on the other hand, bulks of data are maintained by data reduction. As mentioned previously, the major remarkable issue in this combining of data is its redundancy. The causes for this redundancy

Table 2 List of primary attributes

Name of the Attribute	Description
Age	Age of the patient
National Institute of Health Stroke Scale (NIHSS)	0—No stroke
	1—Ischemic
	2—Intracerebral hemorrhage
	3—Subarachnoid hemorrhage
Family History	00—No stroke
	11—Diabetes
	22—Heart disease
	33—Stroke
	44—Hypertension
Smoking	0—Never
	1—Past
	2—Current
Body Mass Index (BMI)	18.5 to 24.9—Normal
	25 to 29.9—Overweight
	30 to 39.9—Obesity
	More than 40—Over obesity

are attribute naming, inconsistency, and in the case of attributes attained from the different groups of attributes. This error can be identified by correlation analysis that uses the chi-square test for initial data and correlation coefficient for numerical data [21]. The primary attributes of the dataset used in this research work are available under Table 2.

3.4 Naïve Bayes

The Naive Bayes analysis is an accumulation of dividing algorithms based on Bayes Theorem. This Bayes theorem locates the restrictive exception of a function eventuating the required exception of another event that has previously been completed. The advantage of this theorem is that it can hold multiple datasets with numerous attributes by increasing the accuracy by making it easier. It is highly recommended with fewer databases that use small training data, highly scalable, maintain uninterrupted and distinct data with impertinent feature unconcerned.

$$P(A|B) = \frac{P(B|A)P(A)}{P(B)} \tag{1}$$

3.5 Support Vector Machine (SVM)

Another model for the prediction of stroke under various categories is through SVM. This model has been implemented based on the statical learning theory which is in use by various platforms for recognition of the images and bioinformatics. This algorithm develops prototypes that establish new units that are assigned to the prevailing groups or can be formed to the new groups. Since SVM is implemented on a supervised machine learning model, this can provide results for retrogressions provocation and categorization. It represents 'n' dimensional dataset for arranging the attributes under a particular point. It is referred to as a value of such particular ordinate. The differences among these categories can be explained by the graph of the hyperplane.

$$K(\text{lowrisk}, \text{highrisk}) = (1 + \text{lowrisk} * \text{highrisk})^d \qquad (2)$$

where 'd' means degree of the polynomial.

3.6 Decision Tree

Based on the categorization and regression models, a decision tree can be developed in the form of a tree structure. Dividing the datasets into simpler subsets, a decision tree is evolved as shown in Fig. 1. Since the decision tree cannot indentify stroke detection initially based on the examination of its stroke prediction utilizing available datasets. These datasets individually fall under various labeled classes. Though it represents the characteristics of unsupervised learning and hence can be considered under supervised learning. The single pruning technology can be used in extending improvisations and in gathering the entire required data for developing a tree. The data used in this tree structure involves continuous, discrete, concise, and simple in inferring and outputs in a human-readable perspective.

Fig. 1 Simple classification of decision tree

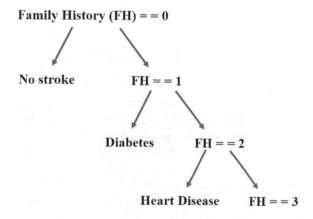

3.7 Logistic Regression

Predictive analysis is entirely based on logistic regression, and it concludes the corre-spondence among independent and dependent attributes of the data. For an instance, a patient with a stroke whose age, blood pressure, and diabetes levels show any impact in the scenario? The output of this possibility maybe 0 or 1 (binary) which is labeled as dependent, and on the other hand, the assuming values are considered as covariant. This technology of regression mechanism can be developed in multiple areas along with machine learning in predicting the type of stroke affected to the patient. Therefore, Sigmoid function is written as

$$\Phi(Z) = \frac{1}{1 + e^{-z}} \tag{3}$$

where $z = b_0 + b_1.\text{age} + b_2. \text{NIHSS} + b_3.\text{family history} + b_4.\text{smoking} + b_5.\text{BMI}$.

3.8 Bagging and Boosting

The mechanism plays a crucial role in gathering assumptions among various machine learning algorithms to conclude the type of stroke. This technology is accurate in finding the results among dependent models. The equal-weighted prediction model is demonstrated in the bagging algorithm by developing an ensemble of models using the learning schemes. The mechanism used in this algorithm is random forest. This algorithm can be maintained under two different stages. Firstly, the development of a random forest algorithm initialized along with assumptions from a random forest classifier. In the next stage, pseudo-code development is carried out for random forest algorithm.

3.9 Random Forest (RF)

Random forest refers to operations performed similarly to that of an ensemble that involves numerous bulks of separate decision trees. Each separate tree specified in the random forest encounters a class of prediction. Such a particular class with assorted votes stands as a prediction of the model. It is an ensemble method of understanding categorization and regression. Every single ensemble acts as a decision tree divider ensuring that the gathering of classifiers is a forest. This mechanism in the training part builds up decision trees and produces the mode of the class or mean value of the prediction of the single trees. Numerous efforts are being made in predicting the stroke which endangers and let the life risk considering these decision trees and have proven that the outcome of these algorithms is showing effective results.

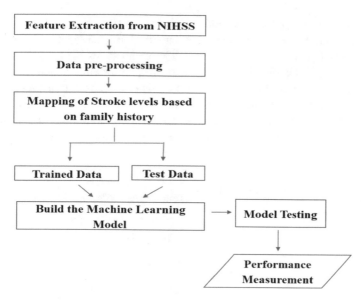

Fig. 2 Model architecture

$$\mathrm{RF}\, f\, i_i = \frac{\sum j \in \text{ all trees norm } f i_i}{T} \tag{4}$$

3.9.1 Proposed Model

This section describes the proposed model architecture as shown in Fig. 2. As a first step, the model initially extracts the features from the NIHSS to find out the stoke levels as ischemic, ICH, and SAH. The identification of primary attributes, data cleaning, and transformation can be done by the data preprocessing technique. Based on the family history, we can map the levels of stroke history that can be diabetes, hypertension, obesity, heart disease, etc., used for early detection of stroke. The available dataset is divided into two phases named as training with 80% subjects and testing with 20% subjects. After training, based on the availability of existing algorithms, the model builds the best accurate machine learning model. The model testing is the stage used for measuring the performance of the model.

3.9.2 Evaluation Metrics

From among the individual records of the patients based on the NIHSS factor, the information has been properly divided into one particular class as represented in a true positive (TP). The left-over data is separated from the patient's record and can be referred to as false positives (FP). In such a particular dataset, the information used is finally concluded as false and is related to different classes are mentionted under false negatives (FN). The amount of data that has been separated and extracted from each class is referred to as a recall. The recall states the stream of the ratio of TP to TP + FN. On the other hand, the proper information which is required to the dataset in each class for representing the data is referred to as precision. This precision ratio signifies TP to TP + FP. From this, the accuracy can be determined in terms of the percentage based upon the proper representation of the information for every individual class which is categorized by data elements under all combined classes.

$$Accuracy = \frac{TP + TN}{Totalelements} \tag{5}$$

$$Precision = \frac{TP}{TP + FP} \tag{6}$$

$$Recall = \frac{TP}{TP + FN} \tag{7}$$

$$F - Score = \frac{2 * Precision * Recall}{Precision + Recall} \tag{8}$$

4 Results and Discussion

In Table 3, we can observe more accuracy gain for the ischemic stroke under the evaluation by the random forest model with sensitivity 86% and specificity 95%.

Table 3 Performance evaluation of random forest with three classes of stroke

Type of stroke	Accuracy	Precision	Recall	Sensitivity	Specificity
Ischemic	92.7	93.3	91.1	86	95
ICH	88.9	89.4	87.3	82	93
SAH	87.3	91.2	90.8	98	92

Table 4 Performance evaluation of Xgboost with three classes of stroke

Type of stroke	Accuracy	Precision	Recall	Sensitivity	Specificity
Ischemic	94.2	94.3	94.9	89	95
ICH	91.3	89.8	91.1	85	96
SAH	89.7	89.6	89.1	90	97

Table 5 Performance evaluation of logistic regression with three classes of stroke

Type of stroke	Accuracy	Precision	Recall	Sensitivity	Specificity
Ischemic	88.9	88.1	90.9	86	93
ICH	90.4	89.3	91.2	82	91
SAH	89.9	90.2	90.4	92	94

Table 6 Performance evaluation of Naïve Bayes with three classes of stroke

Type of stroke	Accuracy	Precision	Recall	Sensitivity	Specificity
Ischemic	88.9	88.2	90.9	90	95
ICH	90.4	89.3	91.2	88	94
SAH	89.9	90.2	90.4	89	96

Table 7 Performance evaluation of SVM with three classes of stroke

Type of stroke	Accuracy	Precision	Recall	Sensitivity	Specificity
Ischemic	94.7	89.1	90.9	84	92
ICH	89.6	87.3	89.1	83	94
SAH	91.3	90.3	90.4	92	95

In Table 4, we can observe the more accuracy gain for the ischemic stroke under the evaluation by the Xgboost machine learning model with sensitivity 89% and specificity 95%.

In Table 5, we can observe the more accuracy gain for the ICH stroke under the evaluation by the logistic regression model with sensitivity 82% and specificity 91%.

In Table 6, we can observe the more accuracy gain for the ICH stroke under the evaluation by the Naïve Bayes model with sensitivity 88% and specificity 94%.

In Table 7, we can observe the more accuracy gain for the ischemic stroke under the evaluation by the SVM model with sensitivity 84% and specificity 92%.

Figure 3 shows the pictorial representation of machine learning models on the three levels of strokes such as ischemic, ICH, and SAH. Out of all existing models, it is observed that the SVM model has attained the highest accuracy with 94.7 for the type of stroke 'ischemic' and also for SAH with 91.3%. For the type of stroke ICH, the Xgboost model has attained the highest accuracy with 91.3%.

4.1 AUC–ROC Curve

The defects in the binary classification can be examined by the graph representing a receiver operator characteristic (ROC). The results from this ROC curve can be determined by eradicating the signal from among the noise. It can also be used in

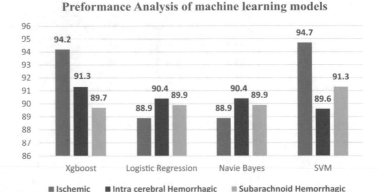

Fig. 3 Performance evaluation of machine learning models on the three classes of stroke

representing the probability that shows TPR and FPR at different frequency values. The portion of the area under the curve (AUC) is represented as the ability factor of an individual that differentiates among the classes and defines the briefing of the ROC curve.

Whenever the value is set to 1, it represents the accurate measure of division which can be referred to as significant model of AUC. The weaker measure of the AUC is set to the value 0 which represents the bad measure of classification. It also can be referred to as the interchanges values of the results obtained. It identifies and states 0 s as 1 s and vice versa. While at the point of 0.5 in AUC, it represents that there is no specific class under the classification model.

To determine the exactness of this model, usually receiving operating characteristic (ROC) curves is widely implemented in this job. Under dichotomic classification, references can be identified by considering a single variable X as a value determined for each example. These instances are divided on the threshold values of T, and if $X > T$, it represents positive else negative. If the referred example is positive which can be followed by X, then $f1(x)$ is the probability density function. Whereas if the examples considered as negative, then $f0(x)$ remains as the probability density function for the value of X. Further, the true positive rates and false positive rates are distinguished and can be determined by Figs. 4, 5, and 6:

$$\text{True Positive Rate}(T): \int_{T}^{\infty} f1(x) \tag{9}$$

$$\text{False Positive Rate}(T): \int_{T}^{\infty} f0(x) \tag{10}$$

Fig. 4 ROC for random forest model

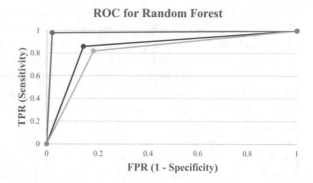

Fig. 5 ROC for Xgboost model

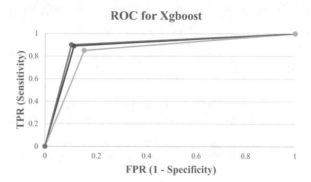

Fig. 6 ROC for SVM model

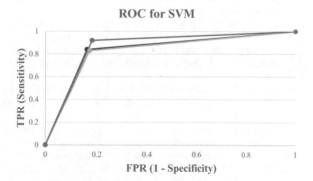

5 Conclusion and Future Scope

The main aim of this research work is to pull out the factors about NIHSS by the machine implementing models and to predict the intensity and criticality of the occurrence of the various types of strokes, namely ischemic stroke, intracerebral hemorrhagic, and subarachnoid hemorrhagic strokes. There are various advantages in detecting and identifying the severity of these strokes. The system has been developed in such a way that it by default divides and predicts the stroke and complexity into four divisions based on the NIHSS features related to the day-to-day activities. Later, the system intimates patient and their family with the type of the stroke and its complex nature with detailed data in present-day life as such the patient can immediately seek medical emergency care and can be taken to the nearest medical centers. Since NIHSS has 18 factors, out of which, only five major factors are utilized for the actual proposed system considering the patient's age which involves getting the medical emergency care depending upon the intensity and depth of the strokes that has occurred. Rather this the technology developments in the future models should represent the RS scores based on the image assumptions and possible opportunities for the fast recovery by reconsidering the multinomial models. To maintain this, a very bulk and enormous database and deep learning techniques should be implemented for better results and accuracy rather than these machine learning models.

References

1. V. Turova, Machine learning models for identifying preterm infants at risk of cerebral hemorrhage. PLOS ONE **15**(1), e0227419 (2020). https://doi.org/10.1371/journal.pone.0227419
2. K. Matsumoto, Impact of a learning health system on acute care and medical complications after intracerebral hemorrhage. Learn. Health Syst. **e10223** (2020). https://doi.org/10.1002/lrh2.10223
3. C.H. Lin, Evaluation of machine learning methods to stroke outcome prediction using a nationwide disease registry. Comput. Methods Program Biomed. **190**, 105381 (2020). https://doi.org/10.1016/j.cmpb.2020.105381
4. J. Heo, Prediction of intracranial aneurysm risk using machine learning. Sci. Rep. **10**(1), 6921 (2020). https://doi.org/10.1038/s41598-020-63906-8
5. E. Klang, Promoting head CT exams in the emergency department triage using a machine learning model. Diagn Neuroradiol **62**(2), 153–160 (2020). https://doi.org/10.1007/s00234-019-02293-y
6. Y. Fang, Programmed cell deaths and potential crosstalk with blood–brain barrier dysfunction after hemorrhagic stroke. Front. Cell. Neurosci. **14**(68) (2020). https://doi.org/10.3389/fncel.2020.00068
7. T.I. Shoily, Detection of stroke disease using machine learning algorithms, in 10th International Conference Computer Communication Network Technology, 1–6 (2019)
8. Y. Xie, Use of Gradient Boosting machine learning to predict patient outcome in acute ischemic stroke on the basis of imaging, demographic, and clinical information. Am J Roentgenol **212**, 415–51 (2019). https://doi.org/10.2214/AJR.18.20260
9. J. Liu, Prediction of hematoma expansion in spontaneous intracerebral hemorrhage using support vector machine. EBio Med **43**, 454–459 (2019). https://doi.org/10.1016/j.ebiom.2019.04.040
10. S. Park, Predicting delayed cerebral ischemia after subarachnoid hemorrhage using physiological time series data. J. Clin. Monitor. Comput. **33**(1), 95–105 (2019). https://doi.org/10.1007/s10877-018-0132-5
11. Y. Yu, Prediction of hemorrhagic transformation severity in acute stroke from source perfusion MRI. IEEE Trans. Biomed. Eng. **65**(9), 2058–2065 (2018)
12. B. James, Predictors of symptomatic intracranial haemorrhage in patients with an ischaemic stroke with neurological deterioration after intravenous thrombolysis. J. Neurol. Neurosurg. Psychiatr. **89**(8), 866–869 (2018). https://doi.org/10.1136/jnnp-2017-317341

13. W.J. Powers, 2018 Guidelines for the early management of patients with acute ischemic stroke: a guideline for healthcare professionals from the American Heart Association. Am. Stroke Assoc. Stroke **49**(3), e46–e99 (2018). https://doi.org/10.1161/STR.0000000000000158

14 M. Monteiro, Using machine learning to improve the prediction of functional outcome in ischemic stroke patients. IEEE/ACM Trans. Comput. Biol. Bioinform. **15**(6), 1953–1959 (2018). https://doi.org/10.1109/tcbb.2018.2811471

15. P. Govindrajan, Classification of stroke disease using machine learning algorithms. Neural Comput. Appl. **32**(3), 817–828 (2020). https://doi.org/10.1007/s00521-019-04041-y

16. R. S. Jeena, Machine Intelligence in stroke prediction. Int. J. Bioinform. Res. Appl. **14**(1/2), 29–48 (2018). https://doi.org/10.1504/ijbra.2018.089192

17. M. Singh, Stroke prediction using artificial intelligence, in 2017 8th Industrial Automation and Electromechanical Engineering Conference, pp e158–e161 (2017). https://doi.org/10.1109/iemecon.2017.8079581

18. B. Vamsi, M. Divya, D. Bhattacharyya, stroke_analysis, mendeley data. **V1** https://doi.org/10.17632/jpb5tds9f6.1

19. A. Wouters, Prediction of outcome in patients with acute ischemic stroke based on initial severity and improvement in the first 24 h. Front. Neurol. **9**, 308 (2018). https://doi.org/10.3389/fneur.2018.00308

20. F.M. Sacks, Dietary fats and cardiovascular disease: a presidential advisory from the American Heart Association. Circulation **136**(3), e1–e23 (2017). https://doi.org/10.1161/CIR.0000000000000510

21. D.L. Tirschwell, Validating administrative data in stroke research. Stroke **33**(10), 2465–2470 (2002). https://doi.org/10.1161/01.str.0000032240.28636.bd

Classification of Brain Tumors Using Deep Learning-Based Neural Networks

Jayavani Vankara, Muddada Murali Krishna, and Srikanth Dasari

Abstract Deep neural networks have grown as the most interesting topic in past few years. This new machine learning technique has been applied in several real-world applications. It can be considered as a machine learning tool of high strength for solving the higher dimension problems. Deep neural network is used in this paper, for organizing the 66 brain MRI datasets into categories of four. We have considered four dissimilar classes; those are glioblastoma, sarcoma, metastatic bronchogenic carcinoma tumors and normal tumors. The classifier has been integrated with the discrete wavelength transform (DWT), which is used as a tool for extracting features, and the principal components analysis (PCA) for dimensionality reduction.

Keywords Brain tumor · Deep learning · Neural network · Classification

1 Introduction

The human brain is considered as most important and composite organ in a human body parts, which has billion of cells. A brain tumor is the resultant of an uncontrolled and unrestricted cell division. It leads to irregular and uncurbed group formation of cells inside and around the brain, which increases the change and cause of abnormal brain activity and destroys the healthy cells [1, 2]. Brain tumors have two classifications, i.e., benign or low-grade tumors called grade 1 and grade 2 and malignant or high-grade tumor which are called grade 3 and grade 4. Malignant tumors are more aggressive than benign tumor. The mass of abnormal cells does not contain cancer

J. Vankara · M. M. Krishna (✉) · S. Dasari
Department of CSE, Dr. Lankapalli Bullayya College of Engineering, Visakhapatnam, Andhra Pradesh, India
e-mail: muralimuddada@lbce.edu.in

J. Vankara
e-mail: vjayavani@lbce.edu.in

S. Dasari
e-mail: dsrikanth@lbce.edu.in

© The Author(s), under exclusive license to Springer Nature Singapore Pte Ltd. 2021
S. K. Saha et al. (eds.), *Smart Technologies in Data Science and Communication*,
Lecture Notes in Networks and Systems 210,
https://doi.org/10.1007/978-981-16-1773-7_3

cells. The rate of growth of benign brain tumors is moderate and does not escalate into other tissue. Malignant brain tumors contain cancer, a cell whose rate of growth is quite rapid and tends not to have clear frontier. Malignant tumors grow rapidly, so they are more menacing and can easily spread to other parts of the brain.

Based on the origination, the damaged tumors in the brain are classified into two types, namely primary tumors and secondary tumors. Primary tumors arise in brain by self and escalate to other parts of the brain, while secondary tumors arise elsewhere and escalate to the brain [3, 4]. Brain MRI (MRI scan), one of the current methods for brain tumors detection, is modeled for understanding the present status in together finding phase and treatment phase. MRI images have played a great impact in the automatic medical imaging technology as it provides transparent and complete information about the brain structure and the abnormalities within it [3, 5, 6]. There are various techniques presented by various researchers using MRI images as they have high resolution and can be scanned and loaded to computer. However, in the previous few decades, support vector machine (SVM) and neutral networks (NN) were extensively used [7]. The modern research has broadened its wings to deep learning and is one of the most functional techniques of machine learning. The deep learning architecture is proficient enough for solving an intricate and multifarious problem without using a large number of nodes, for example, SVM and K-nearest neighbor. This is why they are frequently used in innumerous areas of health informatics like medical analysis of image, bioinformatics and medical informatics [7–9].

The concept of classification of brain tumor is summarized in this paper by using deep learning and MRI images as the input. Finally, its performance is calculated based on various benchmarks. The aim of this technique is to distinguish between different categories of tumors such as sarcoma, glioblastoma, multiforme and metastatic bronchogenic carcinoma tumors by using the MRI images of brain. The set of features that are obtained via DWT are used in this technique. DWT is a routine used for extracting features from the sliced brain MRI images, which are helpful in training the DNN classifier for segregating the various brain tumors. The paper is planned as follows: Sect. 1 gives concise prologue, and Sect. 2 covers a preview on deep learning concept and as well as its architecture. In Sect. 3, the tactics has been described, and experimental results and discussion are fully explained and described.

2 Synopsis of Deep Learning

DL is a subdivision of machine learning (ML). It deals with superior and major datasets. It is based on multiple training levels of neural layers using a set of input features. In this, the higher-level neurons are defined from the neurons in the previous level, and the same lower-level neurons can be utilized as input by various neurons of the next level [10]. The DL framework enhances the conventional neutral network (NN) by attaching several intermediate layers, bounded by the input and the output layers, in the traditional NN architecture for complex and nonlinear relationships [8–10]. CNN is a type of deep learning architecture which has gained recognition in few years because of its ability to solve the complex and stringent problems in relatively shorter period [8, 9]. A convolutional neural network model is a chain of feed forward layers. This type of network is implemented using max or average pooling layers in convolutional neural network. After the last pooling operation, CNN then uses the output of the pooling layer to feed into a fully connected feed forward neural network (FFNN) which converts the 2D outputs of the prior layers into 1D output for classification [10].

One significant feature of CNN models is that they can work without the procedure of extracting features before applying it. The greatest and major disadvantages of CNN architecture are that its processing is slow and hence it is quite time consuming. The next disadvantage considered is, as another significant point, the complexity in training. The training is difficult since it needs a large labeled dataset. The next disadvantage, which can be considered, is its hardware requirement. The hardware requirement is more for processing images with higher pixel counts, for example, 256×256 [11–13]. Deep neural network (DNN) is a type of DL architecture, which is importantly useful for classification or regression. It is basically a FNN in which the data passes from the input neural layer to the output classifier layer through number of hidden intermediate layers which are normally more than two [14, 15]. Figure 1 depicts the architectural structure of CNN where N1 represents the first neural layer containing neurons for input data, NO represents the classifying layer which consists of neurons for the various categories of the output, and Nh1 are the intermediate layers.

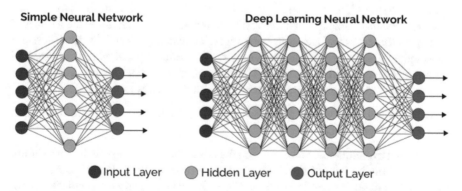

Fig. 1 DNN architecture [16]

3 Problem Statement

The MRI scan image of the brain is categorized into four different classes, which we have considered are glioblastoma, normal, sarcoma and metastatic bronchogenic carcinoma tumors using DNN classifier.

4 Proposed Model

The methodology suggested uses DNN architecture for classification. It basically acts as a classifier to recognize the tumors in MRI images. The complete steps of the tactic can be understood as the following steps:

a. Data acquisition that contains the MRI image of brain.
b. Fuzzy C-means for the segmentation of MRI image.
c. Extraction of the features using discrete wavelet transform (DWT) and dimensionality reduction using PCA.
d. Applying deep neural network (DNN) for classification.

The total modeling of the analyzer consists of three distinct steps, namely data analysis, data acquisition and data visualization. Data visualization is the output phase.

A. Data Acquisition

According to the WHO, a hundred types of brain tumors are known which might vary in source, location, size of the tissues and character. In this technique, the focus will be only on three types of carcinogenic tumors. They being: Glioblastoma: These are classified as grade 4 malignant brain tumors. These tumors are extremely dreadful in nature. These are crucial tumors. These develop from astrocytes, which are star shaped cells and supports the nerve cells. It mainly initiates in the cerebrum.

Sarcoma: It varies in grades from grade 1–grade 4 and starts from the connective tissues like blood vessels.

Metastatic Bronchogenic Carcinoma: This is a secondary malignant tumor in brain. This starts from bronchogenic carcinoma lung tumor and then slowly spreads in the brain. The dataset needed is collected from Website of Harvard Medical School (https://med.harvard.edu/AANLIB/). This set of data consists of 66 MRI images of real human brain, in which 44 are abnormal and 22 are normal. The brain MRIs were in T2 weighted axial plane and 256 * 256 pixels.

B. **Image Segmentation**

Image segmentation is basically non-trivial task and helps in differentiating the normal part of the brain tissues such as white matter (WM) and grey matter (GM)

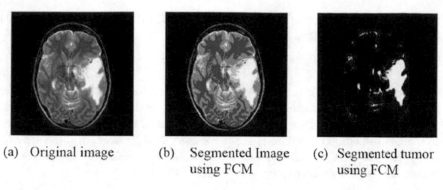

(a) Original image   (b) Segmented Image   (c) Segmented tumor
                                using FCM   using FCM

Fig. 2 Segmented image using C-means algorithm

tissue in MRI scans [12]. This procedure is of high eminence as it acts as the input for the next phase. We have used the technique of fuzzy C which means cumulating technique to subdivide the image since it had good result in the earlier works and good for correlation purpose [13]. Figure 2 shows the segmented image using fuzzy logic algorithm.

C. Feature Extraction and Dimensionality Reduction

Once we are done with the segmentation of the MRI image, the feature extraction followed by dimensionality reduction stage comes into the picture. The features are extracted using discrete wavelength transform (DWT) which extracts the most pertinent features at contrasting aspects and scales as they provide information of the signal by the help of cascaded filter bands or high pass and low pass filters in order to obtain the characteristics in a hierarchical manner [13]. Two levels Discrete Wavelength Transform disintegration of an image is verified in Fig. 3, where the coefficient of low pass and the high pass filters are represented by the functions $h(n)$ and $g(n)$, respectively. There are four sub-band images at each level as a result, i.e., (LL, LH, HH and HL). The estimate component of an image is LL sub-band, whereas LH, HL and HH sub-bands might be considered as the important constituent of the image. [13, 14]. In our approach, we have used three-phase decomposition of Haar

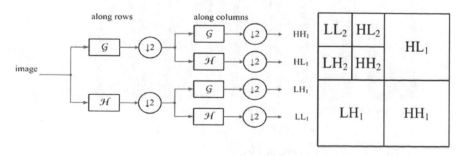

Fig. 3 Representation of level 1 and level 2 decomposition

wavelet that was adopted in our earlier endeavor [13] to extract 32 * 32 = 1024 features for every MRI scan [13].

D. Classification

After the selection and extraction of feature, the deep neural network is used for categorization and is executed on the feature vectors obtained. In order to build and train the deep neural network, having a seven layers of hidden model, the seven-gold cross validation procedure is used. We even used another ML classification algorithm from WEKA (https://www.cs:waikato:ac:nz/ml/weka/) for evaluating and calculating the performance of chosen classifier [13].

5   Experiment Results and Discussions

Two tools were used for the results and experiment.

(1)   The brain MRI dataset was prepared, and MATLAB R2015a was used for first three steps.
(2)   For the evaluation and classification of preferred classifier, WEKA 3.9 tool was used.

The efficiency of the proposed procedure was measured on the benchmark of average categorization rate, F-measure average recall and average area under ROC curve (AUC) of all the four classes, i.e., glioblastoma, normal, sarcoma and metastatic bronchogenic carcinoma tumors. As seen from Fig. 4 and Table 1, deep neural

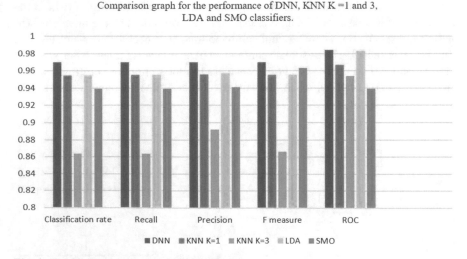

Fig. 4 Graphical comparison for DNN and KNN

Table 1 Performance of the different benchmark algorithms

Algorithm	Classification rate (%)	Recall	Precision	F-measure	AUC (ROC)
DNN	97.77	0.89	0.97	0.97	0.84
KNN K = 1	93.15	0.86	0.856	0.75	0.67
KNN K = 3	76.56	0.84	0.82	0.66	0.54
LDA	92.15	0.55	0.57	0.55	0.83
SMO	92.85	0.39	0.41	0.63	0.39

network classifier provided better and improved results when fused with DWT feature extraction tool when compared to other classifiers.

6 Conclusion and Forthcoming Efforts

This paper has presented a practiced procedure combining DWT and DNN, which classifies the MRI scanned images to three types of tumors of brain. The processing time and size of the images were quite good for implementing the process. It is also tested with large size of images for better understanding of the model developed. The currently considered DNN classifier is working well when compared with the other model considered in the current article.

References

1. A. Kavitha, L. Chitra, R. Kanaga, Brain tumor segmentation using genetic algorithm with svm classifier. Int. J. Adv. Res. Electr. Electron. Instrum. Eng. **5**(3), 10–16 (2016)
2. T. Logeswari, M. Karnan, An improved implementation of brain tumor detection using segmentation based on soft computing. J. Cancer Res. Exp. Oncol. **2**(1), 06–014 (2010)
3. K.G. Khambhata, S.R. Panchal, Multiclass classification of brain tumor in MRI images . Int. J. Innov. Res. Comput. Commun. Eng. **4**(5), 8982–8992 (2016)
4. V. Das, J. Rajan, Techniques for MRI brain tumor detection: a survey. Int. J. Res. Comput. Appl. Inform. Technol. **4**(3), 53–66 (2016)
5. E.I. Zacharaki, S. Wang, S. Chawla, D. Soo Yoo, R. Wolf, E. R. Melhem, C. Davatzikos, Classification of brain tumor type and grade using MRI texture and shape in a machine learning scheme. Magn. Reson. Med. Off. J. Int. Soc. Magn. Reson. Med. **62**(6), 1609–1618 (2009)
6. L. Singh, G. Chetty, D. Sharma, A novel machine learning approach for detecting the brain abnormalities from MRI structural images, in *IAPR International Conference on Pattern Recognition in Bioinformatics*. (Springer, Berlin (2012), pp. 94–105
7. Y. Pan, W. Huang, Z. Lin, W. Zhu, J. Zhou, J. Wong, Z. Ding, Brain tumor grading based on neural networks and convolutional neural networks, in *37th Annual International Conference of the IEEE Engineering in Medicine and Biology Society (EMBC)*. IEEE (2015), pp. 699–702.
8. G. Litjens, T. Kooi, B.E. Bejnordi, A.A.A. Setio, F. Ciompi, M. Ghafoorian, J. Van Der Laak, B. Van Ginneken, C.I. Sanchez, A survey on deep learning in medical image analysis. Med. Image Anal. **42**, 60–88 (2017)

9. D. Ravi, C. Wong, F. Deligianni, M. Berthelot, J. Andreu-Perez, B. Lo, G.-Z. Yang, Deep learning for health informatics. IEEE J. Biomed. Health Inform. **21**(1), 4–21 (2016)
10. S. Tharani, C. Yamini, Classification using convolutional neural network for heart and diabetics datasets. Int. J. Adv. Res. Comput. Commun. Eng. **5**(12), 417–422 (2016)
11. K.B. Ahmed, L.O. Hall, D.B. Goldgof, R. Liu, R.A. Gatenby, Fine-tuning convolutional deep features for MRI based brain tumor classification, in *Medical Imaging 2017: Computer-Aided Diagnosis* (International Society for Optics and Photonics, 2017), p. 101, no. 34, p. 101342E
12. N. Gordillo, E. Montseny, P. Sobrevilla, State of the art survey on MRI brain tumor segmentation. Magn. Reson. Imaging **31**(8), 1426–1438 (2013)
13. H. Mohsen, E. El-Dahshan, E. El-Horbaty, A. Salem, Brain tumor type classification based on support vector machine in magnetic resonance images. Ann. Dunarea De Jos University of Galati, Math. Phys. Theor. Mech. Fascicle II Year IX (XL) **1**, 10–21 (2017)
14. M. Ahmad, M. Hassan, I. Shafi, A. Osman, Classification of tumors in human brain MRI using wavelet and support vector machine. IOSR J. Comput. Eng. **8**(2), 25–31 (2012)
15. N. Asmita Ray, Thirupathi Rao and Debnath Bhattacharyya: Analysis of Breast Cancer Detection and Classification in Early Stage from Digital Mammogram. Asian Int. J. Life Sci. **20**(2), 1–15 (2019)
16. https://cdn-images-1.medium.com/max/800/1*5egrX--WuyrLA7gBEXdg5A.png. Accessed on 5 Jan 2021

Deep Learning for Lung Cancer Prediction: A Study on NSCLC Patients

Madhuri Thimmapuram⬤, Sowjanya Pentakota, and H. Naga Chandrika

Abstract This study explores the application of deep learning in clinical tomography taking into account computerized evaluation of the X-rays and the possible improvement of patient delineation. Our results demonstrate that in-depth learning systems can be used for the definition of death rate risk based on CT descriptions of NSCLC patients' care specifications. This evidence will encourage the forthcoming exam to unravel the clinical and natural foundations of deep learning systems as well as approval for forthcoming knowledge.

Keywords NSCLC · Convolutional neural system · Benchmarking

1 Background

NSCLC (non-small-cell lung cancer) patients frequently show changing clinical classes and results, even inside a similar tumor phase. This investigation investigates deep learning applications in clinical tomography taking into account the computerized evaluation of radiographic attributes and possibly enhancing patient delineation. Our outcomes give proof that deep learning systems might be utilized for the death rate risk definition dependent on care specification CT descriptions from NSCLC patients. This proof inspires upcoming examination into superior unraveling the clinical and natural premise of deep learning systems just as approval in imminent information.

M. Thimmapuram
Assistant Professor, Department of Computer Science and Engineering, Vardhaman College of Engineering (A), Shamshabad, Hyderabad, Telangana State 501218, India

S. Pentakota (✉)
Research Scholar, Vishakhapatnam, Andhra Pradesh, India

H. Naga Chandrika
Research Scholar, Dr. A.P.J. Abdul Kalam University, Indore, Madhya Pradesh, India

2   Techniques and Results

An integrative analysis of performed on 7 free datasets around 5 organizations equaling 1183 NSCLC patients [(survival mean = 1.7 years (series 0.0–11.7), age mean = 68.3 years (series 32.5–93.3)]. Then exterior confirmation was applied in information of CT (calculated tomography), a three-dimensional CNN (convolutional neural system) is utilized to collect the signatures of prognostic for patients undergoing treatment with radiation treatment [$n = 772$, survival mean = 1.3 years [series 0.0–11.7), age mean = 67.9 years (series 32.45–93.38)]. We at that stage a transfer learning methodology to deal with accomplish the equivalent for surgical treatment patients ($n = 389$, survival mean = 3.1 years (series 0.0–8.8), age mean = 69.1 years (series 37.2–88.0) (Table 1).

We initiate that the CNN forecasts were essentially connected with 2-year in general existence from the beginning of radiotherapy (area in accordance with the operational distinctive curve [AUC] = 0.70 [$p < 0.001$, 95% CI 0.63–0.78]) and surgical treatment (AUC = 0.71 [$p < 0.001$, 95% CI 0.60–0.82]) patients. The CNN was additionally ready to essentially define patients into high and low death rate hazard bunches in both the surgery ($p = 0.03$) and radiotherapy ($p < 0.001$) datasets. Furthermore, the CNN was initiate to fundamentally beat arbitrary timberland models that are based upon clinical constraints—comprising sex, age, and cancer hub metastasized phase—just as show high-level heartiness between readers (Spearman's rank correlation = 0.88) and against test-re-testing (intra-class correlation factor = 0.91) varieties. The areas were distinguished with the highly commitment toward forecasts and featured the significance of tumor-adjoining tissue in understanding delineation. So that a superior comprehension of the qualities caught by CNN was improved. Then the fundamental discoveries were presented on the natural premise of the caught phenotypes as being connected to transcriptional forms and cell cycle. Constraints

Table 1 Techniques and results	*Integrative analysis*			
	N	1183		
	Survival mean	1.7 Years	Series	0.0–11.7
	Age mean	68.3 Years	Series	32.5–93.3
	Three-dimensional CNN (convolutional neural system) for patients undergoing treatment with radiotherapy			
	N	772		
	Survival mean	1.3 Years	Series	0.0–11.7
	Age mean	67.9 Years	Series	32.45–93.38
	Transfer learning approach to deal with accomplish the equivalent for surgery patients			
	N	389		
	Survival mean	3.1 Years	Series	0.0–8.8
	Age mean	69.1 Years	Series	37.2–88.0

incorporate the review idea of this investigation just as the dark discovery type of deep learning systems.

3 Data Analysis

Tumors were physically molded and affirmed by a qualified professional reader. With the slice-texture surpassing in-plane goals, all the datasets were re-sampled into homogeneous voxels of unit measurement to make sure likeness, in which 1 voxel relates to 1 mm^3. This has been achieved by applying direct and closest neighbor interruptions for the picture and comments, correspondingly. If various annotations which are disconnected already found, larger volume has been considered for the study.

4 Preprocessing of Data for Deep Learning

Offered full three-dimensional tumor division, both the jumping box and center of mass (CoM) of the tumor explanations were determined. Three-dimensional homogeneous stripes of magnitude $50 \times 50 \times 50$ were separated across every CoM, catching over 60% of the tumor jumping boxes' measurements in the radiotherapeutic training dataset. The stripes have been stabilized later to a 1–0 series utilizing upper and lower Hounsfield unit limits of 3072 and -1023 correspondingly. An expansion factor of 32,000 was applied to the stripes, generating a size for training of nearly 5.89 and 9.39 million information tests for the surgery and radiotherapy datasets, separately. The above-mentioned extensions consist of random interpretations ± 10 pixel resolution in each of the 3 axes, arbitrary revolution at 90° interims along the transverse axis only, and arbitrary spinning along every one of the 3 axes. A magnitude was done progressively throughout the professional development. No adjustment- or assessment-time reinforcement has been implemented.

5 Deep Learning

In this study, a three-dimensional CNN engineering was utilized. The system contains a sum of four three-dimensional convolutional layers of 63, 928, 894, and 509 channels with portion series of $3 \times 3 \times 3$, $3 \times 3 \times 3$, $3 \times 3 \times 3$, $5 \times 5 \times 5$, and separately. Two maximal pooling strata of part series $3 \times 3 \times 3$ have been implemented following the 2nd and 4th convolutional strata. A progression of four completely joined strata—with 13,824, 513, 039, and two units—if significant level thinking prior to the forecast possibilities were determined in the finishing layer of SoftMax categorizer. Training description is as below: We applied the inclination based upon the probability associated with the optimization tool Adam [1] with a worldwide learning rate of 1×10^{-03} with no rot, a bunch size of 16, dropping out [2] of 25% and half on the convolutional and completely associated levels, individually, and a L2 normalization [3] penalizing chance of 1×10^{-05}. To evade the inward covariance, move issue [4], bunch standardization has been implemented over all levels, with the information layer as an exemption. Flawed redressed straight units (defective ReLUs) [5] with alpha $= 0.1$ were the initiation capacity of decision over the whole system before the last SoftMax enactment.

During a training session of the CNN within the therapeutic radiology dataset, we utilized a randomized lattice search investigating diverse hyper-factors involving convolution part size, regularization term, learning rate, batch size, and input patch size. Concerning the overall architecture, we began with a thin system, in which the under-fitting happens, and steadily included levels. The model was streamlined on the changing dataset utilizing initial halting [6]. Through a 1000-age limit, the model with the greatest execution on the tuning dataset was picked. In applying move learning on the surgical treatment training dataset, the quantity of definite levels to tweak was investigated. The ideal environment encompassed calibrating the last characterization layer just whereas maintaining prior levels permanent. With many less parameters to prepare, the cluster size and learning rate were expanded to 24 and 1×10^{-02} correspondingly. Deep learning structure of Google Tensor flow [7] was employed to prepare, test, and tune the CNN.

6 Findings

6.1 Tumor Categorization Utilizing Three-Dimensional Deep Learning Networks

In surveying the capacity of deep learning systems to evaluate the radiological qualities of tumors, we conducted an integrative examination on seven free datasets totaling **1183** patients. It is found and autonomously approved prognostic sign utilizing a CNN for patients being undergoing treatment with radiation treatment ($n = 772$, incorporating 609 with 2-year follow-up for survival study). We at that point utilized an exchange learning way to deal with accomplish the same for patients ($n = 389$, incorporating 367 with 2-year follow-up for survival study). The engineering of the system was intended to get three-dimensional input m^3 encompassing the center of the first-level tumor—in light of clinician-found seed points. The trained system was deployed to foresee generally survival probability 2 years from the start of the particular treatment.

Beginning with the radiation treatment patients, the investigation was part into a finding stage and an self-determining test stage. Inside the finding stage, a three-dimensional CNN was prepared on the Harvard RT dataset (2-year survival expired/alive $= 134/159$, survival mean $= 2.2$ years [series 0.0–11.7], female/

male = 153/140, age mean = 67.9 years [series 32.5–93.3]) utilizing enlargement, although the autonomous Radboud dataset (2-year survival perished/alive = 76/28, survival mean = 0.9 years[series 0.1–8.2], male/female = not accessible, age mean = 65.9 years [series 44.4–85.9]) was utilized to an iterative tune and upgrade the CNN's hyper-constraints just as the tumor three-dimensional input fix sizes till the most excellent forecast score was accomplished. Past this disclosure stage, the prognostic CNN was bolted and tried on the free Maastro dataset (2-year survival expired/alive = 151/60, survival mean = 1.0 years [series 0.0–5.8], female/male 114/269, age mean = 69.0 years [series 34.0–91.7]). The CNN indicated a huge prognostic force in anticipating 2-year survival (AUC = 0.70 [$p < 0.001$, [95% CI 0.63–0.78]). Kaplan–Meier bend investigation was conducted to assess the CNN's exhibition in separating high and low death rate chance gatherings. A critical existence contrast ($p < 0.001$) was seen among the two gatherings on the free Maastro dataset.

So as to build up a prognostic deep learning system for careful patients, we utilized an exchange learning methodology. The last expectation strata of the radiation treatment-trained CNN were tweaked on the Moffitt dataset (2-year survival expired/alive = 50/133, survival mean = 2.8 years [series 0.0–6.3], male/female = 83/100, age mean = not accessible) utilizing enlargement Techniques. The free MUMC dataset (2-year survival perished/alive = 24/64, survival mean = 3.3 years [series 0.2–8.8], male/female = 61/27, age mean = 68.0 years [series 37.2–83.3]) was utilized to the iterative tune and enhance the CNN's hyper-factors just as distinguish the ideal levels for calibrating. Then the CNN was bolted and tried on the autonomous test dataset M-SPORE (2-year survival perished/alive = 17/80, survival mean = 4.5 years [series 0.3–7.8], male/female = 44/53, age mean = 70.0 years [series 46.0–88.0]), where it showed a noteworthy prognostic presentation (AUC = 0.71 [$p < 0.001$, 95% CI 0.60–0.82]). Kaplan–Meier bend examination indicated a noteworthy survival distinction (p = 0.03) among high and lower death rate chance gatherings inside the M-SPORE test dataset.

6.2 Threshold Values Against Clinical Factors and Highlights of Designed Imaging

The deep learning networks were standardized against randomized forest models that are based upon clinical data (sex, age, and TNM phase). Such clinical models accomplished an exhibition of AUC = 0.58 ($p = 0.4$, 95% CI 0.39–0.77) and AUC = 0.55 ($p = 0.21$, 95% CI 0.47–0.64) for the surgery and radiotherapy datasets, separately. Furthermore, the single-dimensional examination proposed that such clinical factors didn't have a huge correlation with existence. Deep learning achieved essentially superior for both treatment kinds. The deep learning systems were likewise contrasted with arbitrary timberland models dependent on designed highlights portraying tumor shape, voxel power data (measurements), and examples (surfaces). The built component models showed a prognostic exhibition of AUC = 0.58 ($p = 0.275$, 95% CI

0.44–0.75) and AUC $= 0.66$ ($p < 0.001$, 95% CI 0.58–0.75) for the surgery and radiotherapy datasets, individually. In spite of the fact that the deep learning systems exhibited improved execution across the built models for both patient gatherings, this distinction was huge for surgical treatment patients ($N = 1000$; stage test, $p = 0.035$) however, was not huge for radiation therapy patients ($N = 1000$; change test, $p = 0.132$). Such outcomes were affirmed with a meta p-value test ($p = 0.06$). At last, the deep learning systems were contrasted with tomography factors generally utilized in clinical practice, to be specific tumor volume and most extreme width. We observed that tumor volume accomplished an exhibition of AUC $= 0.51$ ($p = 0.85$, 95% CI 0.37–0.66) and AUC $= 0.64$ ($p < 0.001$, 95% CI 0.56–0.73) for the surgery and radiotherapy datasets, separately. The deep learning systems were fundamentally improved for the surgical treatment dataset ($p = 0.004$) and non-altogether improved on the radiation treatment dataset ($p = 0.056$), as affirmed with a meta p-value test t ($p < 0.001$).

Lastly, the deep learning systems were in comparison to the imaging factors which are employed in the clinical practice commonly, the parameter was maximal diameter and tumor volume. It is observed that tumor volume attained a result of AUC $= 0.51$ ($p = 0.85$, 95% CI 0.37–0.66) and AUC $= 0.64$ ($p < 0.001$, 95% CI 0.56–0.73) for the surgical procedure datasets along with radiotherapy database correspondingly. The deep learning networks were limit considerably acceptable for the surgical treatment dataset ($p = 0.004$) and the non-considerably acceptable on the radiation treatment dataset ($p = 0.056$), as concluded with a meta p-value test ($p < 0.001$). These findings were similarly noticed for maximal diameter which was taken from the training set.

7   Discussion

In this examination, we surveyed the efficiency of deep learning systems in foreseeing 2-year by and large existence of NSCLC patients from CT information. We prepared a three-dimensional CNN to start to finish on patients undergoing treatment with radiation therapy and utilized an exchange learning methodology for such undergoing treatment with the medical procedure. We exhibited CNN's capacity to fundamentally delineate patients into high and low death rate hazard gatherings, just as its strength in test–retest and between reader changeability situations. Notwithstanding benchmarking standards against include building strategies; we additionally featured locales with the biggest commitments to the caught prognostic sign, both inside and past the tumor volume. At long last, our primer genomic affiliation examines recommended connections between deep learning highlights and transcriptional and cell cycle forms.

This attempt expands upon an assortment of deep learning applications in clinical tomography that has developed while the exceptional prevalent exhibition of CNNs in late image order rivalries [8, 9]. Scarcely any deep learning examinations to the present day have investigated forecast, with highly tending to different errands including division, location, and danger grouping [10]. While highlight description is mechanized in such deep learning draws near, radiomics has basically depended on the mining, determination, and ensuing characterization of predetermined highlights utilizing additional AI strategies comprising randomized woods, shallow neural systems, and bolster vector machines among others [11]. Such techniques have discovered applications in the guess of the nasopharynx carcinoma in MRI [12], the pulmonary cancer in CT [13], and beginning period NSCLC in CT/ PET (positron emission tomography) to give some examples. Thus, in this examination, we standardized the deep learning systems against randomized backwoods models based on built highlights, with the presentation of the arbitrary timberland models being inside recently watched series [9]. These models showed a second-rate execution when contrasted with the deep learning systems, in spite of the fact that this distinction was just noteworthy for surgery patients. These outcomes might be because of the more elevated levels of reflection characteristic in deep learning highlights their designed partners. Furthermore, and as far as information designs, built highlights were removed solely from inside tumor comments. Deep learning inputs, be that as it may, were involved three-dimensional m^3 permitting the system to consider tumor-encompassing tissue. This impact is amplified in the littler tumors undergoing treatment with surgery comparative with their bigger radiation therapy partners, possibly clarifying the essentialness of the surgical treatment findings. Surgical treatment patients are frequently avoided from built radiomics contemplates [14], where there is no prognostic sign were distinguished, with the referred to thinking being the absence of a method of reasoning in foreseeing a tumor reaction dependent on its phenotype on the off chance that it is surgically removed. Our outcomes indicate the prospective efficiency of deep learning systems in defining this particular patient gathering.

We likewise investigated models based on a lot of clinical highlights, involving sex, age, and TNM phase. Such models done ineffectively in both the surgery and radiotherapy datasets, possibly because of the constrained highlights accessible and normal to every one of the six datasets. Imaging highlights normally utilized in the facility, specifically tumor volume and most extreme width, performed moderately well on the radiation therapy datasets, but instead ineffectively on the surgical treatment datasets, as has recently been shown [15]. The two models were beaten by deep learning draws near, despite the fact that the thing that matters was just noteworthy for the surgical treatment datasets. Additional examinations are expected to research the prognostic connection among such highlights and deep learning highlights for radiotherapy patients, particularly given the entrenched connection among survival and tumor volume in this gathering [16]. Such outcomes likewise indicate the prognostic predominance of deep learning highlights for surgical treatment patients.

The endeavor's to recognize striking districts inside images via actuation planning indication at the importance of tumor-encompassing tissue in the quiet definition lines up with endeavors that grandstand the prognostic estimation of tumor area [17] just as the significance of understanding the communications among tumors and their environmental factors as a method for successful cancer anticipation and care [18]. At last, our starter genomic affiliation study grandstands connections between deep learning system expectations and transcriptional, cell cycling, and additional DNA imitation forms, for example, DNA fix or harm reaction. This recommends deep learning high-lights might be motivated by hidden atomic procedures generally identified with the expansion of cells and subsequently movement of tumors. Besides, almost all funda-mentally improved natural procedures had a negative advancement score, showing a converse correlation to the survival expectations. This recommends the quality articulation present in cell multiplying pathways will, in general, be diminished, with superior system scores showing superior survival likelihood. As the correlation among built tomography highlights and natural path shave just been set up [19], our investigation stretches out these correlations to deep learning. Qualities of this investigation incorporate the generally enormous—in cancer imagery terms—set of 1183 NSCLC patients with tuning, training, and testing on free datasets. The datasets were diverse as far as tomography procurement considerations, clinical phase, and the board, therefore displaying clinical realities. This recommends deep learning strategies may, in the end, be adequately vigorous and can be generalized for down to earth application in clinical consideration. Notwithstanding being a non-obtrusive and financially savvy routine clinical test [20], CT imaging gives a moderately steady radiodensity metric normalized across gear merchants and imaging conventions contrasted with additional imagery modalities (e.g., PET and MRI). In contrast with built radiomic techniques that need cut-by-cut tumor comments—a tedious and costly procedure that is exceptionally inclined to between reader inconstancy—our method-ology might generate greater throughput as it just involves a solitary snap seed point assessment generally inside the focal point of the tumor volume. The 2-year survival endpoint used at this point is a pertinent survival limit for NSCLC patients and some that were recently utilized in anticipation endeavors [21]. Our examination alludes to the utility of move learning inside clinical tomography and throughout treatment categories, a discovery that is additionally reinforced by performance measurement standards against start to finish training of the surgical operation training dataset. A few constraints ought to likewise be noted. By plan, the review idea of this investiga-tion obstructed the capacity to check exactly where and how such an instrument can possibly be coordinated into the clinical work process. Thus, the prognostic informa-tion refined into the deep learning systems depends on previous therapy alternatives and conventions and might not be satisfactorily situated to gather a prognostic sign for a patient undergoing treatment with progressively current methods. The haziness of deep learning systems is an extra restriction. Highlight definition, mining, and determination in such methods—a significant wellspring of inconstancy in designed radionics [11] are completely computerized and happen certainly. This happens at a costly expense: interpretation. Therefore, this discovery like systems is exceptionally hard to troubleshoot, seclude the explanation for specific results, and foresee when and where disappointments will occur. Without solid hypothetical support [22], deep learning highlights are anonymous, and the tomography qualities they determine are profoundly dark.

This equivocalness is in severe complexity to the master founded all around char-acterized designed highlights and is frequently exacerbated in visualization issues where the main methods for approval are long-haul mortality follow-up through forthcoming investigations. Also, a superior comprehension of the system hyper-factor space is required, conceivably gave by utilizing different changing datasets inside the revelation stage and before the last test stage. Another constraint lies in the information space. Notwithstanding the previously mentioned inter-reader variability, and test–retest, CT stability, and dataset heterogeneity contemplates executed in this, the systems' affectability to different varieties in clinical factors and image procurement factors, comprising noise record levels, tube current, and reproduction explicit factors and so on, have not been studied. At long last, as the survival times utilized in this investigation are by and large instead of being cancer-explicit, they might be impacted by outer considerations and bring vulnerability into the issue. Offered the permanent info size of the deep learning systems utilized in this investigation, upcoming research headings incorporate investigating grouping system models that acknowledge contributions of synchronous multi-scale goals [23] or variable sizes [24]—a methodology normal to completely convolutional systems utilized in image division. Contributions of fluctuating scales can possibly consider joining the enormous tumors in radiation therapy patients with their generally littler partners in surgical treatment patients into single prognostic system while keeping up heartiness against such variety. As far as interpretation, training neural systems with unraveled shrouded layer portrayals is a functioning territory of research [25]. While our initiation mapping considers offering a subjective proportion of system consideration, a progressively quantitative perception and analysis of system portrayals is required, particularly with applications in the clinical space. Also, a protect against neural systems' vulnerable sides is needed intending to our feeble comprehension of their helplessness to antagonistic assaults [26], and all the more explicitly the affectability of clinical images to certain announced illogical properties of CNNs [27]. At last, ongoing developments in the tomography genomics rouse additional investigations past our starter GSEA study [28]. Once thoroughly assessed in future imminent investigations, deep-learning-based prognostic signatures can feature the particular natural conditions of tumor formation showed by a provided patient and in this way empower more focused on treatment applications that abuse explicit organic characteristics.

The improvement of prognostic biological indicators for NSCLC patients is a functioning region of study, where tumor arranging data is increased with protein-based proof, hereditary, sub-atomic, and radiographic proof [29]. The absence of a really prognostic clinical highest quality level ruins the capacity to precise the standards those biological indicators and additionally burdens the requirement for imminent approval. Whereas TNM arranging is frequently used in the center as the essential methods for NSCLC forecast and medication determination, it is mostly planned as a discrete proportion of tumor degree and a clinical specialized device, notwithstanding being basic and static by structure. Then again, quantitative imaging highlights surmised through deep learning are persistent and high-level-dimension and might be utilized to expand the more elevated level, coarser delineation gave by TNM organizing. In the wake of considering the previously mentioned impediments, a prognostic imaging apparatus may permit the change to a better arrangement empowering the recognizable proof of fitting treatment anticipates the particular level of patient. Several potential applications for those change might be in over-seeing beginning period NSCLC patients, for whom surgical intervention speaks to a remedial backbone though having high repeat dangers [8]. Adjuvant chemotherapy is regularly controlled as a method for diminishing these dangers [30]. Though N and T phases are recognized to be related to a repeat in these patients [31], we realize that patients with comparative clinical attributes can show broad varieties in the occurrence of repeat [32] and survival [33]. A better characterization inside a similar stage may take into account recognizing low and high mortality chance patients. In like manner, generally safe patients might be saved the antagonistic physical and mental impacts just as related expenses of adjuvant chemotherapy, and, on the other hand, progressively severe post-treatment observation of those at high hazard might be arranged. Furthermore, an increasingly nitty–gritty definition might illuminate careful methodologies and procedures, engage high-chance patients with the decision of adjunctive therapy detailed rules governing that are best suited their ideal ways of life, and recognize long-haul recipients from such treatment [34]. Deep learning calculations that gain as a matter of fact offer access to exceptional conditions of knowledge that, now and again, coordinate human insight. Past tomography, multi-modal nature of deep learning guarantees the reconciliation of various equal floods of data traversing genomics [35], pathology, electronic well-being records, online networking, and numerous different modalities into ground-breaking incorporated indicative frameworks [36]. In spite of various detours including the requirement for normalized information assortment strategies, assessment rules, planned approval, and detailing conventions [37], the best foreseen clinical effect of these calculations will be inside exactness medication. This rising methodology takes into account early conclusion and tweaked quite explicit medicines, hence conveying the suitable clinical consideration to the correct patient at the perfect time [38]. Although clinical tomography has consistently given a person's appraisal of afflictions, artificial intelligence calculations dependent on tomography bio-signature's guarantee to precisely delineate patients and empower new research roads for customized medicinal services.

8   An Outline of Author

8.1   Reason for This Research Conducted

The following are the some of the reasons to conduct this study.

- Cancer is considered to be one of the main sources of mortality around the world, with lung cancer have being the second-most regularly analyzed cancer in the two people in the USA.
- Prognosis in patients with lung cancer is basically decided over tumor organizing, which thus depends upon a discrete separation and moderately coarse.

- Radiological clinical images provide patient-and tumor-explicit data that could be utilized to supplement scientific predictive assessment endeavors.
- Latest propels in radionics across uses of man-made reasoning, deep learning and PC vision take into consideration the mining and extracting of various quantifiable highlights from radiological images.

8.2   Findings and Work Carried Out by Author

The following are the some of the work performed by the author.

- We structured an examination arseriesment containing 7 free datasets around 5 organizations making 1183 patients with non-little cell lung cancer depicted with processed magnetic resonance imaging and undergoing treatment by either medical procedure or radiotherapy.
- We assessed the analytical signatures of measurable visualization highlights removed across deep learning systems and surveyed their capacity to separate patients into high and low death rate hazard bunches according to a 2-year in general in order to survive failure.
- In patients undergoing treatment with a medical procedure, deep learning systems essentially beat models dependent on predefined tumor includes just as tumor volume and greatest distance across.
- In expansion to featuring image areas with predictive impact, we assessed the deep learning highlights for heartiness against physical imagery ancient rarities and info changeability, just as associated them with atomic data through quality articulation information.

References

1. D.P. Kingma, J. Ba, *Adam: A Method for Stochastic Optimization* (2014). arXiv:1412.6980
2. N. Srivastava, G.E. Hinton, A. Krizhevsky, I. Sutskever, R. Salakhutdinov, Dropout: a simple way to prevent neural networks from overfitting. J. Mach. Learn. Res. **15**, 1929–1958 (2014)
3. A.Y. Ng, Feature selection, L1 vs. L2 regularization, and rotational invariance, in *Proceedings of the Twenty-First International Conference on Machine Learning* (ACM Digital Library, New York, 2004), p. 78
4. S. Ioffe, C. Szegedy, *Batch Normalization: Accelerating Deep Network Training by Reducing Internal Covariate Shift* (2015). arXiv:1502.03167
5. A.L. Maas, A.Y. Hannun, A.Y. Ng, Rectifier nonlinearities improve neural network acoustic models. 30th International Conference on Machine Learning; 2013 Jun 16–21; Atlanta, GA, US
6. L. Prechelt, Early stopping—but when?, in *Neural Networks: Tricks of the Trade*, ed. G.B. Orr, K.-R. Müller (Springer, Berlin, 1998), pp. 55–69
7. M. Abadi, A. Agarwal, P. Barham, E. Brevdo, Z. Chen, C. Citro, et al., *Tensor Flow: Large-Scale Machine Learning on Heterogeneous Distributed Systems* (2016). arXiv:1603.04467

8. H. Uramoto, F. Tanaka, Recurrence after surgery in patients with NSCLC. Trans. Lung Cancer Res. **3**, 242–249 (2014)
9. H.J.W.L. Aerts, E.R. Velazquez, R.T.H. Leijenaar, C. Parmar, P. Grossmann, S. Carvalho, et al., Decoding tumour phenotype by noninvasive imaging using a quantitative radiomics approach. Nat, Commun. **5**, 4006 (2014)
10. G. Litjens, T. Kooi, B.E. Bejnordi, A.A.A. Setio, F. Ciompi, M. Ghafoorian et al., A survey on deep learning in medical image analysis. Med. Image Anal. **42**, 60–88 (2017)
11. C. Parmar, P. Grossmann, J. Bussink, P. Lambin, H.J.W.L. Aerts, Machine learning methods for qantitative radiomic biomarkers. Sci Rep. **5**, 13087 (2015)
12. B. Zhang, X. He, F. Ouyang, D. Gu, Y. Dong, L. Zhang et al., Radiomic machine-learning classifirs for prognostic biomarkers of advanced nasopharyngeal carcinoma. Cancer Lett. **403**, 21–27 (2017)
13. H. Kim, C.M. Park, B. Keam, S.J. Park, M. Kim, T.M. Kim, et al., The prognostic value of CT radiomic features for patients with pulmonary adenocarcinoma treated with EGFR tyrosine kinase inhibitors. PLoS ONE **12**, e0187500 (2017)
14. J. Lao, Y. Chen, Z.-C. Li, Q. Li, J. Zhang, J. Liu et al., A deep learning-based radiomics model for prediction of survival in glioblastoma multiforme. Sci. Rep. **7**, 10353 (2017)
15. T.P. Coroller, V. Agrawal, V. Narayan, Y. Hou, P. Grossmann, S.W. Lee, et al., Radiomic phenotype features predict pathological response in non-small cell lung cancer. Radiother. Oncol. **119**, 480–486 (2016)
16. J. Zhang, K.A. Gold, H. Lin, S. Swisher, S.M. Lippman, J.J. Lee et al., Relationship between tumor size and survival in non-small cell lung cancer (NSCLC): an analysis of the surveillance, epidemiology, and end results (SEER) registry. J. Clin. Orthod. **30**, 7047 (2012)
17. K. Shien, S. Toyooka, J. Soh, H. Yamamoto, S. Miyoshi, Is tumor location an independent prognostic factor in locally advanced non-small cell lung cancer treated with trimodality therapy? J. Thorac. Dis. **9**, E489–E491 (2017)
18. M. Egeblad, E.S. Nakasone, Z. Werb, Tumors as organs: complex tissues that interface with the entire organism. Dev. Cell. **18**, 884–901 (2010)
19. S.A. Ahrendt, Y. Hu, M. Buta, M.P. McDermott, N. Benoit, S.C. Yang et al., p53 mutations and survival in stage I non-small-cell lung cancer: results of a prospective study. J. Natl. Cancer Inst. **95**, 961–970 (2003)
20. OECD iLibrary, *Health Equipment: Computed Tomography (CT) Scanners.* (OECD iLibrary, 2018) [cited 2018 Nov9]
21. A. Cistaro, N. Quartuccio, A. Mojtahedi, P. Fania, P.L. Filosso, A. Campenni et al., Prediction of 2 years-survival in patients with stage I and II non-small cell lung cancer utilizing 18F-FDG PET/CT SUV quantifica. Radiol. Oncol. **47**, 219–223 (2013)
22. R. Shwartz-Ziv, N. Tishby, *Opening the Black Box of Deep Neural Networks Via Information*
23. M. Ghafoorian, N. Karssemeijer, T. Heskes, M. Bergkamp, J. Wissink, J. Obels et al., Deep multi-scale location-aware 3D convolutional neural networks for automated detection of lacunes of presumed vascular origin. Neuroimage Clin. **14**, 391–399 (2017)
24. J. Long, E. Shelhamer, T. Darrell, Fully convolutional networks for semantic segmentation. 28th IEEE Conference on Computer Vision and Pattern Recognition; 2015 Jun 7–12; Boston, MA, US
25. Q.-S. Zhang, S.-C. Zhu, Visual interpretability for deep learning: a survey. Front. Inf. Technol. Electron Eng. **19**, 27–39 (2018)
26. X. Yuan, P. He, Q. Zhu, X. Li, *Adversarial Examples: Attacks and Defenses for Deep Learning* (2017). arXiv:1712.07107
27. S.G. Finlayson, H.W. Chung, I.S. Kohane, A.L. Beam, *Adversarial Attacks Against Medical Deep Learning Systems* (2018). arXiv:1804.05296
28. H.X. Bai, A.M. Lee, L. Yang, P. Zhang, C. Davatzikos, J.M. Maris, et al., Imaging genomics in cancer research: limitations and promises. Br. J. Radiol. **89**, 20151030 (2016)
29. M.K. Thakur, S.M. Gadgeel, Predictive and prognostic biomarkers in non-small cell lung cancer. Semin. Respir. Crit. Care Med. **37**, 760–770 (2016)

30. Non-Small Cell Lung Cancer Collaborative Group, Chemotherapy in non-small cell lung cancer: a meta analysis using updated data on individual patients from 52 randomised clinical trials. BMJ **311**, 899–909 (1995)
31. X. Wang, A. Janowczyk, Y. Zhou, R. Thawani, P. Fu, K. Schalper, et al., Prediction of recurrence in early stage non-small cell lung cancer using computer extracted nuclear features from digital H&E images. Sci. Rep. **7**, 13543 (2017)
32. J.M. Pepek, J.P. Chino, L.B. Marks, T.A. D'amico, D.S. Yoo, M.W. Onaitis, et al.: How well does the new lung cancer staging system predict for local/regional recurrence after surgery?: A comparison of the TNM 6 and 7 systems. J. Thorac. Oncol. **6**, 757–761 (2011)
33. C.-F. Wu, J.-Y. Fu, C.-J. Yeh, Y.-H. Liu, M.-J. Hsieh, Y.-C. Wu et al., Recurrence risk factors analysis for stage I non-small cell lung cancer. Medicine **94**, e1337 (2015)
34. R. Arriagada, A. Dunant, J.-P. Pignon, B. Bergman, M. Chabowski, D. Grunenwald et al., Long-term results of the international adjuvant lung cancer trial evaluating adjuvant Cisplatin-based chemotherapy in resected lung cancer. J. Clin. Oncol. **28**, 35–42 (2010)
35. J. Ngiam, A. Khosla, M. Kim, J. Nam, H. Lee, A.Y. Ng, Multimodal deep learning. 28th International Conference on Machine Learning; 2011 Jun 28–Jul 2; Bellevue, WA, US
36. C.F. Lundström, H.L. Gilmore, P.R. Ros, Integrated diagnostics: the computational revolution catalyzing cross-disciplinary practices in radiology, pathology, and genomics. Radiology **285**, 12–15 (2017)
37. P. Lambin, R.T.H. Leijenaar, T.M. Deist, J. Peerlings, E.E.C. de Jong, J. van Timmeren et al., Radiomics: the bridge between medical imaging and personalized medicine. Nat. Rev. Clin. Oncol. **14**, 749–762 (2017)
38. European Society of Radiology, Medical imaging in personalised medicine: a white paper of the research committee of the European Society of Radiology (ESR). Insights Imaging. **2**, 621–630 (2011)
39. A. Krizhevsky, I. Sutskever, G.E. Hinton, ImageNet classification with deep convolutional neural networks, in *Advances in Neural Information Processing Systems 25*. ed. by F. Pereira, C.J.C. Burges, L. Bottou, K.Q. Weinberger (Curran Associates, Red Hook, NY, 2012), pp. 1097–1105

Lung Cancer Detection Using Improvised Grad-Cam++ With 3D CNN Class Activation

Eali Stephen Neal Joshua⬥, Midhun Chakkravarthy⬥, and Debnath Bhattacharyya⬥

Abstract Lung Cancer Detection using improvised Grad-Cam++ with 3D CNN with class activation for classification of lung nodules and early detection of lung cancer. So, the question is how can deep learning methods we use to solve high-impact medical problems such as lung cancer detection. And more specifically how can we use 3D convolution neural networks in this specific application for detection of lung cancer. No matter how good your deep learning model is if it's not interpretable to people in the domain it's really hard for them to adopt. So, all models recently they've been known to have very good accuracies especially deep learning models specifically but if the domain expert can't trust the model then it doesn't mean. Gradient weighted class activation mapping or Grad Cam++ to visualize the models decision-making an increased radiologist trust and improve adoption in the field. We have achieved an overall accuracy of 94% on LUNA 16 dataset which was better compared with remaining architectures, as per the literature study we done.

Keywords Lung cancer · Alex Net · CNN · Gradient weighted class activation · 3D CNN · Grad Cam++ · Medical image analysis

1 Introduction

Lung cancer is a leading cause of cancer death among both men and women in the United States and this is around even more than a hundred thousand deaths a year (Stewart Editor, n.d.) [1] in the US alone. So, it's a pretty big scale problem and one of the reasons the five-year survival rate is also only 17%. when we look at why the survival rate is so low what we find is that early detection can really improve the

E. S. N. Joshua (✉) · M. Chakkravarthy
Department of Computer Science and Multimedia, Lincoln University College, 47301 Kuala Lumpur, Malaysia
e-mail: stephen_neal@lincoln.edu.my

D. Bhattacharyya
Department of Computer Science and Engineering, Koneru Lakshmaiah Education Foundation, Vaddeswaram, Guntur 522502, Andhra Pradesh, India

© The Author(s), under exclusive license to Springer Nature Singapore Pte Ltd. 2021
S. K. Saha et al. (eds.), *Smart Technologies in Data Science and Communication*,
Lecture Notes in Networks and Systems 210,
https://doi.org/10.1007/978-981-16-1773-7_5

chances of survival, however many times by the time the lung nodule or the cancer is detected it's already too late in the process for intervention. Good treatment to happen so the idea is that early detection of malignant lung nodules can significantly improve the chances of survival and prognosis.

The problem with this is that detection of lung nodules is quite time-consuming and difficult due to the volume of data involve. There can be end as well as in Terraria all just variants so what this means is that since the CT scan [2] data is so huge it has millions of voxels but long nodules pretty tiny compared to the size of that CT scan data. The problem is that one radiologist might see it and classify this as a nodule while another radiologist might not even see the same thing. There's some subjectivity and in two radiologists variants and then the nature of the problem is that it's kind of hard to find the nodule in the first place it's almost like trying to find a needle in a haystack. we need to do some more background first so computer tomography [2, 3] or CT is widely used for lung cancer screening and the goal is to see as early as possible can we detect that this disease is here an accurate detection is quite important to the diagnosis of lung cancer but a CT scan can have millions of voxels and out of these along nodule is quite small and hard to detect so this is quite the significant challenge for radiologists today and so what happened is that automated methods recently have shown better accuracies or even comparable accuracy to interpretation by radiologists and in addition to this they can reduce subjectivity and into radiologists variant.

2 Related Work

"Classification of pulmonary CT Images by Using Hybrid 3D-Deep Convolutional Neural Network Architecture" author has worked [4] on the LIDC dataset. The researchers have worked on the two-stage detection of Lung Cancer. First the 3D CNN with SoftMax were used for the classification and secondly 3D CNN with Radial basis function were considered, after the experimentation authors have achieved an accuracy of 90.89%. "Automated Lung Nodule Detection and classification Using Deep Learning Combined with Multiple Strategies", has worked on the LIDC-IDRI benchmark dataset researchers had worked on multi strategy approach by taking customized neural network and gradient boosting algorithm for the classification of Lung cancer benign and malignant nodule. Authors had achieved the accuracy of 91%." An Interpretable deep hierarchical semantic convolutional neural network for lung nodule malignancy classification", has implemented using the hierarchical semantic convolutional neural network (HSCNN) on LIDC/IDRI dataset. The network has classified images into two categories, first one is low level semantic images and second is high-level semantic images. Unified architecture was used for performance of the classifier. Authors had achieved the accuracy of 90.1% only the 3DCNN architecture.

"End-to-End Lung Cancer screening with three-dimensional deep learning on low-dose chest computed tomography" [5] has worked on the low standard images and used the 3D CNN architecture for the classification of the images, attained an overall accuracy of 91%. "Multi-Resolution [5] CNN and Knowledge Transfer for candidate classification in Lung nodule Detection" to classify the benign and malignant tumor the author has approached three way architecture. In the stage one to find out the edges, knowledge transfer, to improve the model multi resolution model had been chosen. Secondly side out branches were calculated. And last loss function with objective equations. This was conducted on LIDC/IDRI dataset. The authors have achieved an accuracy of 91.23%.

3 Convolutional Neural Network

A new architecture called a convolutional neural (Lecun-Bengio-95a, n.d.) [6] network so basically what a CNN is it consists of the input and output like any other deep learning method and it'll have layers called convolutional layers and Max pooling layers.

The general idea is that the input will go through a series of convolution [7] and max pooling layers. And then this will map to some fully connected layers at the end which map to an output. Kernel will convolve over the input so it'll start all the way at the left and it'll convolve [8] it'll go to the next layer next layer [9] and then basically what happens is that the dot product is taken between the kernel and the input and this outputs to some feature map like this these convolution neural layers are able to detect certain features [10] in the image (Fig. 1).

Max pooling layers which reduce dimensions [11] allow for better computational speed and reduce overfitting. CNN layer [12] has features of increasing complexity and the first layer learn edges corners things like that and then as you go further the intermediate layers will learn more complex parts of the object and finally the last layers will detect [13] full objects.

In the face example maybe like eyes nose things such as that such as faces and so the idea is that these CNN's will have convolutional layers Max poll layers and finally it'll output to some decision made layer and that's just a general idea of how our 3D CNN are essentially the same thing except your input your input will be 3D data like this instead of being a singular layer it'll be a volume [14] and then your

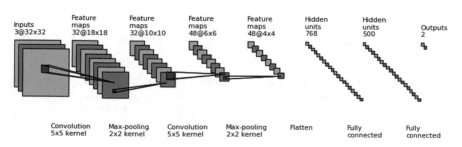

Fig. 1 Convolutional neural network

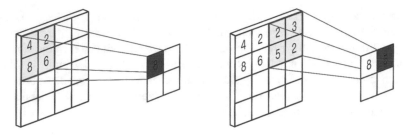

Fig. 2 Showing 2D max pooling

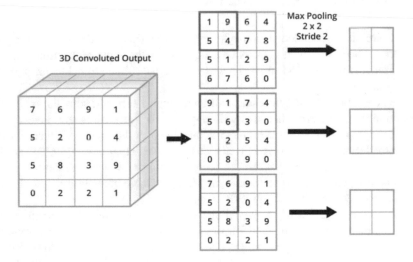

Fig. 3 Showing 3D max pooling

kernels this is the key part your kernels [15] or your filters will also be 3D so what this means is that instead of detecting 2D features such as edges and corners it'll detect the same features but in the 3D fashion which is really critical especially in this lung nodule case because all the data is of 3D nature (Figs. 2 and 3).

4 Proposed Model

The idea is that this 3D data is inputted it goes through some 3D kernels and then it's classified as either healthy or diseased. In this case one module exists in that CT image or long module does not exist so in addition to this like CNN's like other deep neural networks have been black boxes giving users no intuition as to how they're predicting.

4.1 Gradient Weight Class Activation

This study that demonstrates Grad Cam++ techniques [16] for visual no tion on one module [17] classification. The value added here with this study so the objective here is to research and develop 3D [18] CNN to detect lung nodules. CT scan data with better accuracy and higher trust in existing models and this is to ultimately aid in early detection of lung cancer to improve chances of survival and prognosis (Fig. 4).

We prove that a 3D CNN can do better than 2D CNN—detecting lung nodules and also is it possible to derive visual explanations for the internal workings using Grad Cam++ methods. So, we hypothesize that 3D CNN's which can exploit the full 3D nature of the data will have better accuracy. Grad Cam++ algorithm can provide visual explanations for decisions in lung nodule problem by highlighting discriminative [19] regions exactly. The same model type and also can we improve upon that using optimization [20]. This study that demonstrates Grad Cam++ techniques for visual nations on one module classification.

$$Y^a_{\text{Grad−Cam}} = \sum_p K^q_p \frac{1}{z} \sum_\iota 1 \sum_j J^p_{ij} \tag{1}$$

Let us define Jp to be average pooled global output

$$r^p = \frac{1}{z} \sum_i * \sum_j J^p_{ij} \tag{2}$$

Grad-Cam Computes the end scores by

$$p^a_{\text{Grad−Cam++}} = \sum_p^1 *wl^q_p * r^p \tag{3}$$

where wl is the weight connecting the Kth Feature selection. Taking the gradient class score (jp) w.r.t. to feature map we get.

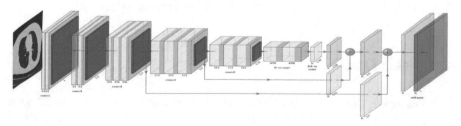

Fig. 4 Proposed model architecture

$$\frac{\delta y k_1^c}{\delta J k_1^p} = \frac{\frac{\delta o}{\delta A o_{ij}^p}}{\frac{\delta J o^p}{\delta A p_{ij}^p}} \tag{4}$$

Taking Partial Derivation (4) w.r.t. A^p, We can see that. $\frac{\delta J^p}{\delta A_{ij}^p}$.Substituting this on (4) we get.

$$e^{-i\omega t} \tag{5}$$

From (3), We get
Summing both sides of (5)

$$\sum_i \sum_j l_p^q = \sum_i \sum_j z \ast \frac{\delta y p^c}{\delta A k_{ij}^p} \tag{7}$$

Therefore

$$g_p^q = \sum_i \sum_j \frac{\delta y n^c}{\delta A g_{ij}^p} \tag{8}$$

Thus, we can say that Grad-Cam is strict generalization of the CAM. This generalization allows the researcher to generate the visual explanation for the domain experts from the 3D CNN based model that will make the process light weight and convolute the process. As, result we will get the exact nodule region in the computer tomography images. We perform a weighted combination of forward activation maps, and follow it by a ReLu to obtain

$$kl_{\text{Grad - Cam++}}^a = \left(\sum_k \alpha j_k^c A p^k \right)$$

So, this is really the value added here with this study so the objective here is to research and develop 3D CNN to detect lung nodules and CT scan data with better accuracy and higher trust in existing models and this is to ultimately aid in early detection of lung cancer to improve chances of survival and prognosis.

4.2 Study Design

The idea here is that to have a golden standard. Inter radiologist variants so the idea to approach this is that by having a panel of radiologists discuss each image and then label it you have a golden standard, for what is actually being seen in that image and then each radiologist's mark lesion does a nodule or non-nodule. The reference standard is that at least three out of four radiologists must identify it as a nodule

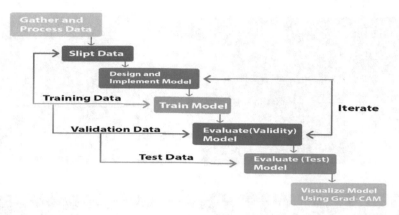

Fig. 5 Proposed study design

greater than eight equals to three millimeters for it to be labelled as a nodule in the data so this is how the data looks basically this was full CT scans that were then chopped up into where exactly the identified as nodular (Fig. 5).

5 Experimental Results and Discussions

5.1 Luna 16 Dataset

Luna 16 [21] dataset which has almost 900 thoracic CT scans. This data set is clean and balanced as already from the start so the scans, with slices greater than 3 millimeters were removed and one of the reasons for this is that the lung nodule is pretty small from the start. If you have slices greater than 30 millimeters you might miss the nodule together so that's some of the pre-processing that was already done and then images were annotated by for experience to radiologists (Fig. 6).

5.2 Split and Pre-process Data

Balanced dataset of 1000 nodule and 1000 non-nodule volumes used. Data divided into three sets for training, validation, and testing. Randomized and split into 1400 volumes for training and 600 for testing. 10% of the training data (140 volumes) used for validation. From the below tables we will summarizes the various architecture model. Table 1 explains about the AlexNet 2D. Table 2 explains about the AlexNet 3D model and last Table 3 summarizes the proposed architecture Network Model.

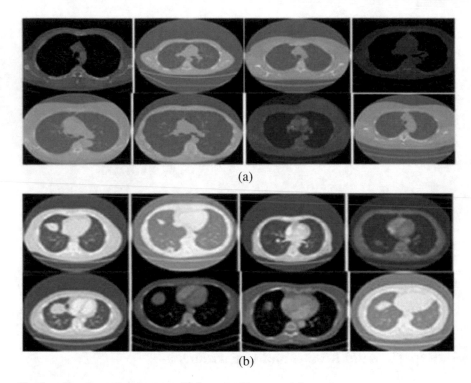

(a)

(b)

Fig. 6 **a** Showing nodule image and **b** Images with non-nodule

5.3 Training Process

Figure 7 explains the Training process of LUNA 16 dataset. Images X from the training dataset fed into the model. Output compared with the training label y, loss is computed, and the model is updated with new parameters. During the iterative training process, a SoftMax activation function was used on the estimates before loss was calculated. Cross entropy was used as the loss function to be optimized. All the models used Adam optimizer with a learning rate of 0.0001 and default parameters $\beta 1 = 0.9$ and $\beta 2 = 0.999$.

5.4 Model Evaluation

Model performance key metrics and then also the visual insights that were generated so this is the AEC so all the CNN's had pretty good AEC is close to one showing that they have a good supper ability and they're able to performed to the detection task well and Alex 3D CNN with an AEC of 0.95 performed better than the 2D CNN with 0.94 and 3D CNN performed the best with an AUC of 0.97 which shows that these

Table 1 AlexNet2D CNN model

Layer	Output shape	Param (#)
Conv 3D_1 (Conv 3D)	(None 24, 32, 32, 16)	102,010
Batch_normalization_v1	(None 24, 32, 32, 16)	64
Max_pooling 3D_1	(None 12, 16, 16, 16)	0
Conv3D_1(Conv 3D)	(None 12, 16, 16, 32)	131,056
Batch_normalization_v1_2	(None 12, 16, 16, 32)	1210
Max_pooling3D_2	(None 06, 10, 10, 312)	0
Conv3D_3	(None 06, 10, 10, 64)	55,360
Batch_normalization_v1_3	(None 06, 10, 10, 64)	256
Conv3D_4	(None 06, 10, 10, 64)	110,656
Batch_normalization_v1_4	(None 06, 10, 10, 64)	256
Conv3D_5 (Conv 2D)	(None 06, 10, 10, 32)	533,210
Batch_normalization_v1_5	(None 06, 10, 10, 32)	1210
Max_pooling2D_3	(None 4, 4, 4, 32)	0
Falatten_1 (Flatten Layer)	(none, 20, 410)	0
Dense_1(Dense layer)	(none, 38)	4,091,000
Batch_normalization_v1_6	(none, 38)	1000
Dropout_1 (Dropout)	(none, 38)	0
Dense_2 (Dense Layer)	(none, 38)	15,048
Batch_normalization_v1_7	(none, 38)	300
dropout_2 (Dropout)	(none, 38)	0
Dense_3 (Dense)	(none, 2)	152

optimizations that are done over iterations and validated are effective in increasing the model classification ability.

I really like to highlight is that this 0.94 recall value it might only be three percent greater than Alex Net 2D CNN but the thing here is that even a point zero one increase in recall is like saving another patients life out of 100 so this recall value is really critical especially if you're maintaining about the same precision because if you're able to increase this recall without sacrificing too much in precision you're basically finding patients that you wouldn't have otherwise found and by early detecting this long module you can approach this cancer differently and you can really make a difference in their life (Tables 4, 5; Figs. 8, 9).

Table 2 AlexNet3D CNN model

Layer	Output shape	Param (#)
Conv 3D_1 (Conv 3D)	(None 32, 32, 32, 16)	102,010
Batch_normalization_v1	(None 32, 32, 32, 16)	64
Max_pooling 3D_1	(None 16, 16, 16, 16)	0
Conv3D_1(Conv 3D)	(None 16, 16, 16, 32)	131,056
Batch_normalization_v1_2	(None 16, 16, 16, 32)	1210
Max_pooling3D_2	(None 10, 10, 10, 312)	0
Conv3D_3	(None 10, 10, 10, 64)	55,360
Batch_normalization_v1_3	(None 10, 10, 10, 64)	256
Conv3D_4	(None 10, 10, 10, 64)	110,656
Batch_normalization_v1_4	(None 10, 10, 10, 64)	256
Conv3D_5 (Conv 2D)	(None 10, 10, 10, 32)	533,210
Batch_normalization_v1_5	(None 10, 10, 10, 32)	1210
Max_pooling2D_3	(None 4, 4, 4, 32)	0
Falatten_1 (Flatten Layer)	(none, 20, 410)	0
Dense_1(Dense layer)	(none, 48)	4,091,000
Batch_normalization_v1_6	(none, 48)	1000
Dropout_1 (Dropout)	(none, 48)	0
Dense_2 (Dense Layer)	(none, 48)	15,048
Batch_normalization_v1_7	(none, 48)	300
dropout_2 (Dropout)	(none, 48)	0
Dense_3 (Dense)	(none, 2)	152

Table 3 Proposed3D CNN model

Layer	Output shape	Param (#)
Conv 3D_1 (Conv 3D)	(None 27, 27, 27, 16)	3472
Batch_normalization_v1	(None 27, 27, 27, 16)	64
Conv3D_1(Conv 3D)	(None 27, 27, 27, 16)	2064
Batch_normalization_v1_2	(None 27, 27, 27, 16)	64
Conv3D_3	(None 27, 27, 27, 16)	16,400
Max_pooling3D_2	(None 10, 10, 10, 312)	0
Batch_normalization_v1_3	(None 23, 23, 23, 16)	64
Max_pooling3D_1	(None 11, 11, 11, 16)	0
Conv3D_4	(None 10, 10, 10, 32)	41,210
Batch_normalization_v1_4	(None 10, 10, 10, 32)	1210
Conv3D_5 (Conv 2D)	(None 9, 9, 9, 32)	10,224
Batch_normalization_v1_5	(None 9, 9, 9, 32)	1210
Max_pooling3D_2	(None 4, 4, 4, 32)	0

(continued)

Table 3 (continued)

Layer	Output shape	Param (#)
Conv3D_6 (Conv 2D)	(none, 3, 3, 3, 64)	164,410
Dense_1(Dense layer)	(none, 48)	4,091,000
Batch_normalization_v1_6	(none, 3, 3, 3, 64)	256
Max_pooling3D_3	(none, 1, 1, 1, 64)	0
Flatten_1(Flatten Layer)	(none, 64)	0
dense_1(Dense Layer)	(none, 256)	16,640
Batch_normalization_v1_7	(none, 256)	1024
dropout_1 (Dropout)	(none, 256)	0
Dense_2 (Dense)	(none, 2)	514

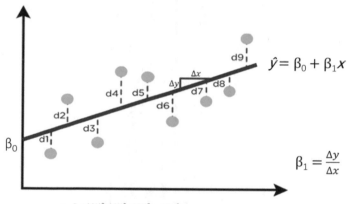

- $D = (d1)^2 + (d2)^2 + (d3)^2 + ... (d9)^2$
- The purple line ($\hat{y} = \beta_0 + \beta_1 x$) minimizes D (the sum of squares)
- β_0 is the y-intercept
- β_1 is the slope of the line ($\Delta y / \Delta x$)

Fig. 7 Training process of proposed architecture

Table 4 Accuracy of CNN with different models

Model	AUC	Accuracy	F-score	Precision	Recall
AlexNet2D-CNN	0.95	88.67	0.88	0.86	0.92
AlexNet3D-CNN	0.96	90.17	0.92	0.88	0.90
Proposed 3D-CNN	0.98	92.17	0.93	0.89	0.96

Table 5 Comparison between existing systems with proposed system on LUNA 16 dataset

LUNA-16 Dataset			Training details			Existing models		Proposed model
	Experimental results							
S. No.	Samples	Training	Testing	Authors	Results	2D AlexNet	3D AlexNet	
1	888	80	20	Zuo.et al. 2019 [3]	Accuracy 88.17%	Accuracy 89.45%	Accuracy 93.23%	Accuracy = 94.17%
2	1018	70	30	Makaju et al. 2018 [2]	Accuracy 89.67%	Accuracy 88.78%	Accuracy 89.13%	Accuracy = 94.17%

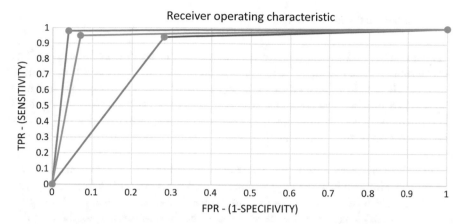

Fig. 8 Showing performance metrices of proposed model with other models

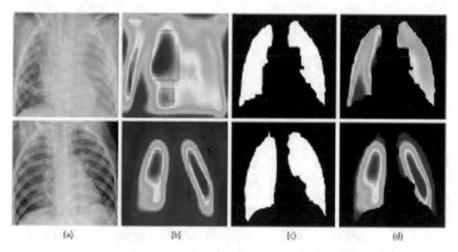

Fig. 9 Visualizing the trained CNN original images in between on the left grad-cam images on the right [22]

6   Conclusions and Future Work

Lightweight model which still had good accuracy in this specific problem as well as downsizing the data and then also if you have problems like in terms of training time one idea is to just decrease data size. One of the big problems about deep neural nets and deep learning approaches is that while they have high accuracies it's really hard for domain experts to accept what they're predicting because they give no intuition and this is one of the general themes that I've noticed like across all of machine learning.

Soft max can get an idea of how sure the model is about its decision and then we can output. The idea here is that it would produce a 3D volume and obviously you can't really see a 3D volume on a slide took one slice of it and displayed it next to a slice of the 3D image 3D volume so a lot of the data we're talking about in this research is 3D data with volume with the z-axis as well but to demonstrate it on a slide. Getting better accuracies in relation to other architectures working for dense Nets working better than ResNet. CT scan data that just not a nodule and figure out whether it's a nodule or not and that's how you approached a false-positive problem.

Acknowledgements This work was supported by Dr. Debnath Bhattacharyya, Professor, Department of Computer Science and Engineering, Koneru Lakshmaiah Education Foundation, K.L. University, Guntur 522502, Andhra-Pradesh, India.

References

1. P. Afshar, A. Mohammadi, K.N. Plataniotis, A. Oikonomou, H. Benali, From handcrafted to deep-learning-based cancer radiomics: Challenges and opportunities. IEEE Signal Process. Mag. **36**(4), 132–160 (2019). https://doi.org/10.1109/MSP.2019.2900993
2. S. Makaju, P.W.C. Prasad, A. Alsadoon, A.K. Singh, A. Elchouemi, Lung cancer detection using CT scan images. Procedia Comput. Sci. **125**(2009), 107–114 (2018). https://doi.org/10.1016/j.procs.2017.12.016
3. A. Chattopadhyay, A. Sarkar, P. Howlader, V.N. Balasubramanian, Grad-CAM++: Generalized gradient-based visual explanations for deep convolutional networks, in *Proceedings—2018 IEEE Winter Conference on Applications of Computer Vision, WACV 2018* (2018), pp. 839–847. https://doi.org/10.1109/WACV.2018.00097
4. H. Polat, H.D. Mehr, Classification of pulmonary CT images by using hybrid 3D-deep convolutional neural network architecture. Appl. Sci. (Switzerland) **9**(5) (2019). https://doi.org/10.3390/app9050940
5. P. Satti, N. Sharma, B. Garg, Min-max average pooling based filter for impulse noise removal. IEEE Signal Process. Lett. **27**, 1475–1479 (2020). https://doi.org/10.1109/LSP.2020.3016868
6. D. Ardila, A.P. Kiraly, S. Bharadwaj, B. Choi, J.J. Reicher, L. Peng, D. Tse, M. Etemadi, W. Ye, G. Corrado, D.P. Naidich, S. Shetty, End-to-end lung cancer screening with three-dimensional deep learning on low-dose chest computed tomography. Nat. Med. **25**(6), 954–961 (2019). https://doi.org/10.1038/s41591-019-0447-x
7. Ö. Günaydin, M. Günay, Ö. Şengel, Comparison of lung cancer detection algorithms. 2019 Scientific Meeting on Electrical-Electronics and Biomedical Engineering and Computer Science, EBBT 2019 (2019). https://doi.org/10.1109/EBBT.2019.8741826

8. T. Kadir, F. Gleeson, Lung cancer prediction using machine learning and advanced imaging techniques. Transl. Lung Cancer Res. **7**(3), 304–312 (2018). https://doi.org/10.21037/tlcr.2018. 05.15

9. M. Kroenke, K. Hirata, A. Gafita, S. Watanabe, S. Okamoto, K. Magota, T. Shiga, Y. Kuge, N. Tamaki, Voxel based comparison and texture analysis of 18 F-FDG and 18 F-FMISO PET of patients with head-and-neck cancer. PLoS ONE **14**(2) (2019). https://doi.org/10.1371/journal. pone.0213111

10. W.W. Labaki, T. Gu, S. Murray, C.R. Hatt, C.J. Galbán, B.D. Ross, C.H. Martinez, J.L. Curtis, E.A. Hoffman, E. Pompe, D.A. Lynch, E.A. Kazerooni, F.J. Martinez, M.L.K. Han, Voxel-wise longitudinal parametric response mapping analysis of chest computed tomography in smokers. Acad. Radiol. **26**(2), 217–223 (2019). https://doi.org/10.1016/j.acra.2018.05.024

11. lecun-bengio-95a. (n.d.).

12. D. Li, B.M. Vilmun, J.F. Carlsen, E. Albrecht-Beste, C.A. Lauridsen, M.B. Nielsen, K.L. Hansen, The performance of deep learning algorithms on automatic pulmonary nodule detection and classification tested on different datasets that are not derived from LIDC-IDRI: a systematic review, in *Diagnostics*, vol. 9, Issue 4. MDPI AG (2019). https://doi.org/10.3390/ diagnostics9040207

13. A. Masood, B. Sheng, P. Li, X. Hou, X. Wei, J. Qin, D. Feng, Computer-assisted decision support system in pulmonary cancer detection and stage classification on CT images. J. Biomed. Inform. **79**, 117–128 (2018). https://doi.org/10.1016/j.jbi.2018.01.005

14. P. Naresh, R. Shettar, et al., Early detection of lung cancer using neural network techniques. J. Eng. Res. Appl. **4**(4), 2248–962278 (2014). www.ijera.com

15. N. Nasrullah, J. Sang, M.S. Alam, M. Mateen, B. Cai, H. Hu, Automated lung nodule detection and classification using deep learning combined with multiple strategies. Sensors (Switzerland) **19**(17) (2019).https://doi.org/10.3390/s19173722

16. K. Roy, S.S. Chaudhury, M. Burman, A. Ganguly, C. Dutta, S. Banik, R. Banik, A comparative study of lung cancer detection using supervised neural network. 2019 International Conference on Opto-Electronics and Applied Optics, Optronix 2019, pp. 1–5 (2019). https://doi.org/10. 1109/OPTRONIX.2019.8862326

17. R.R. Selvaraju, M. Cogswell, A. Das, R. Vedantam, D. Parikh, D. Batra, Grad-CAM: visual explanations from deep networks via gradient-based localization. Int. J. Comput. Vision **128**(2), 336–359 (2020). https://doi.org/10.1007/s11263-019-01228-7

18. S. Shen, S.X. Han, D.R. Aberle, A.A. Bui, W. Hsu, An interpretable deep hierarchical semantic convolutional neural network for lung nodule malignancy classification. Expert Syst. Appl. **128**, 84–95 (2019). https://doi.org/10.1016/j.eswa.2019.01.048

19. D.J. Stewart (ed.), Lung *Cancer Prevention, Management, and Emerging Therapies* (n.d.). www.springer.com/series/7631

20. P. Tripathi, S. Tyagi, M. Nath, A comparative analysis of segmentation techniques for lung cancer detection. Pattern Recogn. Image Anal. **29**(1), 167–173 (2019). https://doi.org/10.1134/ S105466181901019X

21. X. Zhao, S. Qi, B. Zhang, H. Ma, W. Qian, Y. Yao, J. Sun, Deep CNN models for pulmonary nodule classification: model modification, model integration, and transfer learning. J. X-Ray Sci. Technol. **27**(4), 615–629 (2019). https://doi.org/10.3233/XST-180490

22. W. Zuo, F. Zhou, Z. Li, L. Wang, Multi-resolution cnn and knowledge transfer for candidate classification in lung nodule detection. IEEE Access **7**, 32510–32521 (2019). https://doi.org/ 10.1109/ACCESS.2019.2903587

An Automatic Perception of Blood Sucker on Thin Blood Splotch Using Graphical Modeling Methods

D. Sushma, K. V. Satyanarayana, N. Thirupathi Rao,
Debnath Bhattacharyya, and Tai-hoon Kim

Abstract Parasite is a host bacterium known as a plasmodium which lives on a different organism. This parasite is susceptible to malaria, dengue, typhoid, etc. The involvement of parasite in the blood cells will also lead to death of the humans. It is also very important to identify and diagnose the parasite in early blood film images in order to save human life. Therefore, the key slogan of this paper is to identify in less time the parasite in red blood cells using a new image processing technique by blood film images in early phases. *Aim*: In this article, the primary focus is on identifying the blood sucker which occur in red blood cells using thin blood film images in less time using a modern image processing system, in early stage. *Method*: In several steps, the procedure used detects the presence of blood sucker on photographs of blood films. The first step is to obtain the image from an optical microscope laboratory. Using the standard method, the image is then transferred into the grayscale image. The output image which is a grayscale image is transformed into the single-color image i.e., monochrome image which contains the pixel values using the "Binary Threshold method". This monochrome image is then transformed into a matrix format and printed with binary values i.e., zero's and one's. *Conclusion*: The output matrix method will be displayed with the binary values by either one or zero which represents the presence or absence of blood sucker. If all zeroes are displayed in whole image, then no blood sucker presence can be reached

D. Sushma · N. Thirupathi Rao (✉)
Department of Computer Science and Engineering, Vignan's Institute of Information Technology (A), Visakhapatnam 530049, AP, India

K. V. Satyanarayana
Department of Computer Science and Engineering, Raghu Engineering College, Visakhapatnam, AP, India

D. Bhattacharyya
Department of Computer Science and Engineering, K L Deemed to be University, KLEF, Guntur 522502, India

T. Kim
School of Economics and Management,
Beijing Jiaotong University, Chungwon-daero, Beijing, China

© The Author(s), under exclusive license to Springer Nature Singapore Pte Ltd. 2021   71
S. K. Saha et al. (eds.), *Smart Technologies in Data Science and Communication*,
Lecture Notes in Networks and Systems 210,
https://doi.org/10.1007/978-981-16-1773-7_6

in that case, and if any ones are displayed in the blood film images, it may be found that the blood film images contain a blood sucker.

Keywords Blood sucker · Blood film images · Image processing

1 Introduction

Image processing is a way to perform image processing to generate an enlarged image or to collect user information. It is a type of signal processing where images are entered and images or characteristics can be output [1, 2]. Today, photo processing is a technique which is growing quickly. It is a core area of engineering and information technology studies. Two methods, analogue and digital imagery, are used for the processing of photographs. For hard copies such as prints and photographs, analogue image processing can be employed. Now let's have a look at the image processing. The technology is image processing if we are going to detect some object in the image or to perform any type of image operation. The technique of image processing is commonly used in all areas. For instance, this technique is used to identify and detect pathogens from the scanned images.

The key phases involved in image processing are:

1. **Image Procurement**: The first step in image procurement is the acquisition of the photo. This was based on a computer-generated software copy from the electron microscope scan [3].
2. **Filtration**: Filtration is a technique used to alter or improve an image, such as highlighting or eliminating certain characteristics. Filtering the picture involves smoothing, sharpening and strengthening of the edge.
3. **Extraction of function**: Extraction of function is used for extracting image characteristics to recognize significant picture objects. It uses algorithms to identify and separate different desired sections of a digitalized image or video stream.
4. **Segmentation**: Segmenting is a picture separating mechanism into distinct artifacts or sections. Segmenting is typically one of optical image processing's most dynamic activities.

1.1 Blood Sucker

Here blood sucker is also known as parasite. An herbs or group of animals that lives in or on another person, feeding on the parasite organism. The parasite uses its nuclei to sustain its life cycle. It uses the host's control to hold itself. Protozoan, helminth or arthropods caused infections contain parasite conditions. Without a host, a parasite cannot survive, evolve and grow [4–6]. That's why the host never kills, but it is rarely possible to spread any pathogen, and some can be fatal. There are numerous parasites.

Protozoa are of small, single-cell, free-living or parasite type. They can replicate in humans, aid to thrive and also cause serious infections from just one organism to evolve. The protozoa which live in the intestines of humans are usually transmitted via oral fecal route to another individual (e.g., polluted food or water or personal touch). Protozoa in the human blood or tissue travel to other organisms (e.g. a mosquito bite or a sand fly) by arthropod vectors. The helminths are large, multicellular species and are usually noticeable in the naked eye during their adult periods. Helminths, such as protozoans, may be life free or parasite in the organism. In human beings the helminths in their adult state cannot replicate [7, 8].

We reside inside the host. The hotbed, tapeworm and flat worms are included. In the gaps within the host's cells there exists an intercellular worm. Bacteria and viruses are used. Endoparasites are based on a single, vector or carrier organism. The path forwards the endoparasite to the recipient. The mosquito is a host of various malaria-causing organisms, including the protozoan Plasmodium. It is very important to detect all human diseases for the early redemption of human life in biomedical sciences in seconds. In biomedical applications, the analysis of images is important to classify diseases [9–11]. The processing of images may be classified as such camera operations in order to analyze and identify the details using methods used in image therapy. Therefore, image recognition equipment for the retrieval of details in the image is used to take thin images into account.

2 Literature Survey

In the past, numerous algorithms have been developed to automate the detection of parasites by image processing techniques such as threshold segmentation, boundary recognition, cluster processes and Water Shed segmentation, etc.

Yashasvi Puwar, Sirish L. Shah, Gwen Clarke, Areej Almugairi, and Atis Muehlenbachs are the authors proposed a system "Automated and unsupervised detection of malarial parasites in microscopic images". The image-based approach is tested for more than 500 images from 2 separate laboratories. The goal is to distinguish the usage of thin smear photos between positive and negative cases of malaria. Since the procedure is unmonitored, the whole diagnosis cycle requires minimal human involvement. Total malaria tolerance is 100% and the sensitivity ranges from 50 to 88% among all malaria parasite species. To facilitate the diagnostics process, a method based on the digital processing of the Giemsa thin smear image has been developed. The procedure for diagnosis is divided into two parts; listing and identification. In order to automate the enumeration and identification process, the image-based method is presented here; the main advantage being his ability to diagnose without supervision but with a high degree of sensitivity, thereby reducing cases of false negative results.

Amjad Rehman, Naveed abbas, Tanzila Saba, Zahid Mehmood, Toqeer Mahmood, Khawaja Tehseen Ahmed are the authors proposed "Microscopic Malaria Parasitaemia Diagnosis and Grading on Benchmark Datasets". Malaria parasitemia are easily recognized and identified, although far from ideally successful. Falsely reported and abused infants globally have been murdered at tremendous rates. Latest work investigates the definition and degree of artificial malaria parasitaemia in visual pictures of thin blood streams by picture processing and computational vision techniques. Indeed, existing approaches to the computer-based vision-based diagnostic microscopic malaria parasitaemia issue depend partly or morphologically on the new discovery.

Computer Aided System for Red Blood Cell Classification in Blood Smear Image [3] which is proposed by the authors namely Razali Tomari, Wan Nurshazwani Wan Zakaria, Muhammad Mahadi Abdul Jamil, FaridahMohd Nor, Nik Farhan Nik Fuad. Otsu levels, for instance, are used to remove the history of erythrocyte images and filters to minimize noise and unwanted gaps and also to determine whether red blood cell images are regular or irregular. In-vitro identification and counting of red blood cells (RBC), before proper treatment can be proposed, is very important to diagnose blood-related diseases, like malaria and anemia. Pathologists under light microscope perform conventional practice for such a procedure manually. However, manual visual inspection, which leads to variation in RBC identification and counting, is laborious and depends on subjective assessments. In this paper, the RBC process of detection and identification of blood stream images is suggested to be automated with computerized systems. RBCs region is initially extracted from the background by the global method of threshold applied to color images of the green channel.

"Automatic System for Classification of Erythrocytes Infected with Malaria and Identification of Parasites Life Stage" by the authors S. S. Savkarea*, S. P. Naroteb. This paper uses the darkened photos for diagnosis, the RBC segmentation methods, and the overlapping cells, the watershed algorithm and the methods or system for classification of erythrocytes that were contaminated and for the identification of life stages of a parasite.

2.1 Observations from Previous Work

After the discovery of many authors works, the current model was not proposed by any other authors to detect the presence of blood sucker on thin blood film images. There was no matrix model used by previous authors or writers to merge monochrome photos using "Binary Threshold Method" for the identification of the blood sucker presence in thin blood film photos in the RBC's from the different numerous papers mentioned above. Therefore, efforts have been made to consider these observations. In order to determine whether the parasites are present in the blood splotch red blood cells, a new monochrome method was used to treat the image matrix with a new algorithm and findings were addressed in detail in the results section.

3 Proposed Work

In current research, the latest algorithm identifying the blood sucker in red blood cells using thin blood film images. By considering the blood film images the following steps are performed for recognizing the parasite.

3.1 Methodology

The complete process of the new system for assessing the existence of parasites on images of the blood stream can be described in detail.

1. First of all, the thin blood film photos are taken as input to identify the parasite in the red blood cells.
2. By using the standard average method the input photos are converted into greyscale photos in the second step of preprocessing.
3. Then the grey scale photos are transformed into monochrome photo using the "Binary threshold method" which is a segmentation method. Here monochrome image contains two values that are black and white. Black represents the background image which indicates the pixel values as 0, and white represents the foreground image which indicates the pixel value as 1.
4. For each pixel in the monochrome image, the pixel magnitude is computed and the pixels are shown as an output in a matrix-format. 1's and 0's are used in the matrix.
5. Binary values representing zero's and one's in the matrix represents the presence or absence of the blood sucker. Here zero's shows the absence of blood sucker and one's shows the presence of blood sucker.

3.2 Flow Chart

The diagram helps one to understand better the method of identification of parasites, particularly for those who are not aware of the process of image processing (Fig. 1).

4 Results

Two situations were considered in order to better explain the problem and to verify the performance of the existing model considered. In any case, a blood test sample was obtained from the labs and the image was analyzed and attempted to detect the presence of a parasite on the image or not. The accuracy of the detection of the presence of parasites on the blood film images will ensure that the model produced

Fig. 1 Flow chart for
recognizing the parasite

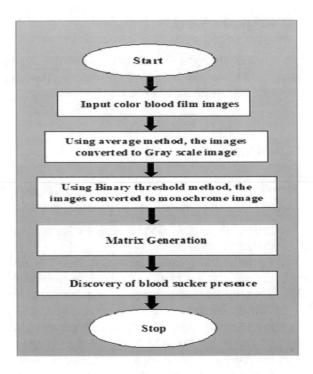

is either accurate or incorrect. In order to verify the same, two photographs were considered, and the experiment was carried out in two situations. The first case concerns the image with the existence of parasites and the second case concerns the image without the presence of the images. The two images were transferred to the original model as input and the feedback from the current model was observed. A matrix type is given for the current model performance. The model is structured to include a matrix of binary values of one or null for the output of the whole model. The whole image is placed or printed as a binary meaning matrix. The position of the parasite is showed in the output matrix, and the spot where no picture or signs of any parasite have been detected is shown as zero formats in the matrix. The output of the whole model will then be in the form of a matrix and, if the parasite exists, the output will be printed as 1's in the matrix model and no parasite trace, then all zeros will be printed.

Test Case 1: Recognizing the blood sucker from an image.

The color thin blood film image is taken as the input image for testing the proposed method in Fig. 2.

Here the input blood film image is converted into grayscale image which is in Fig. 3 using average method.

Here the Grayscale image is taken as input and converted to monochrome image which is in Fig. 4 using "Binary Threshold method" is the next step in the current model.

Fig. 2 Input color image for testing

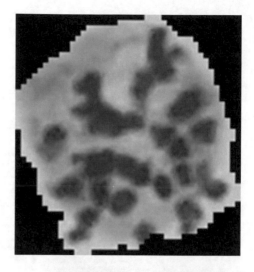

Fig. 3 Converted to gray scale image

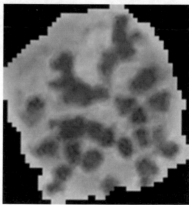

Fig. 4 Converted to monochrome image

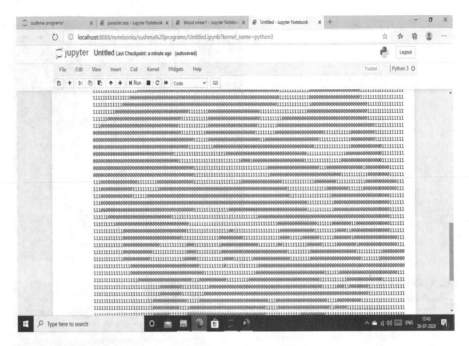

Fig. 5 Matrix generated

The matrix output image was produced from the monochrome image which is taken as input can be seen in Fig. 5. From the above matrix, the zeros indicate the black pixels i.e., blood sucker not presents in the red blood cell images, and one's indicates the white pixels which seems to presence of parasite. So, it shows that the input photo taken for testing gives an accurate result. Figure 5 says that the proposed model is working well for the images with parasites on thin blood film images.

Test Case 2: The second case was the perception of a blood sucker on a photo without a blood sucker, the intended result of this model for this input image is that all values in the matrix should be written with all zeros as the output matrix.

The color thin blood film image is taken as the input image for testing the proposed method in Fig. 6.

Here the input blood film image is converted into grayscale image which is in Fig. 7 using average method.

Here the Grayscale image is taken as input and converted to monochrome image which is in Fig. 8 using "Binary Threshold method" is the next step in the current model.

The matrix output image was produced from the monochrome image which is taken as input can be seen in Fig. 9. By observing the above matrix picture, the image contains only zeros so it indicates the absence of the blood sucker that is no parasite present in the thin blood smear image.

Fig. 6 Input color image for testing

Fig. 7 Converted to gray scale image

Fig. 8 Converted to monochrome image

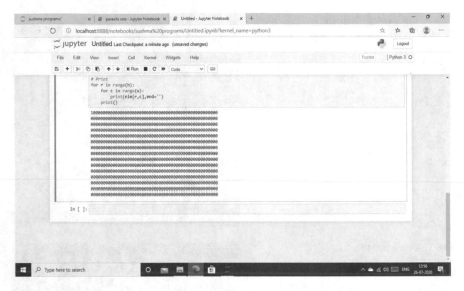

Fig. 9 Matrix generated

5 Conclusions

With image processing methods, the proposed method is in a position to detect the involvement of the parasite in RBC's images in a shorter period. The blood sucker that exists in the blood cells affects the form of the blood cells. Considering this aspect, the matrix produced by the monochrome image that contains zeros when the blood sucker is not present in the image and ones represents the presence of blood sucker may lead to some diseases. Thus, the automated identification of the parasite using the "matrix method" results in low time compared to the other segmentation processes.

References

1. Y. Puwar, S.L. Shah, G. Clarke, A. Almugairi, A. Muehlenbachs, Automated and unsupervised detection of malarial parasites in microscopic images. Malar. J. **10**, 364 (2011)
2. A. Rehman, N. Abbas, T. Saba, Z. Mehmood, T. Mahmood, K. Tehseen Ahmed, Microscopic malaria parasitaemia diagnosis and grading on benchmark datasets. Microsc. Res. Technol. **81**(9), 1042–1058 (2018)
3. R. Tomari, W.N. Wan Zakaria, M.M. Abdul Jamil, F. Nor, N.F. Nik Fuad, Computer aided system for red blood cell classification in blood smear image. International Conference on Robot Pride 2013–2014—Medical and Rehabilitation Robotics and Instrumentation, USA, Elsevier (2013–2014)
4. S.S. Savkarea, S.P. Naroteb, Automatic system for classification of erythrocytes infected with malaria and identification of parasites life stage. 2nd International Conference on Communication, Computing & Security [ICCCS]. Elsevier (2012)

5. V. Naresh Mandhala, D. Bhattacharyya, D. Sushma, Identification of parasite presence on thin blood splotch images. Int. J. Current Res. Rev. Res. **12**(19), 1–8 (2020)
6. A. Bashir, Z.A. Mustafa, I. Abdelhamid, R. Ibrahim, Detection of malaria parasites using digital image processing. International Conference on Communication, Control, Computing and Electronics Engineering (ICCCCEE), IEEE (2017), pp. 1–5
7. G. Lavanya, N. Thirupathi Rao, D. Bhattacharyya, *Automatic Identification of Colloid Cyst in Brain Through MRI/CT Scan Images*. (LNNS Springer, 2020), vol. 105, pp. 45–52
8. R. Anitha, S. Jyothi, V.N. Mandhala, D. Bhattacharyya, T. Kim, Deep learning image processing technique for early detection of Alzheimer's disease. Int. J. Adv. Sci. Technol. **107**, 85–104 (2017)
9. G.R.K. Prasad, N. Siddaiah, P.S. Srinivas Babu, Design and model analysis of circular cantilever sensor for early detection of Parkinson's disease. J. Adv. Res. Dyn. Control Syst. **9**(SP-16), 433–44 (2016)
10. K. Pratuisha, D. Rajeswara Rao, J. Amudhavel, J.V.R. Murthy, A comprehensive study: on artificial-neural network techniques for estimation of coronary-artery disease. J. Adv. Res. Dyn. Control Syst. **9**(SP-12), 1673–1683 (2017)
11. K. Asish Vardhan, N. Thirupathi Rao, S. Nagamallik Raj, G. Sudeepthi, Divya, D. Bhattacharyya, T.-h. Kim, Health advisory system using IoT technology. Int. J. Recent Technol. Eng. **7**(6), 183–187 (2019)

Dual Detection Procedure to Secure Flying Ad Hoc Networks: A Trust-Based Framework

Shikha Gupta⑩**, Neetu Sharma, Rakesh Rathi, and Deepak Gupta**

Abstract FANET is the class of ad hoc networks in which the nodes are named as unmanned aerial vehicles or UAVs or drones. These aerial vehicles communicate with each other using equipped devices. Securing the information from attacks during communication with each other in FANET still needs to be explored in depth because intruders are now the movable nodes present in network and to identify and separate them from legitimate nodes is quite the difficult task. These ad hoc networks must be secured from attacks and threats of bad or malicious nodes, to provide accurate services and information within time. Malicious nodes can break integrity, availability and confidentiality by stealing and modifying information or dropping packets of network. In this article, we discussed about security issues and types of attacks in such ad hoc networks. We are then proposing dual detection procedure to identify the presence and spotting of malicious nodes in FANET. This procedure includes both packet drop and message content-based approach to find the false node in the network. For both the detection procedures, an applied trust model dynamically computes the value of trust for every node involved in communication to evaluate the trusted and non-trusted list. Performance of methodology is found efficient during scalability and dynamicity of network, and result of the same is highlighted in this article with implementation of proposed algorithm.

Keywords Attacks · Threats · Packets · Trust management · Omnetpp · Security · UAVs

S. Gupta (✉) · N. Sharma · R. Rathi · D. Gupta
Department of Computer Science and Engineering, Engineering College Ajmer, Ajmer, India
e-mail: shikhagupta@ecajmer.ac.in

N. Sharma
e-mail: drneetu@ecajmer.ac.in

R. Rathi
e-mail: rakeshrathi@ecajmer.ac.in

D. Gupta
e-mail: deepakgupta@ecajmer.ac.in

© The Author(s), under exclusive license to Springer Nature Singapore Pte Ltd. 2021
S. K. Saha et al. (eds.), *Smart Technologies in Data Science and Communication*,
Lecture Notes in Networks and Systems 210,
https://doi.org/10.1007/978-981-16-1773-7_7

1 Introduction

FANET is the class of ad hoc network in which the nodes are named as unmanned aerial vehicles or UAVs or drones. To complete the defined task, these aerial vehicles are able to communicate with each other. For this, they must be equipped with networking, communication, processing and other related components. UAVs communicate with each other without any infrastructure support or access point. It is required that at least one of FANET node in network is connected to ground controller station (GCS). FANET are the extension of MANET and VANET. UAVs are equipped with communication devices so they can connect to other UAV through wireless communication channel [1]. With the limitation in size of these UAVs, they also restrict the equipped device's capabilities. UAV comes in small and large category as per their size, weight and computing capabilities. Small UAVs have limited resources like storage, processing capabilities and connectivity so they cannot be useful for complex and large missions. In such kinds of scenarios, the large UAVs are used. Therefore, application areas of small UAVs are different from application areas of large UAVs. UAVs are the most important systems used by military since more than 25 years. Other than military, there are lots of areas where FANET communication or UAVs are helpful like search and rescue operations, city monitoring, traffic monitoring and control, wildlife monitoring, automatic sensing in agriculture, goods delivery, weather monitoring, disaster monitoring, etc. [2]. Figure 1 shows a typical FANET scenario.

2 FANET Security

FANETs need to be secured from attacks and threats to provide accurate services and information within time. As FANETs are the sub-classes of MANET and VANET, therefore, attackers are very much aware to the features of FANET and can easily target them. The internal or external attacks can break the confidentiality, integrity and availability of UAVs in flying ad hoc networks. There are various types of attacks in FANETs. Some common attacks according to source and target [3–8] are listed in Fig. 2.

2.1 Attacks According to Source

Source can be internal or external for attacks. Existing attacks according to source type can be classified into internal or external attacks. Figure below shows some of the regular attacks according to source.

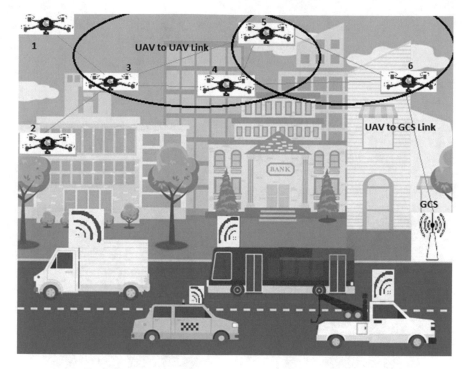

Fig. 1 Flying ad hoc network scenario

2.2 Attacks According to Target

Integrity, availability and confidentiality are basic targets of attacks. These targets are irrespective of internal or external source. Attacker nodes try to identify the loopholes to crack the mentioned targets. According to above targets, attacks can be classified as shown in Fig. 3.

3 Related Work

Chriki et al. [2] mentioned about the increase in importance of FANET in last few years in many fields and applications. Regular connectivity for communication between UAVs and GCS is still challenging. Security is another aspect still open for researches as attackers are very much familiar to FANET because it is sub-class of MANET. Zhang et al. [9] focus the security in UAV system at physical layer, and aim was to send the information successfully to actual receiver in the presence of malicious intermediate. They formulate the approach in non-convex problem and solved by applying iterative block coordinate descent and successive convex optimization methods. Kong et al. [10] worked on UAV-MBN that consists of three categories

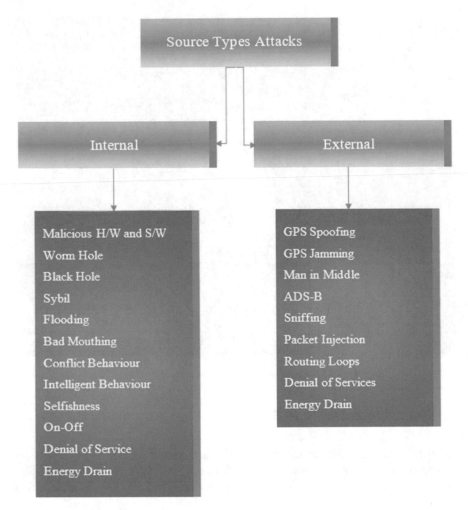

Fig. 2 Attacks categorized according to source type

of nodes. These are (i) regular ground mobile nodes (ii) ground mobile backbones nodes (MBN) and (iii) unmanned aerial vehicles nodes (UAV). Presence of UAV means infrastructure mode, otherwise infrastructureless mode. Switching of modes was performed through intrusion detection system (IDS) mechanisms. Authentication security was based on issuance of certificate as per public key infrastructure (PKI). For crypto public key, authors used de facto standard RSA. Condomines et al. [11] assigned properties to nodes using statistical signature which was based on the concept of wavelets. IDS model was implemented to identify attacks. Each statistical signature provided a unique identification for nodes and traffic. Every signature can be checked or cross-verified from the local database of all signatures of current traffic. Choudhary et al. [12] emphasized security in Internet of drones (IOD) and focused

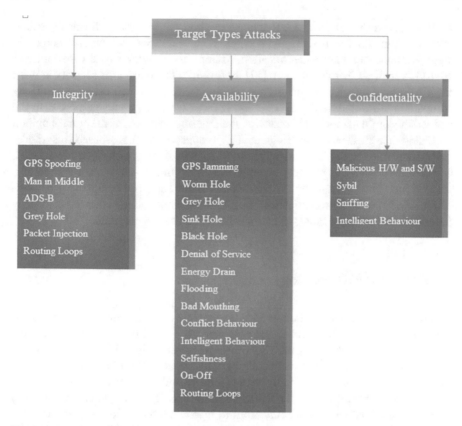

Fig. 3 Attacks categorized according to target

on various categories of vulnerabilities. They categorize the attacks classification according to threats and vulnerabilities. Altawy et al. [13] identified threats related to the system of civilian drones in the national airspace and worked on attacks by malicious node that can cause even crashing or some other security breaks for stealing, modification the information, etc. Also, deployment of IDS can monitor the communication and can perform anomaly detection by identifying abnormal behavior in contrast to predefined protected system behavior. Sharma et al. [14] enhanced the security in FANET by applying a routing mechanism based on trust. Trust value for every node was computed (between 0 and 1) to identify malicious node. Default trust value 0.5 is assigned to each node. On successful forwarding, trust value is increased by 0.001, but when packet is dropped by any node, the corresponding trust value is decreased by 0.001. At particular time when the trust value of particular node reached to 0, it had been assumed that it is a malicious node. Mitchell et al. [15] used intrusion detection techniques for data collection and data analysis. Runtime knowledge-based intrusion detection scheme was used for less false positive rate. Susceptibility to false positive, training and profiling phases are some of disadvantage of knowledge-based intrusion detection scheme. Birnbaum et al. [16] designed monitoring system for

UAV to record the related data. Recursive method applied for performance calculations is named as recursive least squares method (RLSM) which also helped in attacks or hardware failure identification. Attacks may results in hardware failure or control over flight. Birnbaum et al. [17] used profile behavioral for security to warn the UAV about attacks, failure, sensor spoofing, etc. Estimated behavior regularly compared with initial model of behavior. If there were any changes more than the defined threshold in monitored behavior, then the approach warns the control unit on ground to take further action. Singh et al. [18] approached fuzzy-based trust model to calculate uncertainty in behavior of FANET nodes. Nodes were member of good, bad or neutral cluster as per the value of membership functions. Based on properties, performance and environment FANET nodes are categorized using fuzzy classification. Nodes were differentiated from malicious nodes using trust value. Decay function was applied to nodes to evaluate performance degradation.

4 Proposed Methodology

Proposed work identifies malicious node responsible for either dropping of packet or content modification. This combined approach applying trust management among nodes to declare malicious one. Route discovery phase identifies possible routes from source to destination by transmitting RREQ packet to all its neighbors. It is assumed that any malicious node present in network supports in route discovery by forwarding both RREQ and RREP. Destination assigns a unique path_ID (PID) to each discovered path from source to it in its route table including intermediates nodes in path. Destination acknowledges the source to broadcast the message (M) by sending RREP. Source broadcasts message. Destination assigns a unique message_ID (MID) to each message received from different PID. Mapping of MID and PID is marked in mapping table. Dual detection procedure (DDP) applied, i.e., first packet drop-based detection followed by content-based detection to detect malicious node present in network if any.

4.1 Packet Drop-Based Detection (PDD)

In PDD, number of message received (M_r) at destination from different path compared to number of existing PIDs (k) in route table. If any of M_r is not mapped, then corresponding nodes from respective PIDs are placed in the list of untrusted nodes. Nodes common to both trusted and untrusted list will be considered only in trusted list, assuming their valid trust from some another path.

$$N_{valid} \leftarrow N_{trusted} \cap N_{untrusted} \tag{1}$$

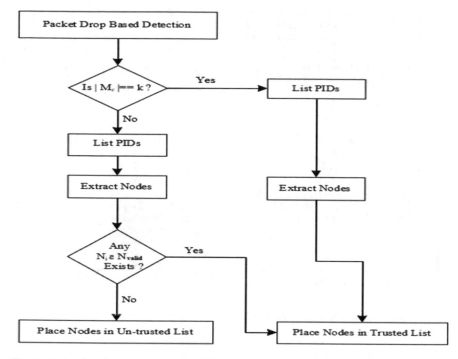

Fig. 4 Packet drop-based detection of malicious nodes

$$N_{\text{untrusted}} \leftarrow N_{\text{untrusted}} - N_{\text{valid}} \tag{2}$$

Figure 4 shows the steps to perform PDD.

4.2 Content-Based Detection (CD)

After PDD, the content of every received message at destination is compared with each other. Unmatched message PIDs listed from mapping table and corresponding nodes are placed in untrusted node list. KMP pattern match algorithm has been used for content matching. Match_M [M_{m}] is vector messages having similar contents with each other.

$$M_{\text{m}} = \text{KMP}_{\text{Match}} \{M_1, M_2, \ldots, M_n\} \quad \text{for } m \leq r \tag{3}$$

$$[N_{\text{untrusted}}] \leftarrow \text{Nodes}\{[M_r] - [M_{\text{m}}]\} \tag{4}$$

Figure 5 shows the steps for detection of paths where malicious node may exist and responsible for content modification.

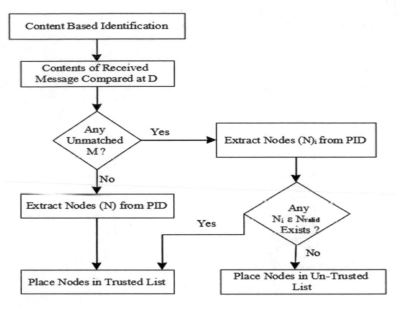

Fig. 5 Content-based detection of malicious nodes

4.3 Trust Model (TM)

Each node has been assigned an initial trust value in the network which can increase
or decrease based on packet forwarding and message modification. For quick identification of malicious node, trust value is decreased by 0.05 while increased by 0.025.
Counter updating is set up for every node in the list of untrusted nodes. When the
trust value reached maximum for node, then it transferred to trusted list, or when it
reached to define threshold, then node processed to declare as malicious.

Algorithm: Trust computation for each node N belongs to PID

```
Initialize Nthreshold to 0
For every transmission (T)
If ( N∈ Nuntrusted)
Ntrust= [Ntrust - 0.05];
else
Ntrust= [Ntrust + 0.025];
WhenNtrust (N) <=Nthreshold
Declare N ; as Malicious
WhenNtrust (N) >= 1
Trusted_List ← N;
```

Table 1 Notation table

Notation	Meaning
M	Message
PID	Unique path ID
MID	Unique message ID for each PID
M_r	Received messages vector at destination
N_{valid}	Valid nodes
$N_{untrusted}$	Nodes in untrusted vector
$N_{trusted}$	Nodes in trusted vector
M_m	Message vector of similar contents
$N_{malicious}$	Malicious nodes

4.4 Malicious Node Information (MNI)

At any instant when a node declared as malicious by trust management, the information is regarding the same to be spread in the network at the earliest so the every other node has information about this malicious node.

$$N_{malicious} \leftarrow N \text{ (Declare by TM)} \tag{5}$$

Remove the malicious node from the list of untrusted nodes and broadcast the MNI to network. Send the information about N to all neighbor's nodes in network. Place N to malicious node list. Notations used are shown below (Table 1).

$$N_{untrusted} \leftarrow N_{untrusted} - N \tag{6}$$

$$\text{Broadcast (MNI)} \tag{7}$$

5 Simulation and Results

This section includes the simulation setup and various parameters considered for proposed methodology followed by the performance result.

5.1 Simulation Setup

To implement the approach, simulation is performed using tool Omnetpp. Table 2 shows the simulation parameters considered for experiment.

Table 2 Simulation parameters

Parameter	Value
Area	$500 \times 500 \text{ m}^2$
Number of nodes	50
Traffic	CBR
Mobility	Random Walk
Node speed	15 m/s
Packet size	1 KB
Transmission range	Up to 400 m
Protocol	TCP
Routing protocol	AODV
Simulation time	300 S

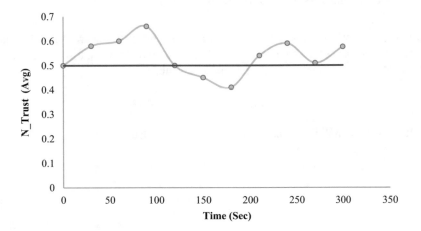

Fig. 6 Network average trust versus time

5.2 Performance Analysis

Analysis of performance of our proposed system is based on the computation of delay of data packets between source and destination and packet drop ratio in the network having malicious nodes. We are assuming that each node in the network is having initial trust of 0.5 value. Network trust at any time instant (t) is given by Eq. 8.

$$N_{\text{Trust}}(t) = \left| \frac{\sum_{i=1}^{N} \text{Trust}(N(i))}{N} \right| \tag{8}$$

Result of above is shown in Fig. 6. This graph shows the behavior of average of total amount of trust in the network for simulation duration in comparison with initial average network trust, i.e., 0.5

Figure 7 shows the effect on delay of packet to destination as the average network trust changes. This result shows that delay is in shrinking phase when the network is trustier.

As discussed in previous sections, the malicious nodes affect the delay because of intermediate processing of data before reaching to final destination. This has been shown in Fig. 8.

Packet drop ratio (PDR) is calculated as a number of dropped packets to number of paths form source to destination (PIDs).

$$PDR = \left| 1 - \frac{\text{Packets Received}}{\text{Packets Transmitted}} \right| \tag{9}$$

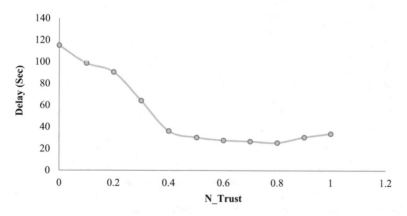

Fig. 7 Effect on delay as network trust varies

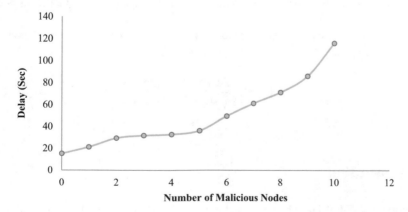

Fig. 8 Effect on delay as network grows with malicious nodes

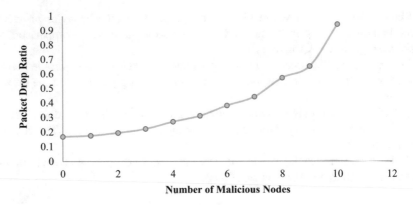

Fig. 9 Packets drop ratio versus number of malicious nodes

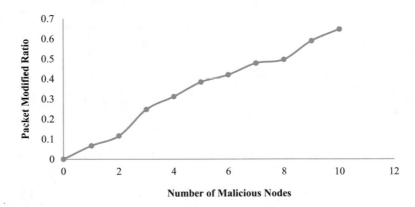

Fig. 10 Packets modified ratio versus number of malicious nodes

Figure 9 shows the result of packet drop ratio with the increase of malicious nodes in the defined network.

Finally, content modification rate has been calculated using number of modified packets identified at the destination (Fig. 10).

$$PMR = \left| \frac{\text{Packets Modified}}{\text{Packets Transmitted}} \right| \tag{10}$$

6 Conclusion

Proposed model of DDP incorporated both packet drop and content-based identification of malicious nodes. Further a direct trust-based computation methodology is applied by trust management system. This dual strategy provides the improvement in malicious node identification, and system performance significantly changes to betterment on removing identified malicious nodes from the network. Hence, overall network performance is improved. Also, the methodology performance was efficient during network dynamicity and scalability. For further research, the model can be extended to compare with another existing models and also on reducing the processing resources.

References

1. I. Bekmezci, O.K. Sahingoz, Ş Temel, Flying ad-hoc networks (FANETs): a survey. Ad Hoc Netw. **11**(3), 1254–1270 (2013)
2. A. Chriki, H. Touati, H. Snoussi, F. Kamoun, FANET: communication, mobility models and security issues. Comput. Netw. **163**, 1–17 (2019)
3. A. Kim, B. Wampler, J. Goppert, I. Hwang, H. Aldridge, Cyber attack vulnerabilities analysis for unmanned aerial vehicles. Infotech@ Aerosp. 2438–2468 (2012)
4. K. Singh, A.K. Verma, Flying ad hoc networks concept and challenges, in *Encyclopedia of Information Science and Technology*, 4th edn. (IGI Global, 2018), pp. 6106–6113
5. R. Altawy, A.M. Youssef, Security, privacy, and safety aspects of civilian drones: a survey. ACM Trans. Cyber-Phys. Syst. **1**(2), 7 (2016)
6. A.Y. Javaid, W. Sun, V.K. Devabhaktuni, M. Alam, Cyber security threat analysis and modeling of an unmanned aerial vehicle system. IEEE International Conference on Technologies for Homeland Security (HST) (2012), pp. 585–590
7. K. Hartmann, C. Steup, The vulnerability of UAVs to cyber attacks-An approach to the risk assessment. 5th International Conference on Cyber Conflict (CyCon) (2013), pp. 1–23
8. K. Mansfield, T. Eveleigh, T.H. Holzer, S. Sarkani, Unmanned aerial vehicle smart device ground control station cyber security threat model. IEEE International Conference on Technologies for Homeland Security (HST) (2013), pp. 722–728
9. G. Zhang, Q. Wu, M. Cui, R. Zhang, Securing UAV communications via trajectory optimization. IEEE Global Communications Conference (GLOBECOM) (2017), pp. 1–6
10. J. Kong, H. Luo, K. Xu, D.L. Gu, M. Gerla, S. Lu, Adaptive security for multilevel ad hoc networks. Wirel. Commun. Mob. Comput. **2**(5), 533–547 (2002)
11. J.P. Condomines, R. Zhang, N. Larrieu, Network intrusion detection system for UAV Ad-hoc communication from methodology design to real test validation. Ad Hoc Netw. **90**, 1–14 (2018)
12. S. Choudhary, K. Gupta, I. You, Internet of Drones (IoD): threats, vulnerability, and security perspectives. 3rd International Symposium on Mobile Internet Security (MobiSec'18) (2018), pp. 1–13
13. R. Altawy, A.M. Youssef, Security, privacy, and safety aspects of civilian drones. ACM Trans. Cyber-Phys. Syst. **1**(2), 1–25 (2016)
14. J. Sharma, Enhancement of security in flying AD-HOC network using a trust based routing mechanism. Int. J. Innov. Technol. Explor. Eng. **9**(1) (2019)
15. R. Mitchell, I.-R. Chen, A survey of intrusion detection techniques for cyber physical systems. ACM Comput. Surv. A1–A27 (2013)
16. Z. Birnbaum, A. Dolgikh, V. Skormin, E. O'Brien, D. Muller, C. Stracquodaine, Unmanned aerial vehicle security using recursive parameter estimation. J. Intell. Rob. Syst. **84**(1–4), 107–120 (2015)
17. Z. Birnbaum, A. Dolgikh, V. Skormin, E. O'Brien, D. Muller, C. Stracquodaine, Unmanned aerial vehicle security using behavioral profiling. International Conference on Unmanned Aircraft Systems (ICUAS) (2015), pp. 1310–1319
18. V. Singh, A fuzzy-based trust model for flying ad hoc networks (FANETs). Int. J. Commun. Syst. **31**(6), 1–19 (2018)

An E-Waste Collection System Based on IoT Using LoRa Open-Source Machine Learning Framework

Puppala Ramya⓪, V. Ramya, and M. Babu Rao⓪

Abstract Nowadays, electronic waste plays a very important role in each and everywhere. It is very extremely wasteful and expensive. The traditional process also has showed its uselessness in the society, as persons do not recycle their electronic waste correctly. With the growth of IoT and AI, the old electronic collection systems could be substituted with real-time sensors for correct observation of electronic waste. The intention of this study is towards advance a waste managing scheme based on long range (LoRa) protocol and TensorFlow machine learning open-source framework. Sensor data is sent by the communication protocol, and object is detected and classified by TensorFlow. The electronic container involves of various sections to separate the waste embracing metal, plastic, paper and normal waste were governed by servo motors. TensorFlow framework is used for actual time object detection and organization of waste data. This object discovery system is skilled with imageries of waste to produce a frozen implication chart used for real-time object detection which is completed over a camera linked to the Raspberry Pi 3 Model B as the key processing unit. In every waste, container placed with ultrasonic sensor for depth of waste identification, and container location is identified by GPS. Long range is a connecting protocol which is utilized to transfer data regarding location produced by GPS and filling depth of container. Radio-frequency identification (RFID) used for personnel identification of electronic waste.

Keywords IoT · TensorFlow · Ultrasonic sensor · RFID

P. Ramya (✉) · V. Ramya
Department of Computer Science & Engineering, Faculty of Engineering and Technology,
Annamalai University, Chidambaram, Tamil Nadu, India
e-mail: mothy274@kluniversity.in

M. Babu Rao
Department of Computer Science & Engineering, Gudlavalleru Engineering College,
Gudlavalleru, Vijayawada, India

© The Author(s), under exclusive license to Springer Nature Singapore Pte Ltd. 2021
S. K. Saha et al. (eds.), *Smart Technologies in Data Science and Communication*,
Lecture Notes in Networks and Systems 210,
https://doi.org/10.1007/978-981-16-1773-7_8

1   Introduction

Internet of Things (IoT) is a message model that imagines an upcoming example
where everyday life substances will be armed with a micro-controller and approximately form of message procedure [1].one of the most important advantage of IoT
is the smart city, i.e. smart persons, smart team and smart skill [2]. IoT can provide
number of different services for the improvement of society [3]. Every modern city
is having waste, and it is very difficult to identify in all the aspects. Long-range
communication is worked on very less power [4–6]. In smart cities, all permissions
have been taken care by higher authorities for improvement, and they used to called
3R (RECYCLE, REDUCE AND REUSE). An instruction on public awareness of
reprocessing the Malaysia activities demonstrations that only 41.8% of the total of
684 members were participated in reprocessing [7]. Additionally, machine learning
has produced advanced answers for lengthily sympathetic human behaviour's [8].
For the better improvement of technologies and for higher accuracy [9].

2   Related Work

For implementation of IoT solution must has 1. Energy-efficient, 2. communication,
3. transmitting information among the media (Table 1).

   However, the present waste systems obtainable through the related work do not
association together a whole waste separation scheme composed with a robust
message net. In this paper, we implement a new technology for better usage of
E-Waste using sensors. Data is transferred through LoRa to the server. Object is
classified by TensorFlow. RFID is used for providing authorization to container.

   Figure 1a 3D system has modelling tool and blender 3D. The electronic section
grips the electrical mechanisms. The waste discovery section has a telescopic stage
that contains the waste provisionally. Next to the similar time, leftover ID is existence
achieved by taking the copy of the leftover and handing out it by the Raspberry Pi
(Fig. 2).

Table 1 Comparison among communication protocols

Protocols	Conn Devices	Current Req	Range	Data Rate	Cost	Structure	Transfer Technique
Bluetooth	255	30MA	10 M	1 MBPS	LOW	STAR	FHSS
ZigBee	>64,000	20MA	100 M	250 KBPS	LOW	MESH	DSSS
Wi-Fi	Depend on IP	100MA	100 M	54 KBPS	MID	P-P	OFDM
LoRa	>5000	17MA	>15KM	50 KBPS	LOW	START	CHIRP

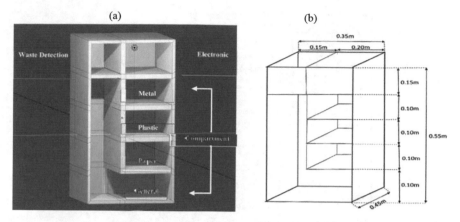

Fig. 1 **a** 3D container, **b** dimension of bin

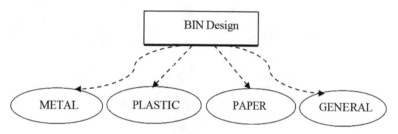

Fig. 2 Waste container is developed with four sections

3 Implementation

LoRa is appropriate to be applied in this scheme since containers remain typically located amid an insufficient rhythm to a few kilometres apart, and LoRa is talented to communicate information from a lengthy distance though overwhelming less energy. Demonstrations the multi-protocol wireless protection linked by the LoRa unit and Arduino-Uno, which performances as a bulge, which performance by way of a gateway (Figs. 3 and 4; Table 2).

Object system is trained depend on training dataset. This type of training is also called as transfer-learning number of sample images are available to train our data. Here, we collected 378 images with a dissimilar location, contextual and illumination condition. Before applying the supervised learning, all the waste images should be labelled. For images, the labelling should be performed based on software labelling. Once all the waste images are collected and labelled to achieve the error rate < 1.0000. Once inference graph will be produced and spread to raspberry pi for detection of thing. The average rate of metal, plastic and paper is 84.7, 95.7 and 86.6%. Once the implication chart is got, it remains spread to Raspberry Pi to achieve waste discovery and documentation. Signifies the process of the container afterward the

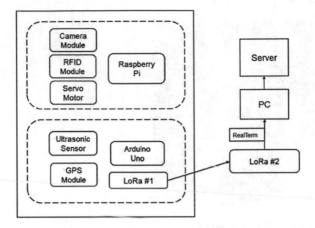

Fig. 3 Block diagram for overall process

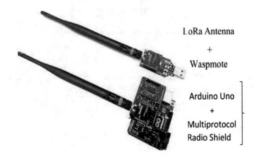

Fig. 4 Block diagram for LoRa

Table 2 Sensor models

Sensor/module	Model
Ultrasonic sensor	HC-SR04
GPS	GY-NE06MN2
RFID	RC522
Camera	PI Camera

documentation of waste. Waste resolve be primary released hooked on the telescopic stage to do waste discovery by means of the implication chart produced before. After the waste lesson is taught, a servo motor will trigger and unlock the telescopic top of the exact waste area. Previously, the determination of the telescopic stage should be phased out in order to allow the waste, with the assistance of significance, to fall into the precise waste portion (Figs. 5 and 6).

Fig. 5 Star topology of
LoRa

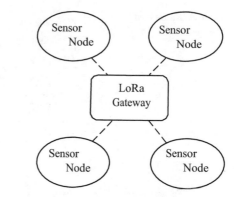

Fig. 6 Steps for object
detection

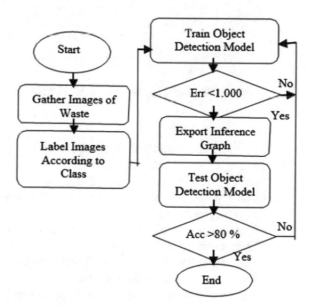

TensorFlow is mainly utilized for detection of object. For object detection, number of steps to be followed; initially, as a human, we need to collect all the waste pictures, once it was collected labelling the images. All waste images can be labelled according to class. Immediately, training will be performed if the error rate less than 1.0000 inference chart will be generated. Once the graph generated test the object detection model. After testing, if we got > 80% accuracy, we will detect an object correctly. Otherwise, repeat the same process (Table 3).

Table 3 Entire cost

Components	Amount	Cost
Arduino-Uno	1	S22
Raspberry Pi	1	S40
Ultrasonic sensor	5	S5
Solar panel	1	S25
Power bank	1	S20
camera	1	S8
RFID	1	S2
GPS	1	S10
Servo motor	5	S8
Bin	FREE	S20
Lora (SPECTRUM COST)		S5
Overall system cost	$180 PER BIN, $380 WITH GATEWAY	

3.1 Algorithm for Detection of E-Waste Management of Bin

1. Droplet waste hooked to the telescopic stage
2. Ultrasonic-Sensor 1 Presence of waste note
3. Take the photo if the waste is available and give it to the raspberry Pi
4. Raspberry-Pi uses the inference map to accomplish waste copy setup
    If, then, (waste = paper)
    Unlock the telescopic lid of the paper section; Exposed retractable step
    Then another (waste = metal)
    Retractable top of exposed metal section; Open retractable stage
    Other waste = rubber, then plastic,
    The retractable lid of the uncovered plastic section; Open telescopic stage
    Otherwise
    Telescopic Step Revealed, The End
5. Lock telescopic lid and folding stage

Signifies the general scheme price of the future scheme. LoRa the whole thing on an uninhibited range. Hence, it originates with not at all range prices. The entry price is lone experienced when apiece dishonourable position. The general scheme price is per bin quantities to $190. The general scheme price is admissible as it goals to decrease the quantity of physical employment work at the reprocessing vegetable, which eventually decreases the price of waste organization. Signifies the yearly working price of dissimilar control unit choices. It is experiential that the primary choice (pretentious the control bank and astral board each has a facility life of 3 centuries and 25 centuries, correspondingly) is inexpensive likened to the additional choice (presumptuous A $0.12 / kWh daily power price). The use of a

Table 4 Annual operations cost of various power module operations

Power compartment	Yearly price
Power bank solar panel	S20/3 years + S25/25 years = s7.67
Universal micro USB Power supply	S0.12/KWH*0.01Kw*24 hr*365 days = $10.51

Table 5 Waste type and its inference time

Waste type	Inference time (MS)
Metal	956
Plastic	951
Paper	973

power bank via an astral board is henceforth considered to be more location-tally friendly and maintainable in the long run (Tables 4 and 5).

4 Outcomes and Discussion of Object Detection System

Figure 7a, b and c has the ideal operational error, localization error and overall error during the whole workout. Organizational error is responsible for estimating the certainty degree of both the grades and upbringings. Localization error is responsible for well-changed container position changes made during object detection. The whole fault is the description error description and the flaw in localization. Presentation authentication is accomplished by hiring the model to categories 300 chance photos of waste after exercising the model for 20,000 epochs to get the perfect graph score. Figure 7d indicates that the map notch has gone through the model's exercise and evaluation level. The optimal one is capable of producing an 86.23% map slash. However, the usual precision of aluminium, plastic and paper is 86.7%, 96.3% and 82.3%, respectively, for existing waste identification. We calculate the drill error and the authentication error to secure the optimal non-over fitted (e) which denotes the error of confirmation and the error of exercise during the exercise process.

5 Conclusion

This paper presented a framework for smart waste organization by smearing devices to show the container rank, LoRa message technique for less control and broad-range information transmission, and TensorFlow-dependent object detection to display excess ID and classification procedures. Thanks to its frivolous countryside, SSDMobilnetV2 is capable of achieving fit in Raspberry Pi 3 Model B, the pre- trained stuff detection device. However, by inflating the amount of training data—in this case, the

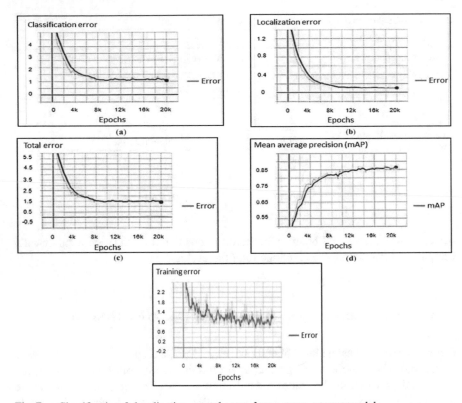

Fig. 7 **a** Classification, **b** localization, **c** total error, **d** map score, **e** terror model

amount of waste imagery—and by inflating the exercise time, the correctness of the model can be higher. The waste isolation is bordered and well synchronized between the object exploration performed by the top of the separate waste segment by Raspberry Pi and the servo motor regulatory. An RFID unit controls the bin's lock-up system. Ultrasonic sensors monitor the filling speed, while the GPS module tracks the scene and the real bin. LoRa operating at the 915 MHz frequency band conveys data on the bin rank of the filling stage, location and real-time from the bin to the LoRa gateway. The implementation of this automated isolation and filtering model in the bin aims to minimise running costs and restore the waste management system. We are ready to transform the area into a smart city at the same time. The waste detection model would be easier in future by inflating the volume of waste imagery in the dataset to inflate the model's elasticity in waste classification. In addition, to define and identify the shortest path to the container for the determination of repair, an automated direction-finding method may be introduced. The existing method of waste organisation will be better with this in mind to bring humanity into an olive green and better lifespan.

References

1. L. Atzori, A. Iera, G. Morabito, The internet of things: a survey. Comput. Netw. **54**(15), 2787–2805 (2010)
2. A. Meijer, M.P.R. Bolívar, Governing the smart city: a review of the literature on smart urban governance. Int. Rev. Adm. Sci. **82**(2), 392–408 (2016)
3. A. Zanella, N. Bui, A. Castellani, L. Vangelista, M. Zorzi, Internet of things for smart cities. IEEE Internet Things J. **1**(1), 22–32 (2014)
4. M. Shahidul Islam, M.T. Islam, M.A. Ullah, G. Kok Beng, N. Amin, N. Misran, A Modified meander line microstrip patch antenna with enhanced bandwidth for 2.4 GHz ISM-band internet of things (IoT) applications, in *IEEE Access*, vol 7 (2019), pp. 127850–127861
5. S.A. Hassan, M. Samsuzzaman, M.J. Hossain, M. Akhtaruzzaman, T. Islam, Compact planar UWB antenna with 3.5/5.8 GHz dual band- notched characteristics for IoT application, in *Proceedings of IEEE International Conference on Telecommunication. Photon. (ICTP)* (2017), pp. 195–199.
6. N. Misran, M.S. Islam, G.K. Beng, N. Amin, M.T. Islam, IoT based health monitoring system with LoRa communication technology, in *Proceedings of International Confernce on Electronics Engineering Information* (ICEEI, Bandung, IN, USA, 2019), pp. 514–517.
7. A. Mohamed, H.A. Rahman, Study on public awareness towards recycling activity in Kota Bharu, Kelantan Malaysia. Adv. Environ. Biol. **8**(15), 19–24 (2014)
8. S. Han, F. Ren, C. Wu, Y. Chen, Q. Du, X. Ye, Using the tensor flow deep neural network to classify mainland China visitor behaviours in Hong Kong from check-in data. ISPRS Int. J. Geo-Inf. **7**(4), 1–20 (2018)
9. K.S.S. Hulyalkar, R. Deshpande, K. Makode, Implementation of smartbin using convolutional neural networks. Int. Res. J. Eng. Technol. **5**(4), 1–7 (2018)
10. M. Shahidul Islam, M.T. Islam, A.F. Almutairi, G.K. Beng, N. Misran, N. Amin, Monitoring of the human body signal through the Internet of Things (IoT) based LoRa wireless network system, Appl. Sci. **9**(9), 1884 (2019)
11. A.S. Sadeq, R. Hassan, S.S. Al-rawi, A.M. Jubair, A.H.M. Aman, A qos approach for Internet of Things (Iot) environment using mqtt protocol, in *Proceedings of International Conference on Cyber Security* (2019), pp. 59–63.
12. M.S. Medvedev, Y.V. Okuntsev, Using Google tensorFlow machine learning library for speech recognition. J. Phys. Conf. **1399** Art. no. 033033 (2019)
13. M.A. Abu, N.H. Indra, A.H.A. Rahman, N.A. Sapiee, I. Ahmad, A study on image classification based on deep learning and tensorflow. Int. J. Eng. Res. Technol. **12**, 563–569 (2019)
14. Z.-Q. Zhao, P. Zheng, S.-T. Xu, X. Wu, Object detection with deep learning: a review. IEEE Trans. Neural Netw. Learn. Syst. **30**(11), 3212–3232 (2019)
15. C.-W. Chen, S.-P. Tseng, T.-W. Kuan, J.-F. Wang, Outpatient text classification using attention-based bidirectional LSTM for robot-assisted servicing in hospital. Information **11**(2), 106 (2020)
16. T.J. Sin, G.K. Chen, K.S. Long, I. Goh, H. Hwang, Current practice of waste management system in Malaysia? Towards sustainable waste management FPTP postgraduate seminar towards. Sustain. Manage. **1106**(1), 1–19 (2013)
17. M.O. Saeed, M.N. Hassan, M.A. Mujeebu, Assessment of municipal solid waste generation and recyclable materials potential in Kuala Lumpur, Malaysia. Waste Manage. **29**(7), 2209–2213 (2009)
18. A. Periathamby, F.S. Hamid, K. Khidzir, Evolution of solid waste management in Malaysia: impacts and implications of the solid waste bill. J. Mater. Cycles Waste Manage. **11**(2), 96–103 (2009)

Deep Learning Techniques for Air Pollution Prediction Using Remote Sensing Data

Bhimavarapu Usharani and M. Sreedevi

Abstract Due to Environmental and climate changes, these days, the whole world is facing air pollution and this problem is becoming further critical issues, which has terribly disturbed human lives as well as well-being. So, there is an immense need to search on air quality forecasting and has continuously considered as a key issue in environment safeguard. This paper covers the study associated with air pollution prediction using deep learning (DL) techniques based on remote sensing data. Major of the research work on air pollution are on the long-term forecasting of outdoor pollutants particulate matters ($PM_{2.5}$ and PM_{10}), ozone and nitrogen oxide. This paper discusses about the air pollution that causes the mortality and morbidity.

Keywords Air pollution forecasting · Climate change · Deep learning · Ozone · Particulate matter · Remote Sensing · Satellite Images · Google Maps etc

1 Introduction

The major crucial issue that we are facing today is poor quality air, especially in the industrial cities. Today's science and technology provide many smart features and putting the entire world in front of us as well as polluted air. There is the need of accurate and early prediction of the gas emissions by creating an enhanced initial warning system by incorporating with an effective air quality checking organization. Air pollution has risen up human beings severe troubles especially in developing countries like India. Air quality is a terrific medical issue though out the worldwide and causes thousands and thousands of premature deaths each year.

There are 2 types of air pollution:

1 Household Air Pollution
2 Ambient air pollution.

B. Usharani (✉) · M. Sreedevi
Department of Computer Science and Engineering, Koneru Lakshmaiah Education Foundation, Green Fields, Vaddeswaram, Andhra Pradesh, India
e-mail: ushareddy@kluniversity.in

© The Author(s), under exclusive license to Springer Nature Singapore Pte Ltd. 2021   107
S. K. Saha et al. (eds.), *Smart Technologies in Data Science and Communication*,
Lecture Notes in Networks and Systems 210,
https://doi.org/10.1007/978-981-16-1773-7_9

Household Air Pollution: In low- and middle-income countries, cooking and heating depends on substances like charcoal, kerosene, wood, animal dung, coal, crop wastes, etc. In 2016, 3.8 million children and women died prematurely due to household air pollution [1].

Ambient air pollution: Occurs from the from both environmental and human causes. Natural disasters like volcano eruption, earthquakes, wildfires, dust storms emit carbon dioxide, Sulphur dioxide, hydrogen which affects the climate and pollutes the air. Fossil fuel burning, deforestation, clear green lands, more industrialization, fertilizers, agricultural waste is the human activities that pollutes the air, water, and land. 91% of the population in the world are facing ambient air pollution and about 4.2 million early deaths in 2016 due to ambient air pollution [1].

The World Air Quality Index was started in 2007 and it provides a worldwide air quality information [2]. By 2019 the world's highest polluted city was Ghaziabad ($PM_{2.5}$), India with an average of 110.2[3]. According to USAQI by 2020, Dubai, United Arab Emirates has the highest USAQI of 174 and Sydney, Australia has the lowest USAQI of 3 [4].

In India, the National Air Quality Index (AQI) was launched in 2014, it measures the quality of the air and it shows the extent and categories of gases dissolved in air. New Delhi has the most terrible air pollution of some capital cities and this air pollution put down around 1.25 million people every year in India. It is estimated that around 2 billion people may get dislodged from their houses by 2100 due to atmospheric rising zones.

There are 573 monitoring stations in India and this checking is carried off by central pollution control board, state pollution control board, pollution control committee, national environmental engineering research institute [5].

This paper intends to give a comprehensive information primarily based on a variety of works on air quality forecast and predicting pointing out the disparate elements of air quality prediction as follows:

1. Describing the overall application and concentrations of the active air quality predicting and forecasting methods.
2. Discuss the task of data on deep learning models, as well as data sets availability and availability.
3. Summarizing frequent approaches of assessing and validating air quality models.

The rest of the paper is arranged as follows:

Different sources of air pollution and their impact are discussed in the Sect. 2. Section 3 discusses about the deep learning and its variants. Section 4 summarizes the previous work that has contributed up to now for air pollution. Section 5 discusses about the properties and types of remote sensing. Performed survey and the report is given in Sect. 6. The future work in this survey is discussed in Section 7. Finally, this article is concluded in the conclusion section.

2 Major Air Pollutants That Cause Mortality and Morbidity

Globally, air pollution is a primary threat, particularly in urban areas. Air pollution is a mixture of particulate matter, greenhouse emission gases, organic compounds, global warming, deforestation, and solar radiance. World Health Organization (WHO) recognizes that 9 out of 10 people are inhaling polluted air and around 7 million people die every year from breathing the polluted air. Air Pollution from ambient and household causes a premature mortality from heart diseases, chronic, pulmonary, lung cancer and mainly respiratory infections [1]. The Air Quality Index (AQI) gives the daily air quality information regarding the pollutants in the air concerning the health.

According to IEA [6], every year around 6.5 million premature losses are related to air pollution. The negative impacts of air pollution affect the functioning of all life stages [7]. Polluted air shows its effects on the diabetes related mortality even in the general population [8]. Ambient and household air pollution shows its impact on mortality and morbidity. Improving air quality improves social equity and the environment and mitigates the traffic injury, obesity, and noise. There is a serious dangerous air pollutant like particulate matter (PM), ozone(O_3), nitrogen dioxide (NO_2) and Sulphur dioxide (SO_2). The most important air pollutants that effects environment and health are given in Table 1.

3 Deep Learning Architectures

The methods that are proposed for predicting air quality are satellites, spectroscopic method, filter based gravimetric, atmospheric aerosol, social network, mobile information. Deep learning methods are using in numerous engineering and scientific areas, bioinformatics and computational mechanics have been among the early application domains of deep learning. Deep learning is a subgroup of machine learning and refers to the request of a set of algorithms called neural networks and their variants. Deep learning neural network is more efficient to identify the satellite imagery. Different types of deep learning neural networks are discussed in Table 2.

3.1 CNN Models

Convolutional neural networks (CNN) used for solving typical computer concept problems. CNN are useful for 1D problems like times series and 3D image classification. The major feature of CNN is weight sharing and able to learn the relevant features from an image and videos. CNN are good feature extractors and outperforms the neural networks (NN) for image recognition tasks. Some CNN models are given

Table 1 Major air pollutants that cause morbidity and mortality

Pollutant	Sources	Health fffects	Environment effects
PM (2.5 or 10)	Construction areas, Smokestacks, fires, and unpaved roads	Reduced lung function, cardiovascular and respiratory diseases, irritated eyes, nose and throat, asthma, heart attacks and Arrhythmias	River acidify, altering the nutrient solidity in coastal waters, reducing the vitamins in soil, harming forests and farm crops, changing the ecosystems
CO	volcanoes as they erupt, natural gases in coal mines, and even from lightning, motor vehicles, industry, unfueled gas heaters, woodburning heaters, and contained in cigarette smoke	Decreasing the quantity of oxygen achieving the body's limbs and tissues. Excessive levels of CO lead to death	Consequences on greenhouse gases, global warming, land, and sea/ocean temperature increases, increasing storms and extreme climate change and weather events
NO$_2$	motor vehicles, manufacturing, unfueled, heaters, and gas stove	Respiratory irritant and has a range of unfavorable fitness results on the respiratory system	Reduces plant growth, effects habitat or species, effects vegetation and crop yields
SO$_2$	ignition at power plants and industrial effects	Sulphur dioxide aggravates the nose, throat and lungs and irritation on the respiratory ailment, and cardiovascular diseases	deforestation, acidify waterways, acid rain, affect vegetation
O$_3$	motor vehicles and industry	chronic bronchitis, damage lungs, breathingproblems	Damages vegetation along with ecosystems. reducing forest growth and crop yields, harsh weather, Slow the plant's growth, damage from insects, effects of different pollutants, changes to water and nutrient cycles

in Table 3. The advantage of CNN model for air quality prediction is that it takes less training time and relatively easy because the number of weights is fewer than the fully connected architectures. CNN models extract all the important temporal and spatial features effectively.

Table 2 Deep learning architectures

Model	Description	Applications
Radial Basis Network	A sigmoid function gives the result between 0 and 1	Function Approximation, Time series, Prediction, Classification, System Control
Deep Feed Forward Network	By adding more hidden layers, reduces over fitting and generalization	Data Compression. Pattern Recognition. Computer Vision, Financial Prediction
Recurrent Neural Network	Variation to feed-forward (FF) networks. The neurons receive an input with delay in time	Time Series Prediction. Speech Recognition. Speech Synthesis. Time Series Anomaly Detection
Long Short-Term Memory	Introduces a memory cell. It remembers data for longer time	Speech Recognition. Writing Recognition
Gated Recurrent Unit	variation of LSTMs. Works better than LSTM	Speech Signal Modelling. Natural Language Processing
Generative Adversarial Network	The GANs used to distinguish between real and synthetic results.	Generate New Human Poses. Photos to Emojis. Face Aging
deep residual network	Prevents degradation of results	Image Classification. Object Detection
Deep belief network	Represent DBNs as a combination of Restricted Boltzmann Machines (RBM) and Autoencoders (AE)	Retrieval of Documents or Images. Nonlinear-Dimensionality Reduction
Deep convolutional network	Works in the reverse process of the CNN	Optical flow estimation
SVM	Mainly for binary classifications	Face Detection. Text Categorization. Handwriting recognition

3.2 Recurrent Neural Network

Recurrent neural network (RNN) [9] is a category of deep learning architecture mostly applied for time series data modelling. It is similarto feed forward network with an extra internal feedback loop that results in a cyclic graph. Some RNN variants are discussed in Table 4. RNN can use for the non -linear time series prediction problems and rea world data are the nonlinear data. So, RNN gives correct results for real world data when compared to traditional models.

3.3 Long Short-term Memory

Long short-term Memory (LSTM) [15] is a kind of recurrent neural network (RNN) which is a good at processing and predicting time series events. LSTM prevents the

Table 3 CNN models

Model	Description	Parameters	Depth
LeNet	Classifies digits and to recognise hand-written numbers d spatial correlation reduces the computation and number of parameters	60,000	5
Alexnet	Introduces regularization concept and gives an idea of deep and wide CNN architecture	60 Million	8
VGG	Introduces effective receptive field	138 Million	19
GoogleNet/Inception-v1	Introduced the Multiscale Filters within the layers, used the concept of global average-pooling at last layer and sparse Connections, use of auxiliary classifiers to improve the convergence rate	4 Million	22
Resnet	Decreased the error rate for deeper networks, Introduced the concept of residual learning	25.6 Million	152
Inception-v3	demoralized asymmetric filters and bottleneck layer to reduce the computational cost of architectures	23.6 Million	159
Xception	Introduces the varying size filters	22.8 Million	126
WideResnet	Increases the width of ResNet and decreses its depth, facilitate feature reuse	36.5 Million	28
Squeeze and Excitation Networks	Introduced the generic block	27.5 Million	152
PyramidalNet	Increases the width progressively per unit	116.4 Million	200

vanishing gradient problem. LSTM variants are discussing in the Table 5. LSTM has the ability of solving the long-term dependencies which general RNN cannot learn and solve. LSTM is commonly used to solve the time series prediction problems.

4 State of Art

Most of the studies on air pollution prediction do depend on mathematical equations or simulation techniques to illustrate the evolution of air pollution [21]. These traditional methods are represented by classic shallow machine learning algorithms. For example, Chen et al. given a unique approach which is based on daily random forests model for $PM_{2.5}$ concentration value prediction [22].

Table 4 RNN variants

Model	Description	Database
QRNN [10]	computationally efficient hybrid of LSTMs and CNNs. Adding another layer overcomes the weakness of the 1-layer s-QRNN	IMDb movie review dataset, PTB dataset, IWSLT data set
BRNN: Bidirectional recurrent neural networks [11]	considers all input sequences in both the past and future for evaluation of the output vector	TIMIT
MDRNN: Multidimensional RNNs [12]	Extending the applicability of RNNs to n-dimensional data	Air Freight, MNIST database
DeepRNN [13]	An alternative deeper design, which leads to several deeper variants of an RNN	Nottingham, JSB Choralesand Muse Data datasets
Pixel RNNs [14]	Discrete probability of the unrefined pixel values and encodes the entire set of dependencies in the image and take account of quick two-dimensional recurrent layers and an valuable use of residual connections in deep recurrent networks	MNIST, CIFAR-10 and ImageNet datasets

Table 5 LSTM variants

Model	Description	Database
Grid LSTM [16]	A network that is arranged in a grid of one or more dimensions. Works on vectors, sequences, or higher dimensional data such as images	Hutter challenge Wikipedia dataset
Associative LSTM [17]	Based on complex-valued vectors and is closely related to Holographic Reduced Representations and LSTM.	Own dataset, English Wikipedia dataset
Siamese LSTM [18]	Enhance the discriminative capacity of the local features	Market-1501, CUHK03, VIPeR
DECAB-LSTM [19]	Extensive basic LSTM by integrating an attention mechanism to study the significant part of a sentence for a known feature for the text classification	1852 biomedical publication abstracts from PubMed journals

Machine learning utilizes a kind of procedure that does not give preciseoutcomes but does give approximation results [23].

In this work, we mainly focus on the applications of deep learning time series forecasting, for air quality prediction. Among then LSTM can learn long term dependencies using several gates suits well the time series forecasting problems.

Gated recurrent unit (GRU) is an enhanced version of recurrent neural network (RNN) [24] and an effective in time series forecasting [25]. For instance, in [26], authors employ 1D convnets and bidirectional GRUs for air pollution forecasting in Beijing, China. State of art air pollution pre using deep learning and satellite images are given in Table 6. We identified that LSTM networks have a definiteinfluence on the prediction of time series signals, so they are appropriate for air quality prediction.

5   Remote Sensing Data

Earth observation (EO) is the process of gathering the dataregarding Earth through remote sensing. EO satellites have observed applications in variousareas, varying from cartography, urban planning, disaster relief, land management to social analysis, military, and climate studies [33].

The remote sensing images are categorized into groups. Among them the main group is open public images, American Landsat and European Sentinel are the important and gives the free remote sensing images. Freely available data from satellite collections like MODIS, Landsat and Sentinel have democratized access to timely satellite imagery of the whole globe. Cloud Provides like AWS and Google Cloud have gone to store satellite data for free of charge. Another group is the commercial providers like Digital Globe, which will offer the remote sensing image with a resolution up to 25cm per pixel where images are accessible twice each day. By using remote sensing data, the air quality in India is shown in Fig. 1.

Satellite data for the most polluted city in India i.e., Delhi air pollutant forecast is presented in Fig. 2.

5.1   Properties of Remote Sensing

Data Satellites record information that the human natural eyes cannot look at. A picture shown by the satellite can have at least 12 layers, and each layer takesadditional data. By joining these layers, one can make indicators that will give an extra understanding regarding what is going on the earth.

1.   Spatial resolution: It can fix the condition of an image and explain how complete an object can be considered by the image. It is a measurement to decide how trivial an object should be in an order for an imagery system to distinguish it.

Table 6 State of the art air pollution prediction using deep learning and satellite images

Project	Description	Algorithm	Metrics	Database
Alfaseeh et al (2020) [20]	predicting Greenhouse Gas (GHG) emissions rate	LSTM	correlation coefficient, R^2, linear fit, and RMSE	1. Toronto road network 2.The agent-based simulator used a calibrated Intelligent Driver Model (IDM) for vehicular movement
Fan et al (2020) [21]	predicting daily diffuse solar radiation	SVM-PSO, SVMBAT (bat algorithm), whale optimization algorithm (SVM-WOA)	root mean square deviation, R^2, MAE, scatter index (SI)	1.The meteorological data National Meteorological Information Centre (NMIC) of China Meteorological Administration (https://www.nmic.cn/) 2. air pollution dataset – ChinaNationalEnvironmentalMonitoring Centre https://www.cnemc.cn/)
Wang et al (2020) [22]	CO emissions at pedestrian crosswalks	MultiVaraite LSTM	RMSE, MAE, R^2	Shuang long Ave. and JiyinAve regions traffic
Alleon et al (2020) [23]	forecast main pollutants harming humans health	convolutional LSTM	accuracy, MSLE loss	14000 monitoring stations throught the world
Baghirli et al (2020) [24]	GREENHOUSE SEGMENTATION	modified U-Net,a fully convolutional neural network architecture	F1, Kappa, AUC, and IOU	Azersky (SPOT-7) optical satellite images collected from 15 regions in Azerbaijan
AlOmar et al (2020) [25]	Forecasting the Ozone concentration	ANN, MLPNN	RMSE, CC, MAE, ARE and RER	London station
Pak et al (2020) [26]	prediction of PM2.5	CNN-LSTM	MAE, RMSE and MAPE	384 monitoring stations of china (https://www.mee.gov.cn/), meteorological data (https://tianqi.2345.com/)
Ma et al (2020) [27]	prediction of PM2.5 concentrations	Lag-FLSTM	RMSE, MAE, and MAPE	4 monitoring stations in Wayne County

(continued)

Table 6 (continued)

Project	Description	Algorithm	Metrics	Database
HongWei et al (2020) [28]	predicting regional ground-level ozone concentration	Bidirectional Recurrent Neural Network (Bi-RNN)	MAE, R^2RMSE,the Pearson Correlation coefficient (r) and the Coefficient of Determination	six regulatory air pollutants (O_3, SO_2, NO_2 $PM_{2.5}$,CO are obtained from 35 air quality monitoring stations from Beijing area and Traffic monitoring stations are also considered
Wenjing et al (2020) [29]	temporal sliding LSTM extended model(TS-LSTME)to predict air quality in 24 hourly historical PM2.5 concentration data	TS-LSTME	Spatiotemporal correlation, RMSE, MAE, R^2, NRMSE	Jing-Jin-Ji area and 13 cities in Beijing, Tianjin and Hebei province Data from China Meteorological Data: https://data.cma.cn/en) PM2.5 monitoring data (https://106.37.208. 235:20035/)
Kaya et al (2020) [30]	hybrid deep learning model called DFS proposed for future PM10 forecasting	Deep Flexible Sequential (DFS)	MAE, RMSE	Real world data aksaray, Alibeyk¨oy, Be̦sikta̦s and Esenler
Weichenthal et al (2020) [31]	to estimating annual average outdoor nitrogen dioxide (NO2) concentrations using ground-level measurements	Convolutional neural networks (CNNs) and Xception model	RMSE, R^2	Flanders from the "CurieuzeNeuzen" ("Curious Noses") project and Google Static Map
Yan et al (2020) [32]	estimate PM2.5	Entity Dense Net	RMSE, R^2	1434 monitoring stations across China (https://srtm.csi.cgi ar.org and Nighttime VIIRS Day/Night (https://ngdc.noaa.gov/eog/viirs/).

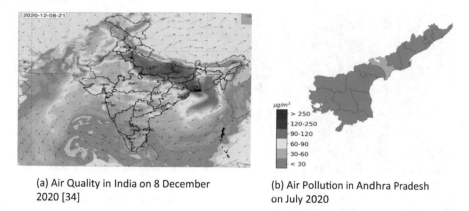

(a) Air Quality in India on 8 December 2020 [34]

(b) Air Pollution in Andhra Pradesh on July 2020

Fig. 1 Air quality in India as on 8 December 2020 [34]

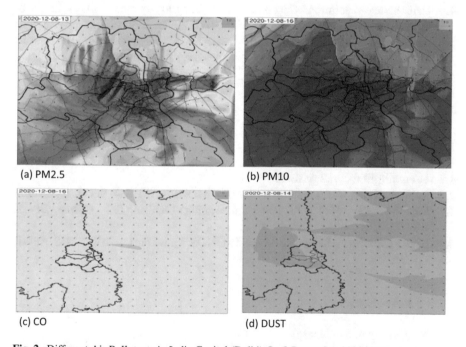

(a) PM2.5

(b) PM10

(c) CO

(d) DUST

Fig. 2 Different Air Pollutants in India Capital (Delhi) On 8 December 2020 [35]

2. Temporal resolution: It depends on their visit time of the satellite. The extent of time varies on the orbital qualities of the platform. Temporal resolution refers to the time among images.

3. Spectral resolution: It refers to the capability of a satellite sensor to assess a particular wavelength of the electromagnetic spectrum. The better the spectral resolution, the clearer the wavelength range for a specific band.

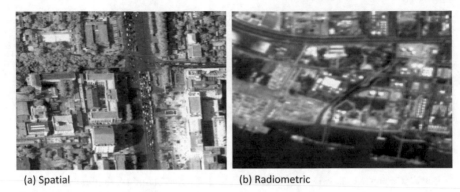

(a) Spatial   (b) Radiometric

Fig. 3 Different resolution of Remote sensing data [36]

4. Radiometric resolution: Capability of an imaging to record many levels of brightness and to active bit depth of the sensor. The radiometric resolution of an imaging system defines its capacity to categorize extremelyminorvariations in energy. Some satellite resolution images are shown in Fig. 3.

5.2 Types of Remote Sensing

5.2.1 Passive Sensors

1. Use sunlight for energy source
2. Low resolution (¿100 meters resolution)- MODIS, AVHRR, SPOT vegetation
3. Moderate Resolution (15–100 m resolution)-Landsat TM/ ETM+, SPOT, ASTER, IRS
4. High Resolution (¡15 m resolution)-IKONOS, Quickbird, OrbViewIRS, SPOT, Corona

5.2.2 Active Sensors

1. Generate their own energy
2. Radar (Radio Detection and Ranging)-Radarsat, ERS, Envisat, Space shuttle
3. Ladar (Light Detection And Ranging)- Mostly airborne platforms for now
4. ICESat is only satellite lidar platform

Different types of satellite images are shown in Fig. 4.

Fig. 4 Images of Sentinel and Landsat Satellites

5.3 *Meteorological Data*

In India, the meteorological data will be provided by the Meteorological centre, Thiruvananthapuram, website (https: //mausam.imd.gov.in/). Daily meteorological data like temperature, wind, relative humidity, rainfall for the Indian capital Delhi [37] is given in Fig. 5.

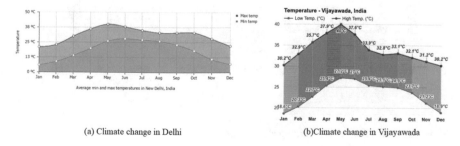

Fig. 5 Metrological Data of Delhi and Vijayawada Cities

6   Survey Report

The goal of this analysis is to gather and analyze the perceptions of communities in India regarding the climate change and the future priorities for action. Different questionnaire was designed and collected information from population of interest regarding climate change. A total of 1000 respondent opinions were collected and presenting those responses in this section. Respondents opinion regarding sufficient technologies is picturized in Fig. 6a respondent's opinion whether to do any special program to mitigate pollutants in climate is shown in Fig. 6b, whether everyone in India have an idea or the awareness to perform climate change services is collected and presented in the Fig. 6c, Indian citizen responsibilities regarding climate awareness has gathered and shown in the Fig. 6d.

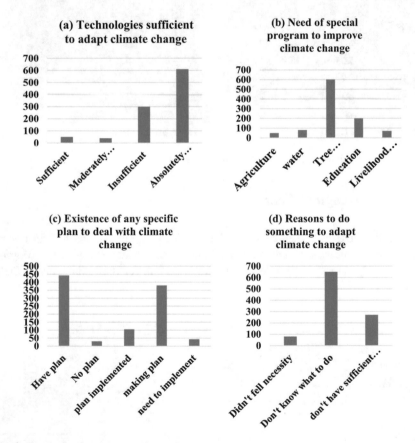

Fig. 6 Respondents opinion regarding climate change in India

7 Future Work

The future work that are identified that need immediate changes:

1. According to the literature, the future work will be using the Distributed Feature Selection (DFS) Approach [38] in deep neural networks
2. According to the literature, the classification and prediction can be improved by MCFS (M-Cluster Feature Selection) [39] approach and correlation-based framework [40] will be our future work to get accurate results
3. Need to elucidate that how the PM exposure affects the cerebrovascular as well as respiratory related morbidities. Need to understand how PM shows its effect on human health.
4. Predicting infectious disease that are caused from the air and water pollution easily by using the sensors [41].
5. Predict how rain effects based on the fossil fuel pollutants and traffic pollutants

8 Conclusion

Air pollution is turn out to be further acute problem in numerous developing countries, which effects the social physical condition and living drastically. The analysis revealed that remote sensing and Geographical improve system (GIS) technology can present valuable data for air pollution and mainly satellite data is suitable for the quantitative assessment of gaseous toxins. Deep learning methods are rapid progressing and a few of them have evolved to be focused in a specific presentation area, this will lead to more exact approximations of air quality all together with measures of uncertainty. Air quality prediction methods mostly use shallow models; but these methods give inadequate results, which motivated us to explore methods of predicting air quality based on deep architecture models.

References

1. WHO Air Pollution. Polluted Air. https://www.who.int/news-room/air-pollution. Online, Accessed on 09 Aug 2020
2. Real-time Air Quality Index. World's air pollution: real-time air quality index. https://waqi.info/. Online, Accessed on 09 Aug 2020
3. World IQAir. World's most polluted cities 2019 . https://www.iqair.com/us/world-most-polluted-cities. Online, Accessed 09 Aug 2020
4. USIQAir. Air quality and pollution city ranking. https://www.iqair.com/us/world-air-quality-ranking. Online; Accessed on 09 Aug 2020
5. cpcbenvis. National Air Quality Monitoring Programme (NAMP) Monitoring Network. https://cpcbenvis.nic.in/airpollution/monetoring.htm, 2020. Online; Accessed 08 Dec 2020
6. IEA. Small increase in energy investment could cut premature deaths from air pollution in half by 2040, says new IEA report . https://www.iea.org/news/small-increase-in-energy-investmen

tcould-cut-premature-deaths-from-air-pollution-in-half-by2040-says-new-iea-report. Online, Accessed on 09 Aug 2020
7. M. Kicinski, G.Vermeir, N. Van Larebeke, E. Den Hond, G. Schoeters, L. Bruckers, I. Sioen, E. Bijnens, H.A. Roels, W. Baeyens, et al., Neurobehavioral performance in adolescents is inversely associated with traffic exposure. Environ. Int. **75**,136–143 (2015)
8. I.C. Eze, M. Imboden, A. Kumar, A. von Eckardstein, D. Stolz, M.W. Gerbase, N. Künzli, M. Pons, F. Kronenberg, C. Schindler, et al., Air pollution and diabetes association: modification by type 2 diabetes genetic risk score. Environ. Int. **94**, 263–271 (2016)
9. M.D. Zeiler, R. Fergus, Visualizing and understanding convolutional networks, in *European Conference on Computer Vision* (Springer, 2014), pp. 818–833
10. S. Xie, R. Girshick, P. Dollár, Z. Tu, K. He, Aggregated residual transformations for deep neural networks, in *Proceedings of the IEEE Conference on Computer Vision and Pattern Recognition* (2017), pp. 1492–1500
11. C. Szegedy, V. Vanhoucke, S. Ioffe, J. Shlens, Z. Wojna, Rethinking the inception architecture for computer vision, in *Proceedings of the IEEE Conference on Computer Vision and Pattern Recognition* (2016), pp. 2818–2826
12. C. Szegedy, S. Ioffe, V. Vanhoucke, A. Alemi, Inception-v4, Inception-resnet and the Impact of Residual Connections on Learning. arXiv preprint arXiv:1602.07261 (2016)
13. F. Chollet, Xception: deep learning with depthwise separable convolutions, in *Proceedings of the IEEE Conference on Computer Vision and Pattern Recognition* (2017), pp. 1251–1258
14. S. Zagoruyko, N. Komodakis, *Wide Residual Networks*. arXiv preprint arXiv:1605.07146 (2016)
15. J. Hu, L. Shen, G. Sun, Squeeze-and-excitation networks, in *Proceedings of the IEEE Conference on Computer Vision and Pattern Recognition* (2018), pp. 7132 7141
16. J. Kim, D. Han, J. Kim, Deep pyramidal residual networks, in *CVPR 2017 IEEE Conference on Computer Vision and Pattern Recognition. IEEE Computer Society and the Computer Vision Foundation (CVF)* (2017)
17. J. Kuen, X. Kong, G. Wang, Y.-P. Tan, Delugenets: deep networks with efficient and flexible cross-layer information inflows, in *Proceedings of the IEEE International Conference on Computer Vision Workshops* (2017), pp. 958–966
18. J. Bradbury, S. Merity, C. Xiong, R. Socher, *Quasi-recurrent Neural Networks*. arXiv preprint arXiv:1611.01576 (2016)
19. M. Schuster, K.K. Paliwal, Bidirectional recurrent neural networks. IEEE Trans. Signal Process. **45**(11):2673–2681 (1997)
20. L. Jiang, X. Sun, F. Mercaldo, A. Santone, Decab-lstm: deep contextualized attentional bidirectional lstm for cancer hallmark classification. Knowl-Based Syst **210**, 106486 (2020)
21. S. Vardoulakis, B.E.A. Fisher, K. Pericleous, N. Gonzalez-Flesca, Modelling air quality in street canyons: a review. Atmosph. Environ. **37**(2), 155–182 (2003)
22. G. Chen, S. Li, L.D. Knibbs, N.A.S. Hamm, W. Cao, T. Li, J. Guo, H. Ren, M.J. Abramson, Y. Guo, A machine learning method to estimate pm2. 5 concentrations across China with remote sensing, meteorological and land use information. Sci. Total Environ. **636**, 52–60 (2018)
23. Q. Wang, Numerical forecast analysis of typical pm2.5 pollution episode over shanghai in autumn. Environ. Monit. China **30**(2), 7–12 (2014)
24. R. Jozefowicz, W. Zaremba, I. Sutskever, An empirical exploration of recurrent network architectures, in *International Conference on Machine Learning* (2015), pp. 2342–2350
25. P.T. Yamak, L. Yujian, P.K. Gadosey, A comparison between arima, lstm, and gru for time series forecasting, in *Proceedings of the 2019 2nd International Conference on Algorithms, Computing and Artificial Intelligence* (2019), pp. 49–55
26. Q. Tao, F. Liu, Y. Li, D. Sidorov, Air pollution forecasting using a deep learning model based on 1d convnets and bidirectional gru. IEEE Access **7**, 76690–76698 (2019)
27. L. Alfaseeh, R. Tu, B. Farooq, M. Hatzopoulou. *Greenhouse Gas Emission Prediction on Road Network Using Deep Sequence Learning*. arXiv preprint arXiv:2004.08286 (2020)
28. J. Fan, W. Lifeng, X. Ma, H. Zhou, F. Zhang, Hybrid support vector machines with heuristic algorithms for prediction of daily diffuse solar radiation in air-polluted regions. Renewable Energy **145**, 2034–2045 (2020)

29. Y Wang, P Liu, C Xu, C Peng, J Wu, A deep learning approach to real-time co concentration prediction at signalized intersection. Atmosph. Pollut. Res. (2020)
30. A. Alléon, G. Jauvion, B. Quennehen, D. Lissmyr, *Plumenet: Large-scale Air Quality Forecasting Using a Convolutional lstm Network*. arXiv preprint arXiv:2006.09204 (2020)
31. O. Baghirli, I. Ibrahimli, T. Mammadzada, *Greenhouse Segmentation on High-Resolution Optical Satellite Imagery Using Deep Learning Techniques*. arXiv preprint arXiv:2007.11222 (2020)
32. M.K. AlOmar, M.M. Hameed, M.A. AlSaadi, Multi hours ahead prediction of surface ozone gas concentration: Robust artificial intelligence approach. Atmosph. Pollut. Res. **11**(9), 1572–1587 (2020)
33. P. Kansakar, F. Hossain, A review of applications of satellite earth observation data for global societal benefit and stewardship of planet earth. Space Policy **36**, 46–54 (2016)
34. ESSO IITM, Air quality early warning system for Delhi Ministry of Earth Sciences, Government of India. https://ews.tropmet.res.in/. Accessed 08 Dec 2020
35. ESSO IITM. Air quality early warning system for Delhi Ministry of Earth Sciences, Government of India. https://ews.tropmet.res.in/index-2.php. Accessed 08 Dec 2020
36. crisp. Images. https://crisp.nus.edu.sg/~research/tutorial/image.htm. Accessed 08 Dec 2020
37. ESSO. 10 days forecast for Delhi. https://ews.tropmet.res.in/10_days_forecast.php. Accessed on 15 Dec 2020
38. S.P. Potharaju, M. Sreedevi, Distributed feature selection (dfs) strategy for microarray gene expression data to improve the classi_cationperformance. Clin. Epidemiol. Global Health, **7**(2), 171–176 (2019)
39. S.P. Potharaju, M. Sreedevi, A novel m-cluster of feature selection approach based on symmetrical uncertainty for increasing classification accuracy of medical datasets. J. Eng. Sci. Technol. Rev. **10**(6) (2017)
40. S.P. Potharaju, M. Sreedevi, Correlation coecient based feature selection framework using graph construction. Gazi University J. Sci. **31**(3) (2018)
41. G. Vijay Kumar, A. Bharadwaja, N. Nikhil Sai, Temperature and heart beat monitoring system using iot, in 2017 *International Conference on Trends in Electronics and Informatics (ICEI)* (IEEE, 2017), pp. 692–695

A Real and Accurate Diabetes Detection Using Voting-Based Machine Learning Approach

Udimudi Satish Varma ⓘ, V. Dhiraj ⓘ, B. Sekhar Babu ⓘ,
V. Dheeraj Varma ⓘ, Gudipati Bharadwaja Sri Karthik ⓘ, and V. Rajesh ⓘ

Abstract Diabetes is a dangerous infection that happens when the pancreas cannot generate insulin or the body could not utilize the insulin appropriately. The pancreas produces a type of hormone named as Insulin. A wide range of food which we eat is additionally separated into glucose. Insulin causes us to get into the human cells and helps us to generate energy. There are generally three types of diabetics—type-1, type-2, and gestational diabetes. According to the International Diabetes Federation (IDF), type-1 diabetes can occur at any stage of your age, but it occurs mostly for teenagers and children's; in type-1, your body delivers very less or no insulin, which means you need to inject the insulin to manage blood glucose levels. Type-2 is most commonly seen in adults and 90% of all diabetes cases are type-2; in type-2, your body does not use insulin properly which it produced. An oral drug and insulin are required to manage blood glucose levels. GDM is one of the type of diabetes which consists of high glucose level during pregnancy and the good part of this is mostly it disappears after pregnancy, if not their children may affect with type-2 in the later stage of their age. 1 among 10 persons is living with these three types of diabetics worldwide so we developed a machine learning classification model to identify the person is affected or not with the highest test accuracy. The machine learning models which we used are logistic regression, linear SVM, K-NN, random forest classifier, GBDT, SVC, and we also used hyperparameter tuning to get the best accuracy from these six respective models. The main aim of this project is to identify the three best machine learning models out of six with highest accuracy and apply the voting classifier by using this, we had acquired 96% of train accuracy and 88% test accuracy.

Keywords Voting classifier · Hyperparameter tuning · Performance metrics · ML classification models

U. Satish Varma · V. Dhiraj (✉) · B. Sekhar Babu · V. Dheeraj Varma · G. B. S. Karthik
Department of Computer Science and Engineering, Koneru Lakshmaiah Education Foundation, Vaddeswaram, AP, India

V. Rajesh
Department of Electronics and Communication Engineering, Koneru Lakshmaiah Education Foundation, Vaddeswaram, AP, India

1 Introduction

The source of diabetes dataset is from UCI Repository University of Washington. The one of the most important parts of machine learning model is to visualize the given dataset. After the visualization, we identified dataset is binary classification type, and we need to classify between the diabetes and non-diabetes, refer Fig. 2. We implemented the pie chart for dependent variable, by these we conclude dataset is a fairly balanced dataset (Fig. 1).

We identified the relationship between the two independent variables using pair plots, refer Fig. 3, and in most of the cases, the glucose level is greater than 100 is identified as diabetes and less than 100 is identified as non-diabetes. We also observed the age group between 30 and 40 has more diabetes when compared to non-diabetes as shown in Fig. 5. Standardizing the dataset is one of the most important to build the machine learning model because without the standardization the accuracy of logistic regression and linear SVM are 67 and 41%, respectively, the drastic improvement is observed, when the dataset is standardized the accuracy of logistic regression and linear SVM are improved from 67 to 74% and 41 to 75%, respectively.

Hypertuning our model helps to get the best accuracy for the machine learning models, we hypertuned our machine learning classification algorithms and we identified the best parameter for the algorithm. We also identified three best algorithms for voting classifier are random forest, GBDT, K-NN, respectively, and we obtained the train accuracy as 96%, test accuracy as 88.7%, F1 score as 0.83, precision score as 0.85, recall score as 0.8.

The performance metrics which we had implemented for evaluating our machine learning models are accuracy, F1 score, precision, recall, and ROC curve.

2 Literacy Survey

Vijayan et al. [1] proposed the AdaBoost algorithm with decision stump for diabetes dataset as base classifier and obtained a best accuracy of 80.72%, when compared to SVM, decision tree classifier, and Naive Bayes. AdaBoost had highest accuracy and performed way better than remaining other classifiers.

Mir et al. [2] have recommended us the best machine learning algorithms. In their research, they suggested us to use support vector machine because it has given the efficient performance when compared to Naive Bayes, simple cart algorithm, and random forest.

Yahyaoui et al [3] proposed an evaluated system on diabetes using the publicly available dataset named as Pima Indians. In the dataset, 500 labeled as diabetes and 268 labeled as non-diabetes. They implemented some mostly used ml classification algorithms like SVM, random forest and they compared with the CNN (deep learning model), and finally they concluded that the best accuracy was obtained for random forest with 83.67%.

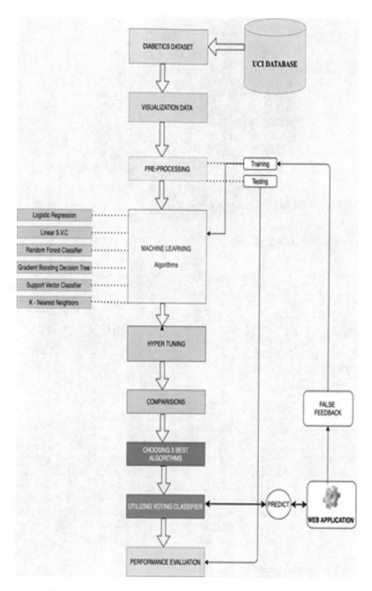

Fig. 1 Flowchart for the methodology

Rubaiat et al. [4] introduced a deep learning approaching by using multiple layer perceptron with feature selection acquired an accuracy as 85.15%. They also compared their result with other experiment models like GRNN, j48graft, hybrid model, MLP, ELM, and fuzzy rules. They clearly shown their result was way far better than others.

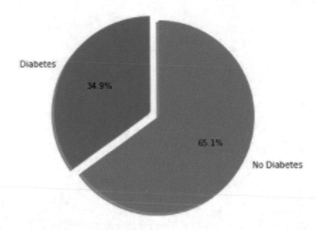

Fig. 2 Pie chart for dependent variable

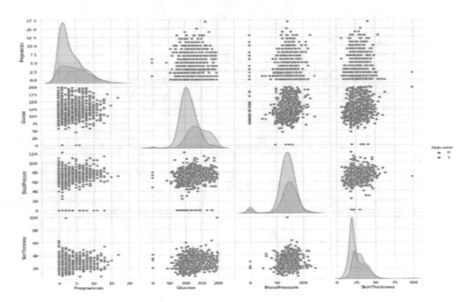

Fig. 3 Pair plot for three features

Agarwal et al. [5] explained the machine learning model helped in health sector, but while coming to diabetes, it does not performed due to huge variations between the respective individuals and lack of proper dataset, and they faced a difficulty in getting the best accuracy. The algorithm they used for diabetes prediction is decision tree classifier, logistic regression, Naive Bayes classifier, SVM, and K-NN. The best accuracy among these classification models is 81.1%. They acquired because of k-fold and cross validation.

Huda et al. [6] explained that due to diabetes the human organ eye has most effect due to diabetes. Sometimes, it occurs complete blindness, they proposed the classification models like decision tree, support vector machine, and logistic regression for detection of diabetes retinopathy in the human eyes. Their proposed model achieved 88% accuracy. Precision as 0.97 and recall as 0.92.

Chakour et al. [7] compared the result of machine learning models like artificial neural network, logistic regression, SVM with their own agent named as control agent. Their own agent retrieves the result of their ml models and after getting result it performs own calculation and acquires the output. They mentioned the classification rate for control agent as 0.97 and sensitivity as 0.98.

Guo et al. [8] proposed the Bayes network and acquired an accuracy 72.3%, after collecting the data for diabetes they implemented the pre-processing techniques, column normalization, and the data was given to the proposed Bayesian model and acquired a best accuracy for detecting the type-2 diabetes.

Khan et al. [9] work has demonstrated that rough sets provide an accuracy comparable to well-known classification techniques such as K-NN, neural networks, or with a principle component analysis. Their accuracy is higher than the remaining models, which they reported, and their result was obtained with less than 100 rules. They acquired an accuracy as 83%.

El Habib Daho et al. [10] implemented a new hybrid learning algorithm using PSO for adjusting function parameters. For their study, they taken the dataset form UCI Repository named as Pima Indians. The result indicates that the implemented model is accepted with a best accuracy.

Pavani et al. [11] they used mostly known ML classification models like DT, random forest, GBDT, K-NN, and Naive Bayes. After implementing the models based on accuracy, precision, recall, F1 score they evaluated the model. In this paper, they obtained the best accuracy 86% for random forest and Naive Bayes.

Mahabub et al. [12] implemented the voting-based classifier and acquired the accuracy 86%. They implemented eleven ML classification algorithms and identified the best three models based on performance metric like F1 score, accuracy, precision, recall and added to voting-based classifier and acquired the best accuracy for Pima Indian dataset.

3 Methodology

The whole work of proposed machine learning model can be seen in below flowchart refer Fig. 1. We also implemented the Web application to take the feedback from the user and add the related data in to the dataset and retrain our model periodically for every 10 days. We had used Python Jupyter notebook and many pre-defined toolkits like numpy, pandas, matplotlib, pyplot, sklearn, and seaborn. The main aim is to find the best three machine learning classification algorithm for voting classifier. In this section, we will explain the best hyperparameter for the ML algorithm. We obtained the best parameter by hypertuning and we will show the feature importance of each model in this section. The classification models used for detecting diabetes are logistic regression, linear SVM, random forest classifier, gradient boosting decision tree (GBDT), K-nearest neighbors classifier, and support vector classifier (RBF kernel).

3.1 Diabetes UCI Repository Dataset

The following dataset is taken from the UCI Repository database. We identified the dataset feature or columns as 9 and the rows as 768. We verified for missing values and we identified few of them using the function is null().any() and the identified null values are replaced with the mean of the column value and we also verified for duplicate values in the dataset and there was no duplicity. The mean of independent variables for pregnancies is 3.845052, for glucose is 120.864531, for blood pressure is 69.10, for skin thickness is 26.496, for insulin is 118.66, for BMI is 31.992, for diabetes pedigree function is 0.4718, and for age is 33.24. The remaining description of these features can be seen in Fig. 4.

	Pregnancies	Glucose	BloodPressure	SkinThickness	Insulin	BMI	DiabetesPedigreeFunction	Age	Outcome
count	768.000000	768.000000	768.000000	768.000000	768.000000	768.000000	768.000000	768.000000	768.000000
mean	3.845052	120.894531	69.105469	26.496094	118.664062	31.992578	0.471876	33.240885	0.348958
std	3.369578	31.972618	19.355807	9.734680	93.701163	7.884160	0.331329	11.760232	0.476951
min	0.000000	0.000000	0.000000	7.000000	14.000000	0.000000	0.078000	21.000000	0.000000
25%	1.000000	99.000000	62.000000	19.000000	68.000000	27.300000	0.243750	24.000000	0.000000
50%	3.000000	117.000000	72.000000	23.000000	100.000000	32.000000	0.372500	29.000000	0.000000
75%	6.000000	140.250000	80.000000	32.000000	127.250000	36.600000	0.626250	41.000000	1.000000
max	17.000000	199.000000	122.000000	99.000000	846.000000	67.100000	2.420000	81.000000	1.000000

Fig. 4 Output for the description of the dataset

Fig. 5 Bar plot for the age parameter

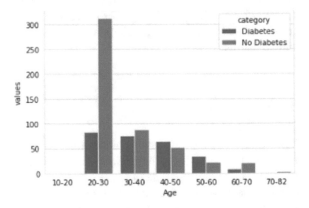

3.2 Visualizing the Dataset

Exploratory data analysis is one of the most important phases to build a robust machine learning classification model. During this visualizing the dataset, we identified few conclusions they are, with pie chart, we identified the dataset is fairly balanced dataset, with pair plot we identified the dataset is not linearly separable and if the glucose level is greater than 100, then it has high probability of getting diabetes, with distplot (combination of both histogram and distribution plot) we identified if age is greater than 30 then there is high chance of getting diabetes and if pregnancies is greater than 3 then there is high chance of getting non-diabetes, with box plot we identified the outlier for blood pressure and insulin, with bar plot we identified few conclusion from the given dataset, we identified the age group between 20 and 30 has less diabetes cases when compared to diabetes and the age group between 30 and 40, and the diabetes case are higher than non-diabetes and the detailed bar graph has shown in Fig. 5.

3.3 Pre-processing and Splitting the Dataset

Pre-processing the dataset is one of the important phases before applying the classification models. In this pre-processing phase, we verify the outlier and replace it with the mean of the column value and we also verify the duplicates values, and it will be removed for improving the results. The standardization the data is must in this pre-processing phase. The standardizing the data means the columns or feature value is set to zero and standard deviation as 1. After this process, we split the dataset into training and testing data, with the train data, we will be training our classification models, and with the testing data, we will be finding the model performance. We split the dataset into 70% as training and 30% as testing data. We handled the outlier by replacing the outlier value with the mean of that column.

3.4 Classifiers Used for Diabetes Detection

We used six classification models for detecting the diabetes or not diabetes. Our aim of this project is to identify the best few machine learning algorithm among logistic regression, linear SVC, random forest classifier, gradient boosting decision tree, K-nearest neighbors, and support vector classier (RBF Kernel based). After identifying the best classifiers, we add the classifier to voting-based classifier. We periodically retrain our models for every 10 days and false feedback we be adding into the excel sheet after authorizing. In this section, we will be clear explaining the hyperparameter of the classifiers and the feature importance of each classifier step by step.

3.4.1   Logistic Regression

For UCI diabetes dataset, the no of features is identified as 9, if we use L1 regularization the sparsity may increase so we had chosen 12 instead of 11.

We used the stochastic gradient descent function to implement logistic regression. If we choose loss as hinge, it works as linear support vector classifier. If we choose loss as log, it works as logistic regression.

In the Probabilistic View

$$W^* = \arg\min \sum_{i=1}^{n} -y_i * \log p_i - (1 - y_i)\log(1 - p_i) + \lambda\|W_i\|12 \qquad (1)$$

where $p_i = \sigma\left(W^T * x_i\right)$, $\|w_i\| \geq 0$.

P_i range varies between 0 and 1.

Y_i is may be 0 or 1 as it is the dependent variable.

In Geometric View

L2 Regularization:

$$W^* = \arg\min \sum_{i=1}^{n} \log(1 + \exp\left(-y_i * W^T * x_i\right) + \lambda\|W_i\|12 \qquad (2)$$

where $W^T = $ transpose of W.

$X_i = i$th value of x values vector.

$Y_i = i$th value of independent value it may be 1 or 0.

λ is the hyperparameter theoretically, but while coming to hypertuning using predefined stochastic gradient descent, it is defined as alpha (α), if the α values is higher, then the regularization will be more stronger.

After the hyperparameter tuning, we identified the alpha (α) value as 0.01 with train accuracy as 76%, test accuracy as 78%, F1 score as 0.65, precision score as 0.73, recall as 0.59, and for area under curve (AUC) refer Fig. 6.

The decreasing order of feature importance for logistic regression is as follows: Glucose > BMI > Pregnancies > DiabetesPredigreeFunction > BloodPressure > Skinthickness > Age > Insulin.

Fig. 6 ROC curve for
logistic regression

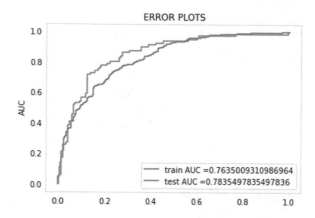

3.4.2 Linear SVM

We used the stochastic gradient descent function to implement linear support vector
machine. If we choose loss as hinge, it works as linear support vector classifier. If
we choose loss as log, it works as logistic regression.

Theoretically:

$$W^*, b^* = \arg_{w,b} \min \sum_{i=1}^{n} \max\left(0, y_i * \left(W^T * x_i + b\right)\right) + \lambda \|W_i\|l_2 \qquad (3)$$

Such that $y_i * \left(W^T * x_i + b\right) \geq 1 - \xi_i$.

$\xi_i \geq 0$, where hyperparameter for SVM is C, $C = 1/\lambda$. We used loss as hinge where
hinge loss $= \max\left(0, 1 - y_i^*\left(W^T * x_i + b\right)\right)$.

ξi is defined as average distance of the miss classified points from the acquired
hyperplane π.

For pre-defined classifier like SGD classifier, we use alpha (α) if the α values is
higher, then the regularization will be more stronger, hyperparameter $C = 1/\alpha$.

After hypertuning the classifier, we identified the best alpha as 0.01 with training
accuracy as 77%, testing accuracy as 77%, precision as 0.73, recall as 0.555, and F1
score as 0.633.

The decreasing order of feature importance for linear support vector machine is
as follows: Glucose > BMI > Pregnancies > DiabetesPredigreeFunction > Insulin >
BloodPressure > Age > Skinthickness.

Fig. 7 ROC curve for
random forest classifier

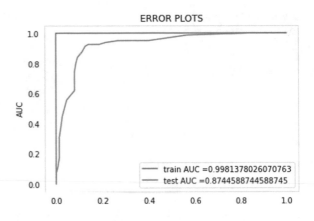

3.4.3 Random Forest Classifier

We used pre-defined function named as random forest classifier, we need to choose the criterion between Gini and entropy, and we chose Gini because the performance of the Gini for these type of dataset is better than entropy.

Gini Index

$$G_i = 1 - \sum_{i=1}^{n} p_i^2 \tag{4}$$

where p_i = probability of the element classified for respective class.

G_i ranges between, $0 \leq G_i \leq 1$.

If $G_i = 0$, then it means the purity of classification.

If $G_i = 1$, then it means the random distribution of element.

The parameter for random forest classifier is n_estimators, after hypertuning the best n_estimators is 25 with train accuracy as 99% and test accuracy as 87%, F1 score as 0.81, precision as 0.833, and recall as 0.80. For ROC curve refer Fig. 7.

As like we mentioned, feature importance for logistic regression and linear SVC and for ensembles model like random forest, we cannot find the feature important.

3.4.4 Gradient Boosting Decision Tree

It is robust to overfitting, the larger value of n_estimators result with better performance. We hypertuned the model and identified n_estimators as 30 and acquired a train accuracy as 94%, test accuracy as 0.8701, F1 score as 0.807, precision as 0.84, recall as 0.777, area under curve (AUC) refer Fig. 8.

Fig. 8 ROC curve for
gradient boosting

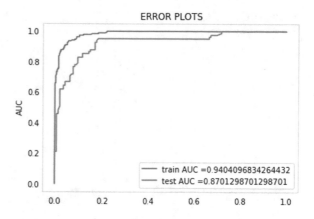

As like we mentioned, feature importance for logistic regression and linear SVC and for ensembles model like random forest, we cannot find the feature important.

3.4.5 Support Vector Classifier (RBF Kernel)

$$\max \alpha_i \sum_{i=1}^{n} \alpha_i - 1/2 \sum_{i=1}^{n} \sum_{j=1}^{n} \alpha_i * \alpha_j * y_i y_j * x_i^{\mathrm{T}} \tag{5}$$

$x_i^{\mathrm{T}} x_j$ can be used as kernel function in our implementation, we used radial basis function (RBF) kernel.

Where $0 \leq \alpha_i \leq C$, C is a hyperparameter for SVC.

If $\alpha_i = 0$ for non-support vectors else greater, than 0 if support vector.

A poor performance of linear SVM for diabetes dataset, so we decided to implement the kernel-based SVM, we used RBF kernel with pre-defined function from sklearn library. Hypertuned the machine learning algorithm with C as 1000, gamma value as 0.01 with a train accuracy as 88%, test accuracy as 80%, F1 score as 0.75, precision as 0.80, and recall as 0.70.

3.4.6 K-Nearest Neighbors

The best algorithm when we identified the outlier data because it has less impact on the outliers, and it only follows the majority voting-based on k-nearest points. The distance is calculated using three ways it can be Euclidean distance, Manhattan distance, and Minkowski distance by default, we used Euclidean distance, where k is a hyperparameter for K-nearest neighbors theoretically based on programmatically, and the hyperparameter is defined as n_jobs.

By default, it uses p value as 2.

Fig. 9 ROC curve for
K-nearest neighbors

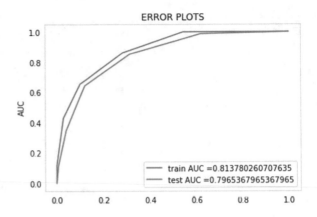

Minkowski Distance

$$\text{dist}(x, y)_p = \left(\sum_{i=1}^{n} \left(|xi - yi|^p \right) \right)^{1/p} \tag{6}$$

dist (A, B) means distance of x_i and y_i.

if $p = 2$ it behaves like a Euclidean distance.

if $p = 1$ it behaves like a Manhattan distance.

We hypertuned the model and identified K as 5 and acquired a train accuracy as 81.3%, test accuracy as 79.6%, F1 score as 0.688, precision as 0.74, recall as 0.64, and for area under curve (AUC) refer Fig. 9.

The decreasing order of feature importance for K-nearest neighbors is as follows: Pregnancies > Glucose > BloodPressure > SkinThickness > Insulin > BMI > DiabetesPredigreeFunction > Age.

4 Identifying the Best Three Models by Their Performance

The performance of the six-machine learning classifier is listed above in Table 1.

Based on the performance, we identified the best algorithms they are (1) random forest, (2) decision tree, (3) SVC, and (4) K-NN.

5 Voting Classifier

This is the proposed model for diabetes dataset form UCI Repository, the two parameters in voting classifier, i.e., hard voting and soft word, the major difference between

Table 1 Output for all the classifiers

S. No.	Algorithms	Train accuracy (%)	Test accuracy	Precision	Recall	F1 score
1	Logistic regression	76	78	0.73	0.59	0.65
2	Linear SVM	77	77	0.73	0.55	0.63
3	Random forest classifier	99	87	0.83	0.802	0.817
4	GBDT	94	87	0.84	0.77	0.807
5	SVC RBF Kernel	85	83	0.802	0.70	0.75
6	K-NN	81	79	0.74	0.64	0.68

both of these are while coming to hard voting it is nothing but majority voting, while coming to soft voting it is the average of results came from regression models. Based on the dataset and our objective, we chosen hard voting for the proposed system, while coming to the performance, we acquired training accuracy as 96% and test accuracy as 88%, F1 score as 0.83, recall as 0.81, and precision as 0.85.

The chosen models for voting classifier are: random forest, GBDT, and K-nearest neighbors.

6 Results

See Figs. 2, 3, 4, 5, 6, 7, 8, and 9.

7 Conclusions

We conclude by choosing the voting classifier by taking the random forest classifier, gradient boost decision tree, and K-NN, and we obtained train accuracy as 96% and test accuracy as 88%. We also hypertuned our classification model to give their best from it.

Future Enhancement

We will be deploying our model to cloud platform using Heroku by using Web application framework named as flask using Python programing language and we will also take the user feedback data and add the data into the dataset and periodically retrain our model for better accuracy.

Acknowledgements We would like to thank the K L Management and the respective faculty who have supported us throughout the work and provided us with all the required facilities.

References

1. V.V. Vijayan, C. Anjali, Prediction and diagnosis of diabetes mellitus—a machine learning approach. 2015 IEEE Recent Advances in Intelligent Computational Systems (RAICS). IEEE (2015), pp. 122–127
2. A. Mir, S.N. Dhage, Diabetes disease prediction using machine learning on big data of healthcare. 2018 Fourth International Conference on Computing Communication Control and Automation (ICCUBEA). IEEE (2018), pp. 1–6
3. A.A. Nayak, K. Pandit, *Prediction and Analysis of Diabetes Mellitus on Different Features using Machine Learning Algorithm* (2019)
4. S.Y. Rubaiat, M.M. Rahman, M.K. Hasan, Important feature selection & accuracy comparisons of different machine learning models for early diabetes detection. 2018 International Conference on Innovation in Engineering and Technology (ICIET). IEEE (2018), pp. 1–6
5. A. Agarwal, A. Saxena, Analysis of machine learning algorithms and obtaining highest accuracy for prediction of diabetes in women. 2019 6th International Conference on Computing for Sustainable Global Development (INDIACom). IEEE (2019), pp. 686–690
6. S.A. Huda, I.J. Ila, S. Sarder, M. Shamsujjoha, M.N.Y. Ali, An improved approach for detection of diabetic retinopathy using feature importance and machine learning algorithms. 2019 7th International Conference on Smart Computing & Communications (ICSCC). IEEE (2019), pp. 1–5
7. I. Chakour, Y. El Mourabit, C. Daoui, M. Baslam, Multi-Agent System based on machine learning for early diagnosis of diabetes. 2020 IEEE 6th International Conference on Optimization and Applications (ICOA). IEEE (2020), pp. 1–6
8. Y. Guo, G. Bai, Y. Hu, Using bayes network for prediction of type-2 diabetes. 2012 International Conference for Internet Technology and Secured Transactions. IEEE (2012), pp. 471–472
9. A. Khan, K. Revett, Data mining the PIMA dataset using rough set theory with a special emphasis on rule reduction, in *8th International Multitopic Conference, 2004. Proceedings of INMIC 2004.* IEEE (2004), pp. 334–339
10. M.E.H. Daho, N. Settouti, M.E.A. Lazouni, M.A. Chikh, Recognition of diabetes disease using a new hybrid learning algorithm for NEFCLASS. 2013 8th International Workshop on Systems, Signal Processing and their Applications (WoSSPA). IEEE (2013), pp. 239–243
11. K. Pavani, P. Anjaiah, N.K. Rao, Y. Deepthi, D. Noel, V. Lokesh, Diabetes prediction using machine learning techniques: a comparative analysis, in *Energy Systems, Drives and Automations.* (Springer, Singapore, 2020), pp. 419–428
12. A. Mahabub, A robust voting approach for diabetes prediction using traditional machine learning techniques. SN Appl. Sci. **1**(12), 1–12 (2019)

Performance Analysis of a Simulation-Based Communication Network Model with NS2 Simulator

G. Rajendra Kumar, B. Dinesh Reddy, N. Thirupathi Rao,
and Debnath Bhattacharyya

Abstract This paper displays a system show for cutting edge cell systems intended to meet the dangerous needs of versatile information while limiting vitality utilisation. The store technique is used in the remote edge of the spread of prominent substance documents, along these lines lessening the substance and the separation between the requestor. Keeping in mind the end goal to improve utilisation of the accessible store space, the reserve system is upgraded and multicast transmission is considered. The numerical after-effects of the following drive demonstrate that mix of reserve and multicast is compelling when there is happening rehashed asked for a couple of substance records show up over the long haul. It can in reality diminish vitality costs within the sight of an extensive interest for postponed resilience content.

Keywords Communication networks · 5G networks · Caching · Multicast · Routing

1 Introduction

We are seeing an uncommon overall development of versatile information movement that is relied upon to proceed at a yearly rate of throughout the following years, achieving exabytes every month. To deal with this "information wave", the rising fifth era 5G frameworks need to enhance the system execution as far as vitality utilisation, throughput and client experienced deferral, and in the meantime improve a utilisation of the system assets, for example, remote data transmission and back-haul interface limit [1]. With the introduction of sensible hand-held gadgets, client demands for versatile broadband are encountering accomplice degree uncommon climb. The strong advancement of information exchange limit hungry applications

G. Rajendra Kumar · B. Dinesh Reddy (✉) · N. Thirupathi Rao
Department of Computer Science & Engineering, Vignan's Institute of Information Technology (A), Visakhapatnam, Andhra Pradesh, India

D. Bhattacharyya
Department of Computer Science and Engineering, K L Deemed to be University, Guntur 522502, India

© The Author(s), under exclusive license to Springer Nature Singapore Pte Ltd. 2021
S. K. Saha et al. (eds.), *Smart Technologies in Data Science and Communication*,
Lecture Notes in Networks and Systems 210,
https://doi.org/10.1007/978-981-16-1773-7_11

like video spouting and transmission report sharing is as of now pushing the breaking points of current cell structures. Inside the next decade, imagined media-rich versatile applications like telecom and 3D optics would require information rates just horrendous with fourth period (4G) systems.

The relentlessly creating enthusiasm for higher information rates and limit require odd theory for taking after period (5G) cell systems. Pleasant trades have such certification. Supportive trades address a substitution grouping of remote correspondence methodology in the midst of which organise hubs empower each other in giving off data to get a handle on special varying characteristics benefits [2]. This new transmission worldview guarantees key execution grabs similar to association recklessness, extraordinary quality, structure limit and transmission change. Pleasant correspondence has been generally considered inside the written work, and stuck terminal exchanging (which incorporates the preparing of low-control base stations to help the correspondence between the supply and thusly the objective) has just been encased inside the 4G future evolution (LTE)—Advanced standard. Mounted terminal exchanging gains upgrades cell structures; however, the aggregate ability of interest may be recognised solely through the use of gadget giving off. The term gadget here implies a cell phone or the other moveable remote gadget with cell property (tablet, compact workstation, et cetera) a client claim. Gadget exchanging makes it practical for gadgets in the midst of a system to execute as transmission exchanges for every interesting and notice a huge unconstrained work arrange [3].

2   Proposed System

Heterogeneous cell systems show that backings are storing and multicast for the administration of the portable clients. Stores can introduce at little cell base stations (SBSs), e.g. picocells and femtocells, focusing to offload movement from the assembled full-scale cell base station (MBS). Estimation examines uncovered up to 66% diminishment in arrange activity by utilising reserving in 3G and 4G systems. In the interim, the remote business started to popularise frameworks that help to reserve. Numerous administrators exploit multicast to effectively use the available data transfer capacity of their systems in conveying a similar substance to different collectors. For instance, multicast regularly utilised for conveying supported substance, e.g. versatile notices in specific areas, downloading news, securities exchange reports, climate and games refreshes. In the interim, multicast has fused in 3GPP details in which the proposed innovation for LTE is called evolved multimedia broadcast and multicast services (eMBMS). This innovation used for different cells where the transmission crosswise over them is synchronous utilising a common bearer recurrence. Henceforth, multicast expends a subset of the radio assets required by a unicast benefit. The rest of the assets can be utilised to help transmissions towards different clients, in this manner upgrading system limit.

3 System Design

Numerous new web-based business applications, including portable sales, will likewise increase unique advantage if remote systems uphold amass correspondence among versatile clients. These two unique ideas converge to decrease theinformation activity in foreign correspondence. HCN displays that backings are reserving and multicast for the administration of the portable clients [4, 5]. Solicitations for a similar substance record produced amid a brief span window are amassed and served through a single multicast transmission. Nearby storing of prevalent records at the small cell base stations (SBS) has been as of late proposed, going for diminishing the movement caused while exchanging asked for content from the centre system to the clients. The reserve plan strategy precisely considers the way that an administrator can serve the solicitations for a similar document that occurs at close-by times using a single multicast transmission. That is, multicasting is the exchange of messages to numerous goals all the while, utilising fewer systems. The data is conveyed to every one of the connections just once, and duplicates made when the connections to the c goals split, therefore making an ideal appropriation way. It decreases unnecessary parcel duplication. A general mix of reserving and multicasting brings about less movement in versatile correspondence and extend vitality effectiveness.

3.1 Favourable Circumstances

1. Multicast administrations do not sit tight for all framework ask. It just reaction the solicitations for a similar record that occurs at close-by times. So, this proposed framework diminished the time delay.
2. This framework for improving the system limit
3. Because of the caching idea, the separation between centre system and the end client gadget had decreased. It fulfils the request on information movement.

Impromptu On-request Distance Vector (AODV) Routing Protocol
The ad hoc on-request distance vector directing convention acquires the great highlights of both DSDV and DSR. The AODV steering convention utilises a responsive way to deal with discovering courses and a proactive approach for distinguishing the latest way. All the more particularly, it discovers courses utilising the course revelation process like DSR and utilisations goal succession numbers to register new courses. The two stages examined in more detail.

3.2   *Route Discovery*

Amid the course disclosure process, the source hub communicates RREQ bundles like DSR. The RREQ parcel contains the source identifier (SId), the goal-identifier (DId), the source grouping number (SSeq), the goal arrangement number (DEQ), the communicate identifier (BId) and TTL fields. At the point when a transitional hub gets an RREQ bundle, it either advances it or readies a route reply (RREP) parcel if it has a substantial course to the goal in its reserve. The (SId, BId) combine is utilised to decide whether a specific RREQ has just been gotten keeping in mind the end goal to dispense with copies. Each middle of the road hub enters the past hub's address and its bid while sending an RREQ parcel. The hub likewise keeps up a clock related to each section to erase an RREQ bundle if the answer is not gotten before it terminates [6–9].

At whatever point a hub gets an RREP bundle, it stores the data of the past hub keeping in mind the end goal to forward the bundle to it as the next jump towards the goal. This goes about as a "forward pointer" to the goal hub. Therefore, every hub keeps up just the following bounce data, not at all like source steering in which all the middle of the road hubs on the course towards the goal are put away. Figure 4.2 demonstrates a case of course disclosure instrument in AODV. Give us a chance to assume that hub 1 needs to send an information parcel to hub seven; however, it does not have a course in its reserve. At that point, it starts a course disclosure process by communicating an RREQ parcel to all its neighbouring hubs.

4   Results

The results of the current model considered are as follows for various nodes which was run on NS 2 simulator. The performance had presented in the form of graphical representation for better understanding of the considered model simulation process (Figs. 1, 2, 3 and 4).

The performance for the scenarios which were considered in the current analysis under the simulation mode had presented in above Figs. 5, 6 and 7. The results were analysed and discussed in detail with the help of graphical representation.

5   Conclusion

We propose a reserving model that can diminish the vitality expenses of broadscale versatile information needs in 5G remote systems. In opposition to conventional reserving plans, common storing procedures essentially convey common substance to close clients, and our reserving technique is all around intended to make additional utilisation of multicast. This is critical because multicast is a worry for the proficient

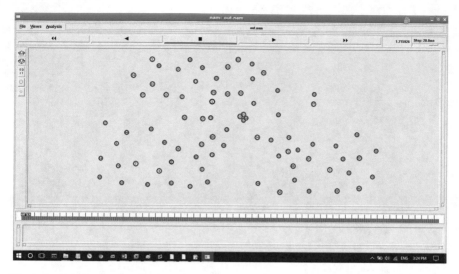

Fig. 1 Starting simulation

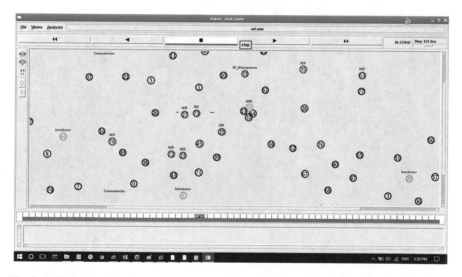

Fig. 2 Initialisation of mobile nodes

conveyance of innovation in an advanced cell arrange. To defeat the NP-hardness nature of the re-get to store issue, we presented a calculation with execution ensures and acquainted a straightforward heuristic calculation with assessing its adequacy via deliberately following the driver's numerical examination. The outcomes demonstrate that portfolio is reserving, and multicast can decrease vitality costs when the interest for deferred resistance content is enormous. At the point when the client endures three minutes of postponement, with the steepness of the substance gets to

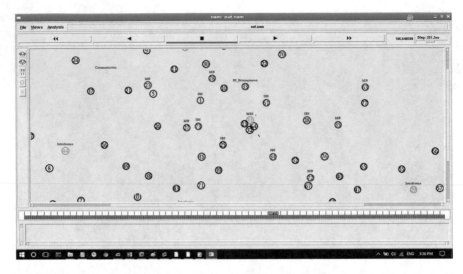

Fig. 3 Macrocell base station updating cache of small base station 34

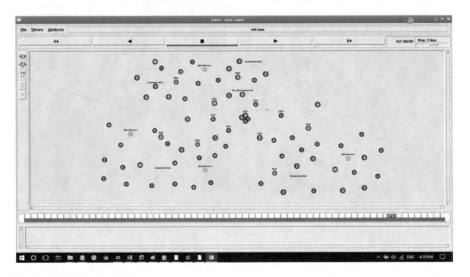

Fig. 4 Simulation ending

show and further increment, the current store program income of 19%. When all said in done, our work can be viewed as an endeavour to methodically join reserving and multicast as methods for enhancing vitality productivity in 5G remote systems.

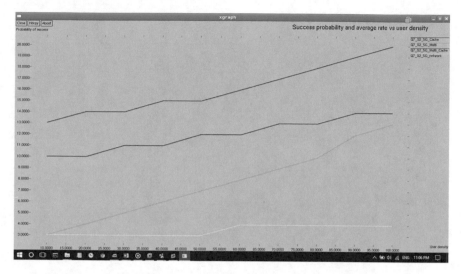

Fig. 5 Comparison of average transmission rate in 5G techniques

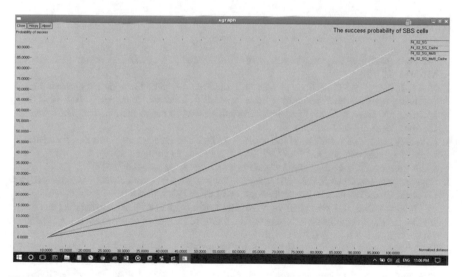

Fig. 6 Cache success probability in SBS's

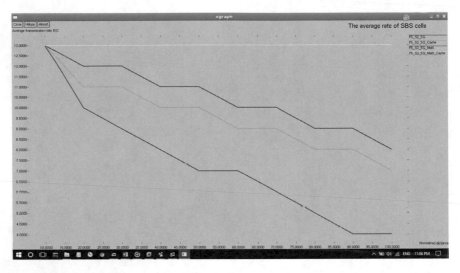

Fig. 7　Average transmission rate in SBS's

References

1. J.G. Andrews, Seven ways that hetnets are a cellular paradigm shift. IEEE Commun. Mag. **51**(3), 136–144 (2013)
2. Y. Xu, J. Wang, Q. Wu, Z. Du, L. Shen, A. Anpalagan, A game theoretic perspective on self-organising optimisation for small cognitive cells. IEEE Commun. Mag. **53**(7), 100–108 (2015)
3. J. Erman, A. Gerber, M.T. Hajiaghayi, To cache or not to cache-the 3G case. IEEE Internet Comput. **15**(2), 27–34 (2011)
4. K.V. Satyanarayana, K.Sudha, N.Thirupathi Rao, D. Bhattacharyya, Analysis of queuing model based cloud data centers, in *SMARTDSC2019, LNNS*, 105 (Springer, 2019), 293–309
5. Light Reading, *NSN Adds ChinaCache Smarts to Liquid Applications* (2014)
6. Saguna, "Saguna Open-RAN", 2015, https://www.saguna.net/products/sagunacods-open-ran. OFweek, China Telecom successfully deployed LTE eMBMS (2014)
7. N. Thirupathi Rao, D. Bhattacharyya, V. Madhusudana Rao, T. Kim, Analyzing the behavior of reactive protocols in wireless sensor networks using TCP mode of connection. Int. J. Eng. Adv. Technol. **8**(4), 1387–1392 (2019)
8. D. Bhattacharyya, N. Thirupathi Rao, S.K. Meeravali, K. Srinivasa Rao, Performance of a parallel communication network having non-homogeneous binomial bulk arrivals with phase type transmission for dynamic resource allocation. J. Green Eng. **10**(9), 7155–7177 (2020)
9. K. Poularakis, G. Iosifidis, V. Sourlas, L. Tassiulas, Multicast-aware caching for small-cell networks, in *IEEE Wireless Communications and Networking Conference (WCNC)* (2014), 2300–2305

A Comparative Analysis on Resource Allocation in Load Balancing Optimization Algorithms

P. S. V. S. Sridhar⬛, Sai Sameera Voleti⬛, Manisha Miriyala⬛,
Nikitha Rajanedi⬛, Srilalitha Kopparapu⬛,
and Venkata Naresh Mandhala⬛

Abstract With the constant involvement of technologies in daily lives, it has been important to establish a system which is more efficient. It is important for individual machines to handle various tasks and schedule them properly. There are numerous technologies that had been created in the recent decades and are unable to comprehend the tremendous capacity of work that is put in the backend. The most under looked algorithms in the modern world is various job management algorithms. With many technical establishments and software that has been created, it can be observed that the most sophisticated features within them was the ability to multitask or carry out various tasks at once. High computational capacity and obtaining fast results is one of the essential demand of the modern world. Two different algorithms specialized in management within a Cloud-Based Architecture are worked. The main focus is upon Heterogeneous Earliest Finish Time (HEFT) and Ant Colony Optimization (ACO) as well. Both the algorithms are established and working along with those algorithms to understand and depict how they both function and work internally. An additional study is done about the drawbacks of HEFT algorithm and benefits of ACO algorithm and a comparative analysis is created. Through this paper, the two algorithms are explained in a clear manner and publish the ideology of why Ant Colony Optimization is more suitable and more efficient than HEFT. A clear illustration and findings are given about how Ant Colony Optimization better than HEFT. To support this assessment, various reasons and statements are given to the conclusion.

Keywords Cloud Computing · Load balancing · Heterogeneous Earliest Finish time · Ant Colony Optimization · Multitasking

P. S. V. S. Sridhar · S. S. Voleti (✉) · M. Miriyala · N. Rajanedi · S. Kopparapu · V. N. Mandhala
Department of Computer Science and Engineering, Koneru Lakshmaiah Education Foundation, Vaddeswaram, Andhra Pradesh, India

1   Introduction

One of the most under viewed algorithms in the modern world is related to job sequencing as well as scheduling. When looked upon from an overview aspect everyone may simply overlook these sophisticated algorithms with the belief that they are not important since their lack of exposure is less within the interface of a system or software. However, in a technical aspect, these are some of the most sophisticated and supreme aspects of any software in addition to hardware device released in the modern market. Although hardware plays a major role in providing computational power to run and process the large loads of traffic as well as tasks which the system is receiving this is only possible with sufficient software to make these possible [1]. Over the past few decades, various seamless algorithms have been created which focus upon working with computer hardware to both multitask as well as process more information in a unique, uniform manner to acquire the required outputs [2]. There are various algorithms installed in systems. But the task scheduling algorithm is far more superior as every algorithm will be further divided into millions of tasks and fed to the system to obtain the required commands. Upon the completion of these tasks, observation is done and these diverse tasks are combined now with the effort of the ACO algorithm to create an output that is shown to the user as the results [3]. As it can be observed, when an algorithm is cleft into millions of tasks it is the responsibility of the task scheduling methods to arrange the tasks suitably and arrange proper computation power and requirements. These play a vital role in ensuring the completion of a task as every task must be allocated the proper computational power within the system's hardware as well as access to several forms of data that may need to be stored or retrieved from within the system [4, 5]. All of these must be achieved in a capable and productive manner to ensure there are no errors upon establishing the task requests. Another important thing which a task scheduling organizer must be specialized upon is to certify that the tasks are correctly allocated with a proper basis to verify the fast processing of every task by having them scheduled by taking significant factors and parameters into accounts [6, 7]. Since a basic idea about the topic is done, the following deals with advantages of proposed system and implementation.

2   Literature Survey

Whenever the traffic increases, the cloud compromises. For the cloud not to compromise scheduling and balancing algorithms are used.

There are countless load balancing algorithms proposed in recent years and each focused on different algorithms and techniques, the techniques of balancing the load in cloud computing, analyzing various algorithms using cloud analyst, and comparing different algorithms like Round Robin, AMLB, and throttled load balancing algorithm. The authors executed all three algorithms in a cloud analyst and explained the comparative contrast between the three algorithms. Supporting these statements, simulated results have also been displayed [8].

A significant way to deliver cloud services in an efficient manner. This paper tells us about the difference between HEFT and ACO and how Ant Colony Optimization is better than Predict Earliest Finish Time. For a job to allocate and monitor, the VMs are checked [9]. Virtual machines and data centers are associated with one another and how they work is scheduled and explains how the resources are allocated. Similarly, this paper proposes a hybrid structure where, though the traffic increases, the Quality of Service should not be down [10].

3 Algorithms Used

This paper provides various environments and scenarios are created for the algorithms with the recently established software CloudSim. CloudSim is an open-source cloud simulation software established by the University of Melbourne. This software has allowed to create and schedule various tasks as well as the creation of different cloud entities such as the data centers and brokers which will play an important role during the execution of the algorithms. In this project, the CloudSim tool is applied upon two algorithms related to task scheduling to compare and depict which may be a better selection of the two. The algorithms implemented are:

I.   Heterogeneous Earliest Finish Time
II. Ant Colony Optimization

3.1 Heterogeneous Earliest Finish Time

The first algorithm that is implemented within the project is the HEFT algorithm. This algorithm has been used by cloud scientists around the world due to its ability to work with heterogeneous systems in an efficient manner [9].

The HEFT algorithms establish two major phases. The first phase of the algorithm is focused upon the scheduling of the distinct types of tasks that are collected by HEFT algorithm. In this phase of the algorithm, it will create a list of all of the obtained tasks and have them sorted with respect to their critical path in descending order (Fig. 1).

It is again later used during the scheduling of the tasks and making sure to complete all of them in a time-efficient manner possible. As the name suggests the main objective of this algorithm is to complete all the tasks in the fastest time possible. Once this first phase is done, it will advance onto the next phase of the algorithm which will play an even more crucial role. In this algorithm, distinct processors are used with various internal structures or software. Due to this, the second phase of this algorithm will be mainly focused upon the selection of an adequate processor to complete the task. The network will be containing several processors to complete

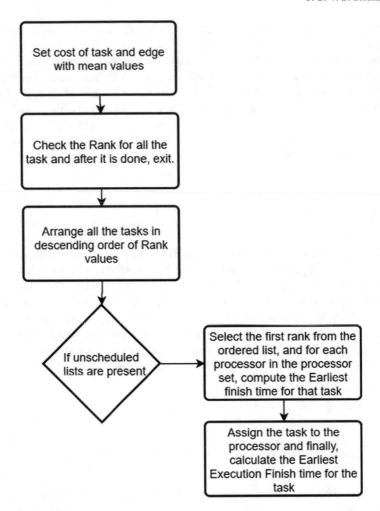

Fig. 1 HEFT algorithm step

the various tasks within the list. It is the responsibility of the proposed algorithm to understand the computational power of each processor and assign tasks to them to ensure that they are not overloaded. Also, make sure to ensure that the time of these tasks assigned to the various processors does not overlap to cause issues within the system. Once both the phases are done, the algorithm will be begun to work upon all of the tasks and have them successfully executed to obtain the desired outputs.

3.2 Ant Colony Optimization (ACO)

The second algorithm implemented in the project is the ant colony-based algorithm. This method of task management is similar to that of the HEFT algorithm; however, its internal structure works differently (Fig. 2).

Instead of creating a list of received tasks and obtaining the most efficient order to execute these tasks, this algorithm will create various graphs with the allotted tasks [11, 12]. Once these graphs have been established, the algorithm will work upon establishing the shortest path possible while taking the computational power of each processor into account at the same time, unlike the HEFT which split these into two different phases. It will further use the various optimization techniques which it has established within itself to further shorten the task execution time.

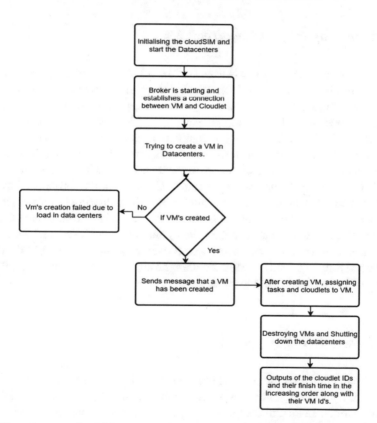

Fig. 2 Execution steps for ACO

4   Methodology

To simulate these algorithms within a cloud simulator, first completely gather the data and establish the required fundamentals within the systems. This project has been apportioned into a total of three different stages which have been crucial to making it possible, and a further explanation about these three stages in detail is given in this section of the paper.

The stages in the methodology are:

(i)   Gathering Project Requirements
(ii)   Testing HEFT Algorithm
(iii)   Testing ACO Algorithm

4.1   Gathering Project Requirements

The first main step in the project is to assemble and set up all of the project requirements. Before beginning to work on running these algorithms through the various simulations, first, gain an understanding of the project requirements and software. The major objective of the project has been to identify which out of these two algorithms (i.e., HEFT and Ant Colony Organization) is more efficient with respect to load balancing and which algorithm displays better results. To accomplish this, gain an understanding of how to assess both algorithms along with the type of environments in which the algorithms would be tested. Once a clear idea is obtained upon the aspects that need to be focused on, the latter is left with gathering the project requirements. CloudSim is used to simulate these algorithms and make sure that CloudSim software was properly installed within the system along with the various libraries and jar files required to compile and execute the algorithms. Once all of these have been successfully obtained and stored within the algorithm, the implementation part will be in the next section of the paper.

4.2   Testing HEFT Algorithm

The environment for the project has been created within the systems, and it is now time to work with the CloudSim software to test the first algorithm. First, initialize all of the various cloud entities functioning as the number of virtual machines, data center, cloudlets, etc. to create the proper testing simulation. Once all of these entities have been properly established and created, work upon the establishment of Heterogeneous Earliest Finish Time and having it executed in the simulation platform within the software. During the simulation, constantly monitor how the algorithm is executing the tasks and observe how well it can manage the load placed upon it. It is keen to note down all the observations so that it can be referred to them later and

Fig. 3 CloudSim
architecture

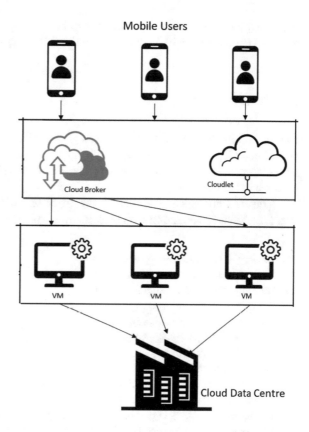

understand how effective it may have been at various stages of the simulation with comparison to the secondary algorithm, i.e., Ant Colony Optimization (Fig. 3).

4.3 Testing Ant Colony Optimization

The final stage of the project will consist of simulating the second algorithm that we wanted to implement. Now that the results of the first algorithm are successfully obtained, the objective while testing the second algorithm is to ensure that ACO and HEFT algorithms are exposed to the same environment within the simulation to ensure that the results are not biased and improper. Make sure that all of the entities used within the first simulation are properly recreated and established in the second simulation for the Ant Colony Organization algorithm as well. Once the simulation has started to take place, it is important to constantly monitor the load handling capacity displayed by the algorithm along with its efficiency. Now, observe and take down the results similarly to those of the first algorithm, and these results can be used later for comparison to establish the best of these two. The results for HEFT and ACO algorithms will be displayed in the following section of paper.

5 Results

In this section, the outputs for both Heterogeneous Earliest Finish Time (HEFT) and Ant Colony Optimization (ACO) are displayed. In HEFT, the number of cloudlets, data centers and, users are fixed in the code. The comparison between these two will be done in the conclusion part.

In Fig. 4, static values are given for the no. of users and data centers. In the initial stage, a global broker along with broker and two data centers are started. Later on,

```
Starting CloudSimExample8...
Initialising...
Starting CloudSim version 3.0
GlobalBroker is starting...
Datacenter_0 is starting...
Datacenter_1 is starting...
Broker_0 is starting...
Entities started.
0.0: Broker_0: Cloud Resource List received with 2 resource(s)
0.0: Broker_0: Trying to Create VM #0 in Datacenter_0
0.0: Broker_0: Trying to Create VM #1 in Datacenter_0
0.0: Broker_0: Trying to Create VM #2 in Datacenter_0
0.0: Broker_0: Trying to Create VM #3 in Datacenter_0
0.0: Broker_0: Trying to Create VM #4 in Datacenter_0
0.1: Broker_0: VM #0 has been created in Datacenter #3, Host #0
0.1: Broker_0: VM #1 has been created in Datacenter #3, Host #0
0.1: Broker_0: VM #2 has been created in Datacenter #3, Host #0
0.1: Broker_0: VM #3 has been created in Datacenter #3, Host #1
0.1: Broker_0: VM #4 has been created in Datacenter #3, Host #0
0.1: Broker_0: Sending cloudlet 0 to VM #0
0.1: Broker_0: Sending cloudlet 1 to VM #1
0.1: Broker_0: Sending cloudlet 2 to VM #2
0.1: Broker_0: Sending cloudlet 3 to VM #3
0.1: Broker_0: Sending cloudlet 4 to VM #4
0.1: Broker_0: Sending cloudlet 5 to VM #0
0.1: Broker_0: Sending cloudlet 6 to VM #1
0.1: Broker_0: Sending cloudlet 7 to VM #2
0.1: Broker_0: Sending cloudlet 8 to VM #3
0.1: Broker_0: Sending cloudlet 9 to VM #4
```

Fig. 4 Execution of HEFT algorithm

VMs are created in the data centers, and a cloudlet is assigned to each VM. Cloudlets are not randomly assigned to VMs; instead, they are assigned in order. After receiving the cloudlets, the finish time (the time which the cloudlet took to go to the VM) is calculated, VMs are destroyed and the Broker is shut down.

In Fig. 5, cloudlet IDs along with their respective VM IDs are displayed with finish time in ascending order. In the out, it is observed that each data center is assigned with more than one task. Till every task in the data center is done, all the remaining tasks should wait. This will increase the waiting time, since there are more than 2–3 tasks in the VMs (created in data center), the load is not balanced. This is the main drawback in HEFT algorithm.

In Fig. 6, Ant Colony Algorithm is started and in this algorithm, number of data centers and cloudlets are dynamically assigned. In this stage, cloud simulation entities for Ant Colony Optimization are initialized.

In Fig. 7, same as the HEFT algorithm, the start of execution of Ant Colony Optimization is observed. All the cloud network is initialized, cloudlets are already defined and started, data centers are started and a communication is established between them. The VMs are created inside the data center and are started. In this,

```
========== OUTPUT ==========
```

Cloudlet ID	STATUS	Data center ID	VM ID	Time	Start Time	Finish Time
0	SUCCESS	3	0	320	0.1	320.1
5	SUCCESS	3	0	320	0.1	320.1
1	SUCCESS	3	1	320	0.1	320.1
6	SUCCESS	3	1	320	0.1	320.1
2	SUCCESS	3	2	320	0.1	320.1
7	SUCCESS	3	2	320	0.1	320.1
4	SUCCESS	3	4	320	0.1	320.1
9	SUCCESS	3	4	320	0.1	320.1
3	SUCCESS	3	3	320	0.1	320.1
8	SUCCESS	3	3	320	0.1	320.1
101	SUCCESS	3	101	160	200.2	360.2
103	SUCCESS	3	103	160	200.2	360.2
105	SUCCESS	3	105	160	200.2	360.2
107	SUCCESS	3	107	160	200.2	360.2
109	SUCCESS	3	109	160	200.2	360.2
111	SUCCESS	3	111	160	200.2	360.2
100	SUCCESS	3	100	160	200.2	360.2
102	SUCCESS	3	102	160	200.2	360.2
104	SUCCESS	3	104	160	200.2	360.2
106	SUCCESS	3	106	160	200.2	360.2
108	SUCCESS	3	108	160	200.2	360.2
110	SUCCESS	3	110	160	200.2	360.2
112	SUCCESS	3	112	160	200.2	360.2
113	SUCCESS	4	113	160	200.2	360.2
114	SUCCESS	4	114	160	200.2	360.2

Fig. 5 HEFT algorithm output

Fig. 6 Initializing cloud
simulation entities for ACO

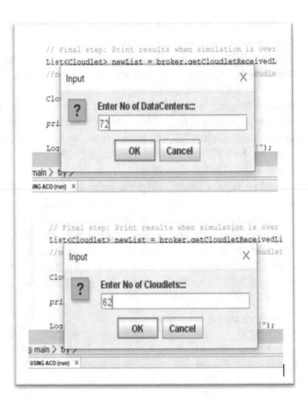

each cloudlet is assigned to a VM uniquely and not in order. VM IDs are generated based on the number of data centers present. As each VM is connected to a data center, the cloudlet that is linked to the VM executes individually on its connected individual data center. Since, each cloudlet is assigned to different data centers, there will be no load on data centers and all the load is equally balanced. There will be no waiting time for the remaining tasks too.

In Fig. 8, after processing all the VMs (i.e., getting request and response from the cloudlets), finally, it is observed which cloudlet is efficiently responding to VM. The cloudlets which are more efficient are displayed in the output along with their VM and Data center ID's.

6   Analysis

In CloudSim, a data center with a host is created and cloudlet is executed. Cloudlets mean a particular task. CloudSim is creating a particular entity called as Cloud Information Service (CIS). CIS is a kind of registry that contains the resources that are available to the cloud. After CIS, the next step is to create a data center. After creation is required to register once it is created. Each data center has one host and

```
0.1: Broker_0: Sending cloudlet 61 to VM #3
0.1: Broker_0: Sending cloudlet 62 to VM #41
0.1: Broker_0: Sending cloudlet 63 to VM #28
0.1: Broker_0: Sending cloudlet 64 to VM #47
0.1: Broker_0: Sending cloudlet 65 to VM #12
0.1: Broker_0: Sending cloudlet 66 to VM #39
0.1: Broker_0: Sending cloudlet 67 to VM #44
0.1: Broker_0: Sending cloudlet 68 to VM #28
0.1: Broker_0: Sending cloudlet 69 to VM #20
0.1: Broker_0: Sending cloudlet 70 to VM #56
0.1: Broker_0: Sending cloudlet 71 to VM #42
0.10400000000000001: Broker_0: Cloudlet 61 received
0.10400000000000001: Broker_0: Cloudlet 54 received
0.10400000000000001: Broker_0: Cloudlet 60 received
0.10400000000000001: Broker_0: Cloudlet 55 received
0.10400000000000001: Broker_0: Cloudlet 66 received
0.10400000000000001: Broker_0: Cloudlet 53 received
0.10400000000000001: Broker_0: Cloudlet 67 received
0.10400000000000001: Broker_0: Cloudlet 64 received
0.10400000000000001: Broker_0: Cloudlet 4 received
0.10400000000000001: Broker_0: Cloudlet 41 received
0.10400000000000001: Broker_0: Cloudlet 70 received
0.10400000000000001: Broker_0: Cloudlet 1 received
0.10400000000000001: Broker_0: Cloudlet 30 received
0.10400000000000001: Broker_0: Cloudlet 2 received
```

Fig. 7 Execution of Ant Colony Optimization

each host has some hardware configurations like RAM, MIPS, Bandwidth, etc. Once done with registering the data center, a broker needs to submit tasks to that particular data center. This broker is a data center broker class responsible for submitting the task to the particular data center.

A broker is an entity at the initial state talks to Cloud Information System (CIS) and retrieves the data that has been registered in the data center with the CIS. Data center broker has some characteristics and it has some tasks called cloudlets. Cloudlets may be one or a set of cloudlets. This set of cloudlets will be submitted to the broker. Once the broker has the details of the virtual machines stored in the data centers, this directly interacts with the data center and assign these cloudlets to some of the VMs which are running on the host.

Cloudlet ID	STATUS	Data center ID	VM ID	Time	Start Time	Finish Time
55	SUCCESS	09	09	00	00.1	00.1
64	SUCCESS	17	17	00	00.1	00.1
73	SUCCESS	35	35	00	00.1	00.1
63	SUCCESS	45	45	00	00.1	00.1
48	SUCCESS	52	52	00	00.1	00.1
31	SUCCESS	54	54	00	00.1	00.1
33	SUCCESS	56	56	00	00.1	00.1
02	SUCCESS	60	60	00	00.1	00.1
01	SUCCESS	61	61	00	00.1	00.1
34	SUCCESS	63	63	00	00.1	00.1
46	SUCCESS	64	64	00	00.1	00.1
07	SUCCESS	65	65	00	00.1	00.1
56	SUCCESS	66	66	00	00.1	00.1
03	SUCCESS	67	67	00	00.1	00.1
49	SUCCESS	69	69	00	00.1	00.1
52	SUCCESS	70	70	00	00.1	00.1
45	SUCCESS	71	71	00	00.1	00.1
24	SUCCESS	72	72	00	00.1	00.1
11	SUCCESS	75	75	00	00.1	00.1

Fig. 8 Output of Ant Colony Optimization

After observing screenshots, the following are the advantages of ACO when compared to HEFT Algorithm:

- In the HEFT algorithm, there is no proper job scheduling while in ACO, there is much importance given to job scheduling and load balancing.
- HEFT only takes care of the current task and evaluates only that particular task, which might lead to poor decisions in some cases. Coming to ACO, along with the current task, the remaining tasks are also evaluated.
- In HEFT, there is no load balancing and the waiting time is more for the tasks that are assigned to VMs. This drawback is corrected using the ACO algorithm where each task is assigned to a unique data center where there is minimal waiting time and since the resources are properly allocated, the load is also properly balanced.
- Heterogeneous Earliest Finish Time is only used in static applications while Ant Colony Optimization Algorithm is used in dynamic applications and in a more efficient manner. An example of the dynamic application of Ant Colony Optimization is the Travelling Salesman Problem.

7 Conclusion

There are two dissimilar load balancing optimization algorithms worked and executed to compare and contrast their capabilities. It is proved that selection of resources for the HEFT algorithm on the frontier of making processor selection is rather lacking. This flaw is caused due to its tendency to select the processor of the current task and not taking into contemplation of those which have been scheduled for the future. Furthermore, HEFT has one more drastic foible which deals with the methodology in which the tasks are handled within the VM's as well as data centers which has led to a very poor load balancing factor. When a comparison between the HEFT and Ant Colony Organization algorithm is established, it is analyzed that the latter algorithm does far better when it comes to the balancing of the load. This is due to the potentiality of the ACO algorithm to bestow a unique and individual VM as well as a data center. Unlike the primary algorithm (i.e., HEFT) is unable to accomplish this and allocates many tasks to a single entity. In this paper, the second algorithm that is implemented ACO. It is seen that this algorithm links various tasks with their own unique cloud entities in various factors to ensure that the execution is both efficient as well and the load balancing is properly accounted for. Due to these several outcomes, the final result that has been observed is that the second algorithm, i.e., ACO algorithm can carry out far more profoundly when compared to the HEFT algorithm due to its internal capabilities of handling loads without having to depend upon outside sources. Though Ant Scheduling Algorithm may have come out on top when compared with HEFT, many softwares still take advantage of these two algorithms within their structures to ensure the efficient functioning of their product as a whole. As these are only the basic algorithms available in the modern world, many new task scheduling algorithms and techniques might be created and obtained in the future which may even further distance themselves from these two in terms of capabilities.

References

1. L. Wang, G. Von Laszewski, A. Younge, X. He, M. Kunze, J. Tao, C. Fu, Cloud computing: a perspective study. New Gener. Comput. **28**(2), 137–146 (2010)
2. K. Nishant, P. Sharma, V. Krishna, C. Gupta, K.P. Singh, R. Rastogi, Load balancing of nodes in cloud using ant colony optimization, in *2012 UKSim 14th International Conference on Computer Modelling and Simulation* (IEEE, 2012), pp. 3–8
3. V.N. Volkova, L.V. Chemenkaya, E.N. Desyatirikova, M. Hajali, A. Khodar, A. Osama, Load balancing in cloud computing, in *2018 IEEE Conference of Russian Young Researchers in Electrical and Electronic Engineering (EIConRus)* (IEEE, 2018), pp. 387–390
4. R. Gao, J. Wu, Dynamic load balancing strategy for cloud computing with ant colony optimization. Fut. Internet **7**(4), 465–483 (2015)
5. S. Dash, A. Panigrahi, N.R. Sabat, Performance analysis of load balancing algorithm in cloud computing. Int. J. Innov. Res. Technol. **6**(6), 11 (2019)
6. Z. Liu, X. Wang, A PSO-based algorithm for load balancing in virtual machines of cloud computing environment, in International Conference in Swarm Intelligence (Springer, Berlin, Heidelberg, 2012), pp. 142–147
7. Y. Shao, Q. Yang, Y. Gu, Y. Pan, Y. Zhou, Z. Zhou, A dynamic virtual machine resource consolidation strategy based on a gray model and improved discrete particle swarm optimization. IEEE Access **8**, 228639–228654 (2020)
8. K.M. Sim, W.H. Sun, Ant colony optimization for routing and load-balancing: survey and new directions. IEEE Trans. Syst. Man Cybernet. Part A Syst. Hum. **33**(5), 560–572 (2003)
9. T. Zhang, S. Wang, G. Ji, A comprehensive survey on particle swarm optimization algorithm and its applications. Mathe. Problems Eng. (2015)
10. S.N. Sivanandam, P. Visalakshi, Dynamic task scheduling with load balancing using parallel orthogonal particle swarm optimisation. Int. J. Bio-Inspired Comput. **1**(4), 276–286 (2009)

11. R Mishra, A., Jaiswal, Ant colony optimization: A solution of load balancing in cloud. Int. J. Web Semant. Technol. **3**(2), 33 (2012)
12. L. Liu, G. Feng, A novel ant colony based QoS-aware routing algorithm for MANETs, in *International Conference on Natural Computation*. (Springer, Berlin, Heidelberg, 2005), pp. 457–466

Artificial Intelligence-Based Vehicle Recognition Model to Control the Congestion Using Smart Traffic Signals

Aleemullakhan Pathan🆔, Nakka Thirupathi Rao🆔,
Mummana Satyanarayana, and Debnath Bhattacharyya🆔

Abstract The present traffic system is a timer-based system which works upon a constant timing. If one side of a traffic junction is having fewer vehicles compared to the other side, still the timer runs the same for both the sides. In our paper, we overcome this problem by using a vehicle recognition model. This AI model runs over the images of all the sides of a traffic junction and recognizes the number of vehicles and the type of vehicles. Depending upon the count and type of vehicles, the timer is reset every time, and the time for every run is calculated by the model. In addition to this, our vehicle recognition (AI) model also focuses on detection of ambulances, so the side with these kinds of vehicles is given first priority over the other sides. This model detects individual vehicles and categorizes them into one of these classes instead of counting vehicles.

Keywords Traffic system · Vehicle recognition · Ambulance · AI model

1 Introduction

Traffic lights are communicating appliances located at roadway junctions, foot traveller crossways and other localities to control the flow of traffic. Traffic lights substitute light-emitting diodes (LEDs) standard colours (red, yellow and green) by illumination following a universal colour code. Normally at least one direction of traffic at a junction has the green light at any moment in the cycle. In some zones, all signals at a junction show red at the same time, to clear any traffic at the junction. The

A. Pathan (✉) · N. T. Rao
Department of Computer Science and Engineering, Vignan's Institute of Information Technology, Visakhapatnam 530049, India

M. Satyanarayana · D. Bhattacharyya
Department of Computer Science and Engineering, K L Deemed to be University, KLEF, Guntur 522502, India
e-mail: Mummana@raghuenggcollege.in

D. Bhattacharyya
e-mail: debnathb@kluniversity.in

© The Author(s), under exclusive license to Springer Nature Singapore Pte Ltd. 2021 161
S. K. Saha et al. (eds.), *Smart Technologies in Data Science and Communication*,
Lecture Notes in Networks and Systems 210,
https://doi.org/10.1007/978-981-16-1773-7_13

current traffic lights can be mainly categorized into two types: manual (MNL) and intelligence (INL); again INL can be categorized into two types such as fixed time control and variable time control. No one of the above-mentioned systems can act as a traffic policeman in the junction efficiently because it cannot detect an emergency vehicle or the number of vehicles.

1.1 Picture Processing

Picture processing is a procedure to execute tasks related to picture and to obtain an enlarged picture or to obtain some useful information from it. It is a kind of signal transmitting in which input signal is an image and output signal may be an image or properties or attributes associated with that image. Picture processing primarily incorporates the following three steps: choosing the image via image accession instruments, evaluating and operating the image. Output can be changed based upon image analysis. Picture processing also involves two types of outcome count such as a huge amount of vehicles and a binary value. The outcome count denoting the existence of ambulances depends on which we will take the needed decision by setting the precedence for each side of the crossway.

1.2 Smart Traffic Signal

A smart traffic signal is a signal which sets the timer based on the number and type of vehicles at a junction. Smart traffic signals also help for smoother flow of the traffic. This smart traffic signal will act as a policeman in the junction and gives the foremost priority to them in case of any emergency. There are two types of existing traffic signals such as:

1.   MNL: MNL stands for manual traffic signal. In this type of traffic signal, the timer is set manually by the traffic personnel present in the junction.
2.   INL: INL stands for intelligence traffic signal which is an automated one. This traffic signal also includes two versions.

Fixed time control: The name itself indicates that the time will be fixed (a fixed duration of red light and green light).

Variable time control: This method is also known as vehicle actuated control. This method is used extensively at disassociated junctions in many countries.

2 Literature Review

A research paper by Lin [1] presents novel focal loss which provides training on a sparse set of hard examples and prevents the vast number of easy negatives from overwhelming the detector during training. To assess the advantage of focal loss and to design and train, a simple dense detector known as RetinaNet is used. When trained with the focal loss, RetinaNet is suitable to correlate the fastness of previous one-stage detector while exceeding the speed of all the present two-stage detectors.

A research paper by Sohn [2] represents a unique artificial intelligence that utilizes only video images of junctions to denote its traffic state instead of using handcrafted features. In simulation assessments using real junction, successive aerial video frames thoroughly communicated the traffic state of an independent four-legged junction, and an image-based RL model outperformed both the real operation of fixed signals and a completely actuated operation.

A research paper by Jomaa [3] presents new artificial intelligence techniques (AIT) and simulation model (SM). An integrated model includes a neural network, fuzzy logic, genetic algorithm and simulation model. The integrated model is also used to adjust traffic lights timing to optimize traffic flow in corresponding traffic lights systems to minimize the traffic jam by controlling traffic lights.

A research paper by Zhang [4] represents a new RL algorithm for partially detected intelligent traffic signal control (PD-ITSC) systems. The execution of this system is examined over various car flows, detection rates and typologies of the road network. This method is capable of effectively decreasing the average waiting time of the vehicles at junctions, even with a low detection rate, thus reducing the travel time of vehicles.

A research paper by Mishra [5] presents the establishment of a mixed technique of artificial intelligence (AI), and computer vision can be desirable to develop an authenticate and scalable traffic system. This method is used to identify the vehicle, its speed, an emergency (in the case of the ambulance), and two closest signals will synchronously share the information to make a reliable decision.

A research paper by Girshick [6] illustrates a simple and scalable detection algorithm that refines mean average precision by above 30% relative to the previous best result on VOC 2012—achieving a mean average precision of 53.3%. Convolutional neural networks with high capacity are used to localize and segment objects and supervised pretraining for an auxiliary task, followed by domain-specific fine-tuning.

A research paper by Płaczek [7] presents a self-organizing traffic signal system for an urban road network. The essential elements of this system are agents that control traffic signals at junctions. The interspace microscopic traffic model is used to estimate effects of its possible control actions in a short-time horizon. The recreation of experiments has shown that the proposed strategy results in an improved performance, particularly for non-uniform traffic streams.

A research paper by Chinyere [8] represents an intelligent system to monitor and control road traffic in a Nigerian city. A composite methodology acquired by the intersection of the structured systems analysis and design methodology and the fuzzy logic-based design methodology was deployed to develop and implement the system. The following fuzzy logic-based system for congestion control was reproduced and tested using a popular intersection in a Nigerian city. The current system removed some of the problems identified in the current traffic monitoring and control systems.

A research paper by Zaid [9] presents intelligent transportation system including smart way to control traffic light time based on number of cars in each traffic light; this paper develops an automatic algorithm to control traffic light time based on artificial intelligent techniques and image for cars on traffic lights; this algorithm is validated by comparing its results with manual results.

A research paper by Xue and Dong [9] illustrates a generic real-time optimal contraflow control method. The introduced method integrates two important functional components: (1) an intelligent system with artificial neural network and fuzzy pattern recognition to accurately estimate the current traffic demands and predict the coming traffic demands and (2) a mixed-variable, multilevel, constrained optimization to identify the optimal control parameters.

A research paper by Africa [10] presents a computer vision that has provided a means to reduce manual labour in parking lots by replacing them with sensors and scanners which would take over. This would be beneficial to the company as this technology would require minimal maintenance. Through image processing and enhancement, the license plates of these cars could be taken in real time and this data could be used to generate the parking bills given the time-in and a time-out of the vehicle.

A research paper by Akinboro [11] represents a mobile traffic management system that provides users with traffic information on congested roads using weighted sensors. A prototype of the system was implemented using Java SE Development Kit 8 and Google Maps. The model was simulated, and the performance was assessed using response time, delay and throughput; mobile devices are capable of assisting road users in faster decision-making by providing real-time traffic information and recommending alternative routes.

A research paper by Ng [12] that illustrates a study of effective optimization technologies in controlling traffic signals is conducted which aims to relieve the congestion problem and increase road efficiency according to the specific needs of Hong Kong. In this paper, a new traffic light system using machine learning with object detection and analysing by the evolutionary algorithm that aims to perform a real-time strategic signal switching arrangement to traffic lights at the intersection was designed to reduce the waiting time of pedestrians and vehicles and provide better travelling experience to road users.

A research paper by de Souza [13] introduces a traffic management system composed of a set of application and management tools to improve the overall traffic efficiency and safety of the transportation systems. Furthermore, to overcome such issues, traffic management systems gather information from heterogeneous sources, exploit such information to identify hazards that may potentially degrade the traffic efficiency and then provide services to control them.

A research paper by Shelke [14] illustrates sensor nodes which monitor traffic information and transmit it to a Dynamic Traffic Management Center (DTMC). It dynamically determines priority of the road segment as critical, high, medium and low using fuzzy logic. Packets are transmitted to DTMC using congestion-aware routing algorithms. The results are simulated in network simulator version 2 and demonstrate that the proposed approach clears emergency vehicles through the intersection with minimum waiting time and optimizes average waiting time and number of vehicles passing through a junction.

3 Proposed Work

To solve the problems in the present traffic system, we have proposed a new model known as smart traffic system which takes the images from the different roads of the junction. Then depending on that image data traffic signals will be assigned. The first priority is given to the side which is having an ambulance and then the side with more number of vehicles (weight). This system overcomes the problem of present systems by giving first preference to emergency vehicles. In addition to the priority, it also estimates the time for which a particular signal should remain, and then based on that data, the next image is taken after that time, thus reducing the burden on the system.

3.1 Algorithms

In this paper, we have focused on simple algorithms which usually perform well in classification tasks, namely:

1 RetinaNet algorithm.
2 Region-convolutional neural network (R-CNN) algorithm.

3.1.1 RetinaNet Algorithm

RetinaNet algorithm consists of a one-stage detector, focal loss and a lower loss. The residual network (ResNet) and feature pyramid network (FPN) act as backbone for feature extraction, for task-specific subnetworks, for classification and bounding box regression. Residual network is used for deep feature extraction. Feature pyramid network (FPN) is used for constructing a rich multi-scale feature pyramid from one single resolution image (Fig. 1).

3.1.2 Region-Convolutional Neural Network Algorithm

Region-convolutional neural network is used to locate objects in an image (object detection). It also includes a sliding window approach to detect the object. By using

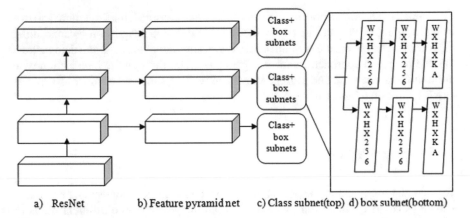

a) ResNet   b) Feature pyramid net   c) Class subnet(top) d) box subnet(bottom)

Fig. 1 Functioning of RetinaNet algorithm

this method, we can pick up the whole image with different-sized rectangles, whereas small images by using a brute-force method. We used this algorithm to choose region proposals. In addition to this, region proposals will create nearly 2000 different regions. In the next step, we have taken each region proposal and we have created a feature vector which represents this image in a much smaller dimension using a convolutional neural network (CNN). They use the AlexNet as a feature extractor. Finally, we have classified each feature vector with the SVMs for each object class (Fig. 2).

The below steps illustrate the mechanism of the smart traffic system that are as follows.

Steps:

1. First loaded the image of the smart traffic system as jpeg or png.
2. Applied the RetinaNet algorithm to process the image.
3. Then counted the number of vehicles.
4. Applied the region-CNN algorithm to process the image.

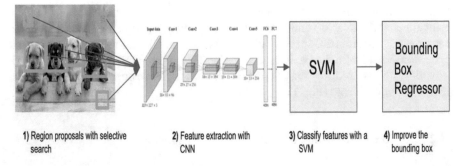

1) Region proposals with selective   2) Feature extraction with   3) Classify features with a   4) Improve the
   search   CNN   SVM   bounding box

Fig. 2 Functioning of region-CNN algorithm

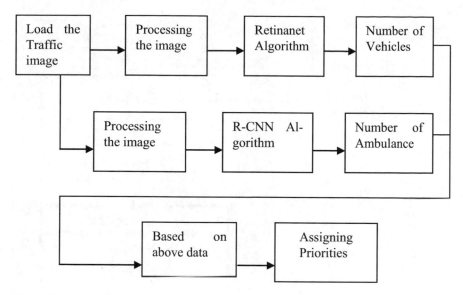

Fig. 3 Structure of the smart traffic network

5. Then found the number of ambulances.
6. Finally, decision has been taken based on the priorities by using the above data.

The structure of the smart traffic network's mechanism is shown in Fig. 3.
The below diagram illustrates the flowchart of the smart traffic network.

3.2 Working Mechanism

To implement the machine learning algorithms, we have followed the below two phases:
1. Construction phase and 2. operational phase.

3.2.1 Construction Phase

In construction phase, we have the following steps:

a. Extracting the required images from the COCO dataset.
b. Loading the dataset.
c. Training the model.
d. Classification of image with Python (Fig. 4).

Fig. 4 Flowchart of the
smart traffic system

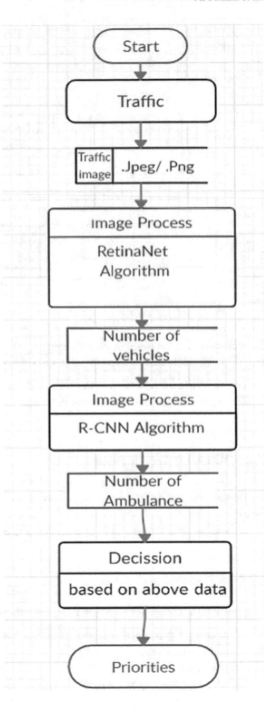

a. Extracting the required images from the COCO dataset
In this step, by writing a small Python script the images with required labels can be extracted. Also, by using the metadata about bounding boxes, training images can be obtained.
b. Loading the dataset
In this step, images are loaded for training using the operating system module provided in Python.
c. Training the model
In this step, by using the sampled image data and label, a single model can be developed which can recognize vehicles in an image. The training takes place in several runs and after tweaking the image data to required pixels, thus improving the accuracy of the model.
d. Image classification with Python

In this step, classification of images is done with Python using machine learning.
There are many models for machine learning, and each model has its own strengths and weaknesses.

3.2.2 Operational Phase

In operational phase, we have the following steps:

a. Perform predictions and b. performance calculation

a. **Perform predictions**:
In this method, the trained model is used for making predictions. The trained model is loaded as a file with ".h5 extension". It contains all the details of the pretrained model. At each run, four images from traffic camera are loaded and the following predictions are made:

  i. Whether the image contains an ambulance.
  ii. Count of vehicles present in the image.

b. **Performance calculation**:
In this method, the accuracy is calculated for each prediction and is combined as an average for the whole system. The accuracy of each run is calculated by factor of vehicles that are identified by the prediction model.
The accuracy is calculated by using the formula such as Accuracy = Average (number of correct predictions/total number of predictions).

4   Results and Discussion

The results of the proposed smart traffic system are categorized as follows:

Output when no image is provided: This method shows the output when no image is provided which is illustrated in Fig. 5.

1. Output when no ambulances are found in images: This method shows the output when no ambulances are found in images which is illustrated in Fig. 6.
2. Output when ambulances are found at both sides of the junction: This method shows the output when ambulances are found at both sides of the junction which is illustrated in Fig. 7.
3   Output when ambulance is struck in the traffic: This method shows the output when ambulance is struck in the traffic which is illustrated in Fig. 8.

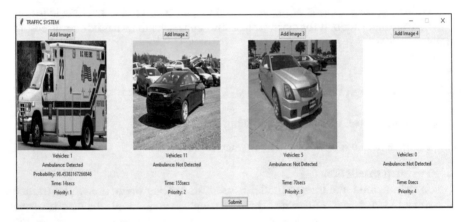

Fig. 5 Output when no image is provided

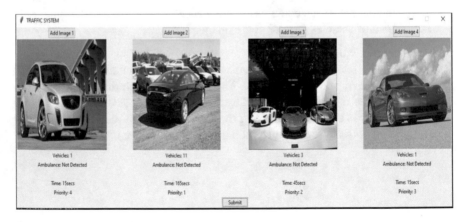

Fig. 6 When no ambulances are found in images

Fig. 7 When ambulances are found at both sides of the junction

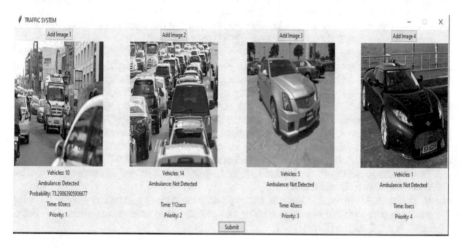

Fig. 8 When ambulance is struck in the traffic

5 Conclusion

The use of traffic signals is used to save time and also the lives of the people. The present system lacks the sense of time management as it runs in the same manner neglecting some of the exceptions such as an ambulance present on the red side of the signal and one side having more number of vehicles when compared to others. Our system overcomes this problem by using the technique of image processing thus giving the first priority if an ambulance is present or if not then it takes the decision based on the number of vehicles. If these traffic lights are implemented all over the traffic junctions, then there will not be any need for the traffic policeman to be physically present over at the junction. Instead, we can use that workforce in some other place, thus saving the time of both government officials and the public on the road.

References

1. T.-Y. Lin, P. Goyal, R. Girshick, K. He, P. Dollar, Focal loss for dense object detection, in IEEE International Conference on Computer Vision (ICCV) (2017)
2. K. Sohn, H. Jeon, J. Lee, Artificial intelligence for traffic signal control based solely on video images. J. Intell. Transp. Syst. (2017)
3. K.A.R. Jomaa, An artificial intelligence techniques and simulation model to control a traffic jam system in Malaysia. Asian J. Bus. Manage. **04**(01) (2016). ISSN: 2321–2802
4. R. Zhang, A. Ishikawa, W. Wang, B. Striner, O.K. Tonguz, Using reinforcement learning with partial vehicle detection for intelligent traffic signal control, in *IEEE Transactions on Intelligent Transportation Systems* (2020)
5. S. Mishra, V. Birchha, A literature survey on an improved smart traffic signal using computer vision and artificial intelligence. Int. J. Innov. Res. Sci. Eng. Technol. **8**(6) (2019)
6. R. Girshick, J. Donahue, T. Darrell, J. Malik, Rich feature hierarchies for accurate object detection and semantic segmentation, in *2014 IEEE Conference on Computer Vision and Pattern Recognition*
7. B. Płaczek, A self-organizing system for urban traffic control based on predictive interval microscopic model. Eng. Appl. Artif. Intell. **34**, 75–84 (2014)
8. O.U. Chinyere, O.O. Francisca, O.E. Amano, Design and simulation of an intelligent traffic control system. Int. J. Adv. Eng. Technol. (2011) ISSN: 2231–1963
9. A.A. Zaid, Y. Suhweil, M. Al Yaman, Smart controlling for traffic light time, in *2017 IEEE Jordan Conference on Applied Electrical Engineering and Computing Technologies (AEECT)*
10. A.D.M. Africa, A.M.S. Alejo, G.L.M. Bulaong, S.M.R. Santos, J.S.K. Uy, Computer vision on a parking management and vehicle inventory system. Int J Emerg Trends Eng Res **8**(2) (2020)
11. S.A. Akinboro, J.A. Adeyiga, A. Omotosho, A. O Akinwumi, Mobile road traffic management system using weighted sensors. Int. J. Interact. Mobile Technol. **11**(5) (2017)
12. S.-C. Ng, C.-P. Kwok, An intelligent traffic light system using object detection and evolutionary algorithm for alleviating traffic congestion in Hong Kong. Int. J. Comput. Intell. Syst.
13. A.M. de Souza, C.A.R.L. Brennand, R.S. Yokoyama, E.A Donato, E.R.M. Madeira, L.A. Villas, Traffic management systems: A classification, review, challenges, and future perspectives. Int. J. Distrib. Sensor Netw. **13** (2017)
14. M. Shelke, A. Malhotra, P.N. Mahalle, Fuzzy priority based intelligent traffic congestion control and emergency vehicle managementusing congestion-aware routing algorithm. J. Ambient Intell. Hum. Comput. (2019)

A Literature Review on: Handwritten Character Recognition Using Machine Learning Algorithms

Sachin S. Shinde and Debnath Bhattacharyya

Abstract Due to wide range of applications such as banking, postal, digital libraries, handwritten character realization is precise sizzling. The maturation comedian in the area of handwritten character processing is application forms processing, digitizing old articles, postal code processing and bank transaction processing as well as many various others applications. The device's handwritten recognition interprets the handwritten characters or phrases of the user into a format understandable by the computer machine. For efficient recognition, many machine learning and deep learning approaches have been planned. We have demonstrated a thorough study of handwritten character recognition phases and various strategies and methods for machine learning and deep learning in this paper. The primary oblique of this paper is to ascertain efficient and trustworthy motion to handwritten character recognition. For character approval, several machines have been used to learn algorithms such as support vector machine, Naive Bayes, KNN, Bayes Net, random forest, logistic regression, linear regression and random tree.

Keywords Handwritten character recognition · Machine learning · Classification algorithms

1 Introduction

2 Overview of Handwritten Character Recognition

Handwritten identification of characters is an area of artificial intelligence science, computer vision and pattern approval. Using his/her intellect and acquisition, humans can easily understand distinct handwritings. Handwritten recognition is a characteristic task because there exists a variety of writing ways. Due to the comparable

S. S. Shinde (✉) · D. Bhattacharyya
Department of Computer Science and Engineering, Koneru Lakshmaiah Education Foundation,
Vaddeswaram, Guntur, Andhra Pradesh 522502, India

© The Author(s), under exclusive license to Springer Nature Singapore Pte Ltd. 2021 173
S. K. Saha et al. (eds.), *Smart Technologies in Data Science and Communication*,
Lecture Notes in Networks and Systems 210,
https://doi.org/10.1007/978-981-16-1773-7_14

situation, the computer machine program does not find good accuracy for the handwritten character recognition task. There are many of literature reviews that focuses on English, Bangla, Marathi, Devanagari, Oriya, Chinese, Latin and Arabic languages [1].

One of the classical applications of pattern recognition is handwritten character recognition (HCR). In general terms, HCR is the method of classifying characters according to the predefined attribute groups from the handwritten texts entered. Handwritten character recognition applications range across the broad domain, such as the processing of digital documents such as data entry mining information, cheque, loan applications, credit cards, tax, character identification, handwritten record digitization, data entry application form reading and dependent, recognizes the unknown language and translates it into a known. Some classifiers such as support vector machine (SVM), Naive Bayes, convolutional neural network (CNN), K-nearest neighbours (KNN), linear regression, logistic regression and random forest (RF) perform the classification task to determine which classifier in this experiment has the highest accuracy rate. In on the web and disconnected methodologies, manually written character acknowledgment frameworks get static contributions to disconnected HCR frameworks that effectively pull in interest in the examination. That implies digitized text records or checked manually written content picture copies [2]. For recognizable proof, the online transcribed character acknowledgment gadget acquires live penmanship. Here, with the utilization of an extraordinary pen, an individual composes on the advanced gadget, and that information is utilized as a live feed for the framework. The vital distinction between the two frameworks is that there is one extra boundary in the online framework, which is time with information [3]. What is more, the strokes, pace, pen-up and pen-down information are likewise included as boundaries [2].

The paper is organized as: Sect. 2 includes phases of handwritten character.

Recognition and a summary overview is given. Section 3 briefly explains different methods for machine learning and deep learning. The character recognition method is described in Sect. 4 and finally in Sect. 5 comparative reviews of some machine learning algorithms, along with some final comments and possible scope in this area.

3 Phases of Handwritten Character Recognition

In Fig. 1, the phases entangled in the operation of the handwritten attribute approval system are correctly represented, including image acquisition, pre-processing, segmentation, extraction of features and classification phases.

Image Acquisition
It is the operation of some source to obtain input data; here, we will acquire handwritten input data for the method of attribute approval. Online and offline systems have been developed based on image or data attainment. MNIST is a very common

Fig. 1 Phases of HCR system

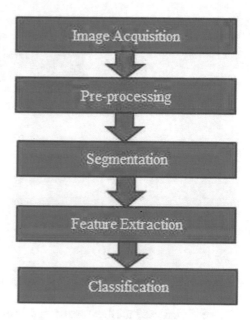

data set for the numerical data used in [4]. Pen-based handwritten digits data set recognition, etc., is some other data sets. Researchers use their data set for the recognition method when there is no standard data set available [5].

Pre-processing

Pre-processing is achieved to improve the image to be used for further processing [3], to improve the consistency of the input data and to make it more suitable for the next step of the recognition system. Different approaches that are carried out in this step are grayscale conversion, binary conversion, noise reduction, binarization, normalization, etc. The simple pre-processing operations on the image are illustrated in Fig. 2 [3].

Grayscale shift, binary figuring and then a outcry reduction technique are completed on the input data in the pre-processing step. The investigator used edge detection for segmentation when assuming the findings in [6] after the grayscale and binary conversion. Thresholding and Otsu's algorithm are commonly used to transform a grayscale image to a binary image.

Segmentation

The method of segmenting the input text data image to the line and then to the several character is segmentation. It motions from the data image the outcast portion or noise. Two forms of segmentation, external and internal, are available. The paragraphs, lines and words are segmented by external segmentation. Internal segmentation, on the other hand, is the segmentation of individual characters from the input text data [1]. For segmentation, different algorithms are available. The basic methods of line segmentation are histogram profiles and linked component analysis. After

Fig. 2 Pre-processed images [3]

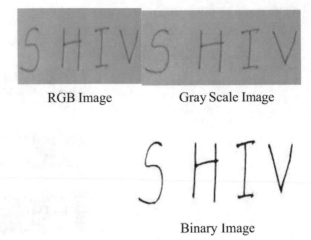

RGB Image   Gray Scale Image

Binary Image

histogram-based segmentation, Figs. 3 and 4 illustrate the line and character looks like.

The device used attribute space espial for the words and the bar chart approach for the attribute and other sign. In [3], the authors use the enchained box proficiency for character division. The resize activity is done on completely sectioned pictures for uniform size following productive division.

Feature Extraction

It is the method by which different and very useful information about an object or a group of objects is collected, so we can identify new unexplored content by twinned them based on that collected information. The feature is that the raw data is robustly represented. The "Feature Extraction" process can be described as the process in

Fig. 3 Segmented line based on histogram [7]

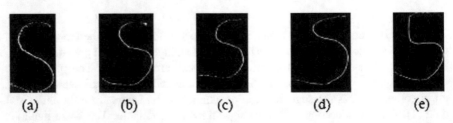

(a)   (b)   (c)   (d)   (e)

Fig. 4 Segmented characters based on histogram [6]

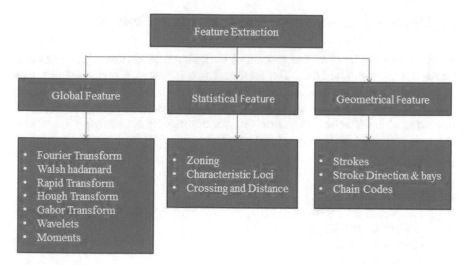

Fig. 5 Feature extraction methods

which the essential information on the centred subject that exists in the image is extracted.

Figure 5 shows various feature extraction methods. There are methods like zone-based, mathematical, structural, chain code histogram, sliding window, gradient feature, hybrid, etc., and they are the some of important feature extraction methods [7]. There are primarily two chain code paths, the 4-neighbourhood and the 8-neighbourhood.

Classification

The predefined class in the classification is allocated to an undefined sample. The digits are categorized and recognized accordingly as the features are extracted. The method of assortment or approval is basically beneficial for decision-making process, as this new character suits the class or looks like. This implies that characters are marked and assigned to marking in the classification stage. Classification information productivity consistently relies upon great extraction and choice of highlights of information. There are different order strategies accessible and every one of them is basically founded on picture preparing and man-made reasoning procedures. There are different classification approaches based on image processing, like as statistical techniques, template matching, structural technique and neural networks, fuzzy logic, genetic techniques based on soft computing.

Machine learning (ML) has used different techniques in the family of hand-written character recognition system (HCR) like as support vector machine, artificial neural networks (ANN), Naive Bayes, nearest neighbour algorithms, decision trees, neurofuzzy, etc.

4   Machine Learning and Deep Learning Technique

Machine learning involves the process of designing a prediction algorithm based on experience or trained data. The important part is learning, and it requires data in the concerned domain after that prediction network organizes itself according to error. The current scenario has attained high complexity because the same field has attracted the attention of researchers. Various models are evolving in machine learning, and some of them are as follows [1]:

(a)   Decision trees
(b)   Nearest neighbour
(c)   Random forest
(d)   Artificial neural network
(e)   Logistic regression
(f)   Linear regression
(g)   Apriori algorithm
(h)   Support vector machine
(i)   K-means clustering
(j)   Naive Bayes classifier
(k)   Neural network

Deep extreme learning machine algorithm divine by the human brain and it is commute a hierarchical level of artificial neural networks to handle the process of machine learning and it has attained pace due to various advancements of hardware and at the same time, algorithmic research that has been done on deep network information processing. Some of the essential algorithms of deep learning are:

a.   Recurrent neural network
b.   Auto-encoder
c.   Restricted Boltzmann machine
d.   Convolutional neural network
e.   Deep belief network
f.   Deep neural network
g.   Deep extreme learning machine
h.   Localized deep extreme learning machine

5   Character Recognition System

There is a variety of challenges in the handwritten character recognition system. Phases of the handwritten recognition system are shown in Fig. 6. There are two basic classifications in character recognition system such as online character recognition and offline character recognition. Online character recognition involves a digital pen and tablet. Offline recognition includes handwritten and printed character.

Two wide categories can be split into optical character recognition such as handwritten character recognition (HCR) and printed character recognition (PCR). As

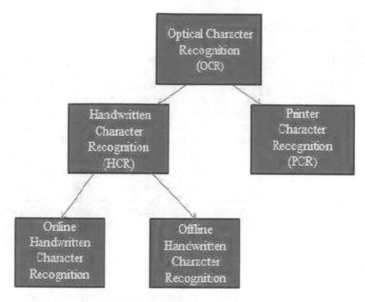

Fig. 6 Phases of optical character recognition

compared to HCR, PCR is comparatively simpler. In addition, HCR is widely spread across the offline recognition and online recognition groups. Figure 6 shows the optical character recognition system hierarchy, online recognition is a mechanism in real time in which characters are recognized as the text is written by the user. In offline identification, the characters of documents already published are remembered. Generally, in offline identification, instruments such as optical scanners or photographs (using a camera) are used. Offline recognition is harder, fewer precise, and has a low degree of recognition because we do not have users' pen strokes here. To build this offline handwritten character recognition framework, various tools such as Python, Android, OpenCV and TensorFlow [8] are used. Also, we have pen strokes of the user in online recognition and therefore it is simpler, more precise as well as the recognition rate is higher.

A number of varieties have handwritten characters. For written terms, segmentation and non-segmentation are involved. Further steps entail choosing features. Optimization can be used to accelerate the classification process. Subsequently, a classification algorithm for reading characteristics is needed. Finally, for desired tasks, a trained model is used for.

6 Classification Algorithms

Table 1 represents comparative analysis of some of the machine learning algorithms used to handwritten character recognition.

Table 1 Comparative analysis of machine learning classification algorithms

Sr. no.	Algorithm	Description
1	Naive Bayes classifier	The Naive Bayes classifier is also called "probabilistic classifier" which offers a simple techniques of signifying and learning probabilistic information with a straightforward semantics and classification algorithm-based mathematically on the theorem of Bayes with an assumption of predictor independence. There is no single algorithm for training such a classifier, but common principles-based family or algorithm. The likelihood is independently encouraged by these properties
2	Nearest neighbour	In the nearest neighbour, the KNN algorithm is used for supervised machine learning that defines a number K that is the nearest neighbour to the data point to be categorized. In K-nearest neighbour algorithm, a labelled data set is adopting as the input and the calculation is kept over as necessary for the output. KNN is also known as example-based or case-based learning
3	Logistic regression	Logistic regression algorithms are important classification algorithm for supervised learning used to examine the association of categorical-independent variable as well as to estimate the likelihood of the target variable To evaluate the result where there are one or more independent variables, the statistical method analyses the data set. The primary objective of logistic regression algorithm is to look for a perfectly suitable model that explains the association between unassociated variables and independent variables
4	Decision trees	The decision tree classifier is a simple and commonly used algorithm classification technique, and it also belongs to family of supervised learning techniques. The data set is divided sequentially or incrementally into subsets and further subsets into smaller subsets for creating associated tree structure. The tree's uppermost node is called as the root node, which plays an important role as a decision tree's best predictor node. In the decision tree, both numerical as well as categorical data can be handled easily
5	Random forest	It is a supervised learning algorithm which is used for both classifcation and regression techniques

7   Conclusion and Future Work

We have examined different techniques/methods used by various researchers for handwritten character recognition in this paper. In our everyday lives, handwritten character recognition is very useful because it covers a wide variety of useful applications. In this region, various researchers suggested their works and successfully achieved a good accuracy rate as well as very few researchers concentrated on the time complexity and handled it.

The handwritten character recognition system may be replaced with deep learning in the future to increase the accuracy of character outcomes. The use of multiple methods of extraction of features helps to increase the precision rate. It should also be notified that the larger the data set taken tends to increase the performance, providing the accuracy needed.

References

1. A. Sing, A. Bist A wide scale survey on handwritten digit recognition using machine learning. Int. J. Comput Sci. Eng. 124–134 (2019)
2. R. Sethi, I. Kaushik, Handwritten Digit Recognition using Machine Learning, in *9th IEEE International Conference on Communication Systems and Network Technologies* (2020), 49–54
3. S.R. Patel, J. Jha, Handwritten character recognition using machine learning approach—a survey, in *International Conference on Electrical, Electronics, Signals, Communication and Optimization (EESCO)* (2015)
4. J. Memon, M. Sami, R. Ahmed Khan, Mueenuddin, Handwritten optical character recognition (OCR): a comprehensive systematic literature review (SLR). IEEE Access (2020), 142642–142668
5. W. Song, S. Uchida, M. Liwicki, Comparative study of part-based hand-written character recognition methods, in *2011 International Conference on Document Analysis and Recognition* (IEEE, 2011), 814–818
6. A. Tahir, A. Pervaiz, Hand written character recognitoin using SVM. Pacific Int. J. 39–43 (2020)
7. B.M.Vinijit, M.K. Bhojak, S. Kumar, G. Chalak, A review on hand-written character recognition methods and techniques, In *International Conference on Communication and Signal Processing* (2020), 1224–1228
8. R. Vaidya, D. Trivedi, S. Satra, M. Pimpale, Handwritten digit recognition using deep learning, in *2nd International Conference on Inventive Communication and Computational Technologies (ICICCT)* (2018), 772–775
9. P. Bojja, N.S.S.T. Velpuri, G.K. Pandala, S.D. Lalitha Rao Sharma Polavarapu, P.R. Kumari, Handwritten text recognition usingmachine learning techniques in applications of NLP. Int. J. Innov. Technol. Exploring Eng. (IJITEE) 1394–1397 (2019)
10. S.M. Shamime, M.B.A. Miah, Handwritten digit recognition using machine learning. Global J. Comput. Sci. Technol. D Neural Artificial Intell. (2018)
11. R. Sharma, B. Kaushik, N. Gondhi, Character recognition using machine learning and deep learning a survey, in *2020 International Conference on Emerging Smart Computing and Informatics (ESCI)* (IEEE, 2020), 341–345
12. P.Thangamariappan, J.C.Miraclin, J. Pamila, Handwritten recognition by using machine learning approach. Int. J. Eng. Appl. Sci. Technol. 564–567 (2020)
13. D. Prabha Devi, R. Ramya, P.S. Dinesh, C. Palanisamy, G. Sathish Kumar, Design and simulation of handwritten recognition system. Mater Today Proc Elsevier (2019)
14. K. Peyamani, M. Soryani, From machine generated to handwritten character recognition; a deep learning approach, in 3rd *International Conference on Pattern Recognition and Image Analysis (IPRIA)* (IEEE, 2017), 243–247
15. L. Abhishek,Optical character recognition using ensemble of SVM, MLP and extra trees classifier, in *2020 International Conference for Emerging Technology (INCET) (INCET)* (IEEE 2020),1–4
16. A. Sharma, S. Khare, S. Chavan, A Review on handwritten character recognition. Int. J. Comput. Sci. Technol. (IJCST) 71–75 (2017)
17. A. Ramzi, Online arabic handwritten character recognition using online-offline feature extraction and back-propagation neural network, in 1st international conference on advanced technologies for signal and image processing (ATSIP) (IEEE, 2014), 350–355

18. C. Patel, R. Patel, P. Patel, Handwritten character recognition using neural network. Int. J. Sci. Eng. Res. (IJSER) 1–6 (2011)
19. S.S. Rosyda, T.W. Purboyo, A review of various handwriting recognition methods. Int. J. Appl. Eng. Res. (IJAERV) 1155–1164 (2018)
20. A. Purohit, S. S. Chauhan, A literature survey on handwritten character recognition. Int. J. Comput. Sci. Inf. Technol. (IJCSIT) 1–5 (2016)
21. S.G. Dedgaonkar, A.A. Chandavale, A.M. Sapkal, Survey of method for chaacter recognition. Int. J. Eng. Innov. Technol. (IJEIT) 180–189 (2012)
22. P. Sharma, R.K. Pamula, Handwritten text recognition using machine learning techniques in applications of NLP. Int. J. Innov. Technol. Exploring Eng. (IJITEE),1394–1397 (2019)

Tweet Data Analysis on COVID-19 Outbreak

V. Laxmi Narasamma, M. Sreedevi, and G. Vijay Kumar

Abstract COVID-19 outbreak as a pandemic in the month of March 2020 announced by World Health Organization and this pandemic virus was originated from Dragon country China, it is a big pandemic virus has been attacking through out of the world and killing lakhs of people. While this pandemic has persevered to have impact on the lives of many thousands of people, various countries have depended on complete lockdown. During this pandemic circumstance, i.e., in lockdown, individuals around the globe have taken long range interpersonal communication destinations or applications to convey their emotions and discover a way to remain quiet themselves as down. In this work, fleeting supposition investigation coronavirus tweets after some time April 16 to April 13 are appeared. The extremity score of each tweet with VADER calculation in NLP is predominantly engaged in this work. The scores are isolated dependent on compound extremity esteem at last tweets are distinguished into three significant assessment classes like positive, negative and impartial and by demonstrating feeling score plot, fleeting assumptions after some time the information perception is additionally accomplished in this work. This work also focuses data visualization by showing sentiment score plot.

Keywords Corona · COVID-19 · Pandemic · Sentiment score · Lockdown · Polarity · VADER

V. Laxmi Narasamma (✉) · M. Sreedevi · G. Vijay Kumar
Department of Computer Science and Engineering, Koneru Lakshmaiah Education Foundation, Vaddeswaram, Guntur, Andhra Pradesh 522502, India

M. Sreedevi
e-mail: msreedevi_27@kluniversity.in

G. Vijay Kumar
e-mail: gvijay_73@kluniversity.in

© The Author(s), under exclusive license to Springer Nature Singapore Pte Ltd. 2021   183
S. K. Saha et al. (eds.), *Smart Technologies in Data Science and Communication*,
Lecture Notes in Networks and Systems 210,
https://doi.org/10.1007/978-981-16-1773-7_15

1   Introduction

Coronavirus pandemic virus disease (COVID-19) outbreak was initially identified
in Wuhan city of China nation, in month of December 2019, and has outbreak in
throughout of the world. Initially, in the month of March, heavy effect was in Italy,
Spain, US countries later on slowly this is been spreader over more than 200 countries worldwide. Within two months, this outbreak has disturbed world people life
and life became uncertainty regarding the future life as first time introduced in the
world. In World history, many epidemic situation periods have been observed. This
epidemics concept has grown in these latest years. Because of the virus spread rapidly
to global. Nowadays, these kinds of situations raises high economic down fall of all
the countries and as well at the personal level. In this pandemic, people have to
express their psychological feelings regarding this economic crisis. The examination
of computerized clients' conduct has seen that they are generally assembling data
from the Internet or person to person communication locales, for example, (Facebook,
Twitter, microblogs…).

To gather conduct, data numerous organizations have contingent upon Web-based
media locales. Twitter is utilized by numerous scientists. In light of the organization of microsites, tweets are taken and for scholarly examination purposes. Twitter
comprises a significant wellspring of chronicle information, permitting its fare for
research. We likewise played out an information investigation concentrate on coronavirus by breaking down Twitter presents in connection on this pandemic sickness,
presented between Walk 2020 on May 2020. The results of analytical researches
would help the outcomes could help shaping identification for the history of the
outbreak and especially how it was apprehended by the world population. In this
research work, tweets have been collected from April 16 2020 to 30th 2020, and are
related to COVID-19 in some or the other way.

This analysis has been done to analyze how the people around of the world dealing
with the situation emotionally. The tweets have been gathered, preprocessed, and then
used for text data analysis with NLP techniques. The main result of this research
work is to analyze emotional words or word cloud used by tweeters around the
world related to COVID-19 with word frequency and identifying sentiment polarity
scores of three emotion classes such as positive, negative and neutral. This work also
analyzed temporal sentiment over time on tweets.

2   Literature Review

Coronavirus (COVID-19) attacked since December 2019 from dragon country China.
Till now, only few publications are done on this topic. This entire literature related
work is on COVID-19 articles and NLP techniques.

Habiba Drias (May 2020) in this paper Twitter tweets to express their feelings
on the COVID-19. Between February 27, 2020 and March 25, 2020, they used a

data collection of nearly six hundred thousand tweets those contain hashtags like #COVID and #coronavirus were posted. In this work, outcome of work is shown, i.e., the results with FP-Growth algorithm. To discover the mostly frequent patterns, FP-Growth algorithm is imposed on the tweets and also association rules are used to know the insights of tweets related to coronavirus [1].

Samuel (May 2020) has done work on public sentiment analysis with machine learning approaches. In this work, they analyzed short text tweets and long text tweets. On short text tweets, they imposed Naïve Bayes algorithm and computed accuracy up to 91% and long text tweets are imposed with logistic regression and got accuracy of 74% [2].

In order to get the emotion analysis on COVID-19, Jelodar et al. (April 2020) have used LSTM recurrent neural network. To obtain the concerns of COVID-19 in order to make confirmation making, they have showed the importance of public emotions [3].

Kaila et al. (March 2020) studied information flow on social network Twitter during corona lockdown period. In this latent Dirichlet allocation, post-processing used for Twitter data analysis. This work analysis addressed mostly frequent accurate and relevant topics related to COVID-19 [4].

Akash D Dubey (April 2020) analyzed the preprocessed tweets for different sentiments such as fear, sadness, and disgust. Analyzed tweets from 11 March to 31 March. They have used NRC emotion lexicon model for sentiment analysis of 12 different country tweets [5].

To analyze several tweets in matter of mental health, Li et al. (June 2020) have done a work with usage natural language processing (NLP) techniques. To classify each tweet into the following, they have trained deep learning models, like emotions: disgust, fear, trust, anticipation, surprise, joy, sadness, and anger. They build the Emotion-COVID-19-Tweet (EmoCT) dataset by manually labeling 1000 English tweets for the training purpose. Furthermore, to discover the reasons that are affecting fear and sadness, we propose and compare two methods [6].

3   Research Methodology

Twitter is a microblogging site that allows tweet text of 140 characters length and initially now it is extended to 280 characters in length. Tweet although it is a small textual information but it gives many clues to both end users and data analysts. The recent observations said that over hundreds of millions tweets per day Twitter receives. It is difficult and challenging task to analyze such kind of huge and big data. Tweets are posted on specific events, particular topic of desire like sports, education, medical, and music.

4   Related Work

This research work focuses on the tweets posted over time April 16, 2020 to April 30, 2020 during corona outbreak and analysis done with methodology as shown in Fig. 1. The tweets are collected from famous widely used research site kaggle.com. The collected tweets over time are represented in the form of temporal plot. The collected tweets been preprocessed with natural language processing techniques (NLP) such as stop word removal, punctuation removal, white space, links removal, and converting into lower case. The preprocessed tweets are verified with word cloud and verified that no special characters and white spaces.

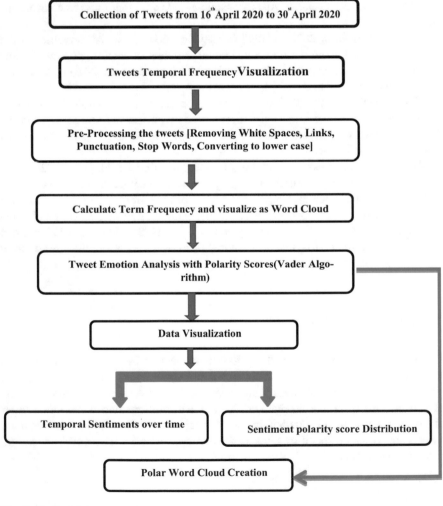

Fig. 1 Methodology

On preprocessed tweets, term frequency is calculated and visualized word frequency and verified with word cloud. This word cloud is formed with popular words of corona pandemic such as COVID-19, Corona, Lockdown, Pandamic, Stay-home, Staysafe, and Cases. Finally, emotion analyzed based on polarity scores for these polarity scores are computed using VADER. VADER is a dictionary and rule-based sentiment analysis tool. VADER not only informs about how positive or negative a sentiment but also informs us of the positivity and negativity score. This entire analysis is done with usage of VADER. This is more efficient than the usage of traditional bag of words collection technique.

4.1   VADER in NLP

VADER is referred to as Valence Aware Dictionary and Entiment Reasoner. It is a rule-based sentiment analysis library and lexicon. With VADER, you can be up and run the performance of sentiment classification very fast even if you do not have negative and positive text samples to train a classifier or need to write a code to search for words in a sentiment lexicon. VADER is also computationally efficient when compared to other machine learning and deep learning approaches. VADER algorithm consists of an inbuilt training library that means we cannot train our model separately. VADER algorithm uses subjectivity and polarity concept. Polarity concept defines positive and negative words as well as it defines the range of polarity which is in between 0 and 1. Polarity, also known as orientation, is the emotion expressed in the sentence. It can be positive, negative or neutral. Subjectivity is when text is an explanatory article which must be analyzed in context. Subjectivity is in between -1 to $+1$ [7–9].

The aftereffects of VADER examination are empowering as well as astounding. The results feature the tremendous benefits that can be accomplished by the utilization of VADER in instances of microblogging locales, wherein the content information is an intricate blend of an assortment of text. In this methodology dependent on the vocabularies of opinion related words, the VADER is a kind of notion investigation whether it is negative or positive identified with every one of the words in the dictionary, and as a rule, how certain or negative. To work out whether these words are positive or antagonistic, the engineers of these methodologies need to get a lot of individuals to physically rate them. It is clearly really costly and tedious. To be exceptionally exact the dictionary needs to have great treatment of the words in your content of intrigue; else, it would not be precise. When there is a solid match between the vocabulary and the content this methodology is exact, and furthermore rapidly returns results even on a lot of text. The engineers of VADER utilized Amazon's Mechanical Turk to get the majority of their appraisals which is a brisk and modest approach to get their ratings! It verifies whether any of the words in the content are available in the vocabulary when VADER investigations a piece of text.

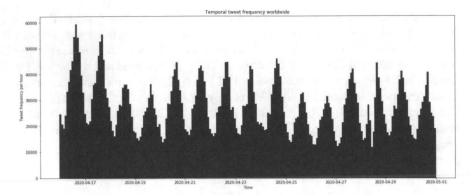

Fig. 2 Temporal frequency of tweets

5 Results and Findings

To handle the research work on COVID-19 tweets analyzing, first step is, this work carried out the gathering of a dataset. The attributes like its tweet title and tweet description, text, tweet text-related keywords or hashtags, tweet publisher, tweet date of publication are described by dataset which contains metadata. From entire dataset, only text attribute is considered to analyze the tweet because tweet attribute only holds tweeter's tweet.

After reading the tweet dataset, 15 days tweets of April 2020 are merged together and temporal frequency over time is shown in Fig. 2.

Word cloud library introduced recently in Python library module that offers an option to analyze and observe the text data through visualization in the form of tags or words, where the importance of a word is explained by its frequency. In recent years, text data has grown exponentially. This data need to be analyzed and to be represented visually (Figs. 3 and 4).

This work analyzed each tweet with sentiment polarity scores. Approximate the sentiment (polarity) of text by sentence (Fig. 5).

Classifying the scores based on Compound Polarity Value. This result value is a value that counts the result values which are normalized between values −1 and +1 (Figs. 6, 7, 8 and 9).

6 Conclusion

The main aim of this research work to analyze the opinions of the people during the pandemic coronavirus lockdown. In this study, people around the world tweeted on COVID-19 with different emotions. As this research work uses polarity scores given NLTK's widely used VADER tool that gives different polarities extreme negative, negative, neutral, positive and extreme positive. By considering compound polarity

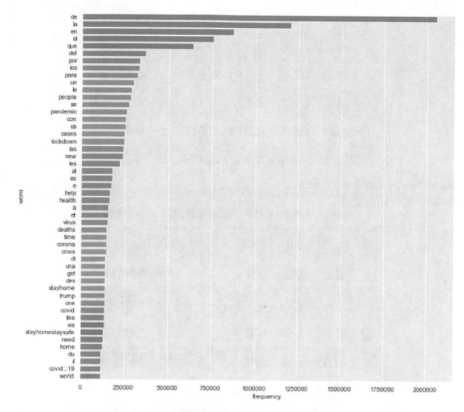

Fig. 3 Word frequency

Fig. 4 Word cloud

	neg	neu	pos	compound
5964648	0.217	0.493	0.289	0.2960
5964649	0.381	0.619	0.000	-0.8225
5964650	0.000	0.807	0.193	0.4767
5964651	0.000	0.323	0.677	0.6369
5964652	0.182	0.640	0.177	-0.0258

Fig. 5 Polarity scores of each tweet

	neg	neu	pos	compound	val
0	0.0	1.0	0.0	0.0	neutral
1	0.0	1.0	0.0	0.0	neutral
2	0.0	1.0	0.0	0.0	neutral
3	0.0	1.0	0.0	0.0	neutral
4	0.0	1.0	0.0	0.0	neutral

Fig. 6 Compound polarity score

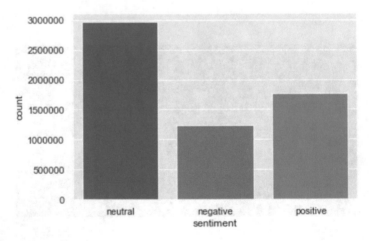

Fig. 7 Sentiment graph

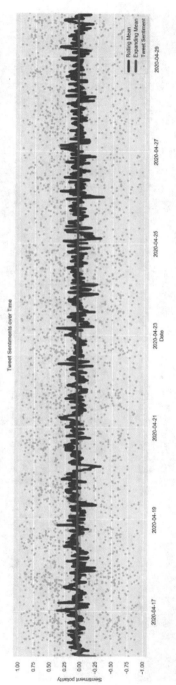

Fig. 8 Temporal sentiments over time

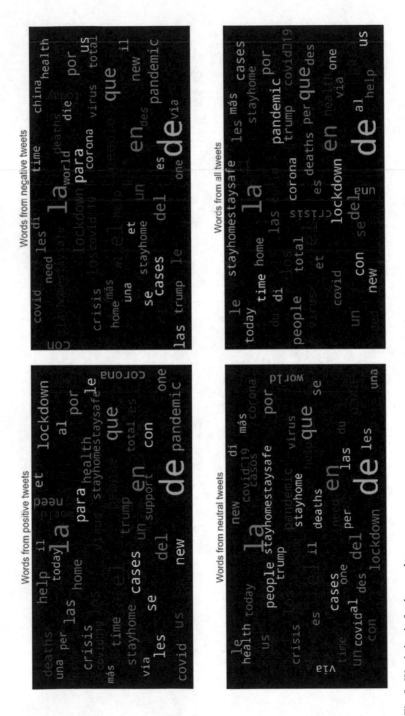

Fig. 9 Word cloud of polar words

score of VADER frame work, all the tweets finally classified to three emotion classes neutral, negative and positive. All the results are found in quick and efficient manner by using VADER framework. This research work is applicable to any dataset that contains tweet text. At present, the Twitter considered the most preferred medium for information spread and highly effective during pandemics and is proved again from the current coronavirus outbreaks. Everybody can depend on Twitter for spreading information and controlling panic among public at large mainly governments, health authorities, and institutions like WHO.

References

1. H. Drias, *Mining Twitter Data on COVID-19 for Sentiment Analysis and Frequent Patterns Discovery* (Algerian Academy of Science and Technology, University of Science and Technology HouariBoumediène, Algiers). medRxiv
2. J. Samuel, G.G.M. Nawaz Ali, M.M. Rahman, E. Esawi, Y. Samuel, *COVID-19 Public Sentiment Insights and Machine Learning for Tweets Classification*
3. H. Jelodar,Y. Wang, R. Orji, H. Huang, *Deep Sentiment Classification and Topic Discovery on Novel Coronavirus or COVID-19 Online Discussions: NLP Using LSTM Recurrent Neural Network Approach.* arXiv:2004.11695v1 [cs.IR] (2020)
4. R.P. Kaila, A.V. Krishna Prasad, Informational flow on twitter—corona virus outbreak—topic modelling approach. Int. J. Adv. Res. Eng. Technol. (IJARET)
5. A.D. Dubey *Twitter Sentiment Analysis during COVID19 Outbreak.* https://papers.ssrn.com/sol3/papers.cfm?abstract_id=3572023
6. I. Li, Y. Li, T. Li, S. Alvarez-Napagao, D. Garcia, What are We *Depressed about When We Talk about COVID19: Mental Health Analysis on Tweets Using Natural Language Processing.* https://www.researchgate.net/publication/340884066_What_are_We_Depressed_about_When_We_Talk_about_COVID19_Mental_Health_Analysis_on_Tweets_Using_Natural_Language_Processing (2002)
7. M. Yadav, S. Raskar, V. Waman, S.B. Chaudhari, Sentimental analysis on audio and video using Vader algorithm. Int. Res. J. Eng. Technol. (IRJET) **06**(06) (2019)
8. S. Baruah, S. Pal an analytical approach to customer sentiments using NLP techniques and building a brand recommender based on popularity Score. Indian J. Comput. Sci. (2019)
9. B.K. Bhavitha, A.P. Rodrigues, N.N. Chiplunkar Comparative study of machine learning techniques in sentimental analysis, in *International Conference on Inventive Communication and Computational Technologies* (ICICCT 2017)

An IoT Model-Based Traffic Lights Controlling System

B. Dinesh Reddy, Rajendra Kumar Ganiya, N. Thirupathi Rao, and Hye-jin Kim

Abstract Several problems of pandemic situations arise in various countries and also around the world in recent years. In recent days, the situation of traffic was going very great in number due to these pandemics. Most of these pandemics are occurring due to the viral spread with the contacts or nearby people. As a result, most of the people are interested to purchase their own vehicles that may be either two-wheeler vehicles or any four-wheeler vehicles. As a result, the traffic problems in the cities are growing at a faster level, and also, the pollution levels are getting increased. Hence, in the current article, an attempt had been made to analyse the traffic on the roads to control the congestions on the city roads at regular intervals of time. The current model was developed on the Python platform, for understanding the performance of the current model, four different cases of input images were selected and tested with the current developed model, and the results were discussed in the results section.

Keywords Traffic · Pandemic situations · Usage of vehicles · Two wheeler · Four wheeler · Lights · Traffic vehicles · Raspberry Pi · IoT model

1 Introduction

Several cities in India are getting more and more in number in recent days [1, 2]. The utilization of vehicles and the number of people staying in cities also increase in large number. As the number of vehicles are increasing in these developing cities, the traffic in those cities is growing in large number from time to time, and as a result, a greater number of traffic jams and traffic congestions occur in the cities. Due to this, people are suffering a lot [3–5]. The journey times from various locations are

B. Dinesh Reddy (✉) · R. K. Ganiya · N. Thirupathi Rao
Department of Computer Science and Engineering, Vignan's Institute of Information Technology (A), Vishakhapatnam, Andhra Pradesh, India

H. Kim
Kookmin University, 77 Jeongneung-RO, Seongbuk-GU, Seoul 02707, Republic of Korea

also getting increased day today. As a result of it, the people are losing their valuable times, valuable money and in some other areas also. Some of the cities already started the concept of odd and even number of vehicles concept in some major cities. One day they will allow all odd-numbered vehicles to move in the cities, and on the other day, the even-numbered vehicles will move on the city roads [6–8].

In the current considered model, an attempt has been made to identify the number of vehicles on road lanes at various signal points on city roads and trying to identify the density of vehicles and the number of vehicles count and based on the count of the vehicles; the traffic signals are controlled automatically with an automated program-based unit which was designed and developed by Python programming and the Raspberry Pi unit [9, 10]. A single unit consists of all the components of a computer, its related units were made, and the working of it was tested with various cases of inputs. These units can be considered as a single unit computer or the controller with which the traffic signals can be controlled and maintained correctly.

2 Methodology Used in the Current Model

The current work had followed the following steps of methodology for solving the problem.

1. Design the proposed system based on a certained scope.
2. Build breadboard circuit.
3. Install and prepare Raspberri Pi single-board computer.
4. Integrate our Java swing applications with Pi4J library for the above circuit.
5. Download and install the OpenCV library.
6. Implement the designed system using the OpenCV library and Java swings to use the cohesion and the countier algorithms to approximately count the number of vehicles in the given frame and accordingly control the time out of green light on the board.
7. Test the system with different images.

3 Existing System

As the vehicle numbers usage had increased a lot in day-to-day life, the congestion occurrence on roads had became common. To solve this problem of congestion on roads, several authors had worked and developed several models. Some of them had worked on density of vehicles such that to control the traffic lights to reduce the congestion.

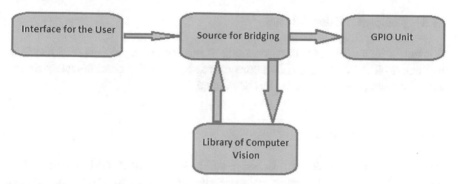

Fig. 1 Modules are working in the project

4 Proposed System

But, the concept of counting the vehicles on road and based on the number of vehicles, controlling the traffic lights on city roads had not done much work. Hence, an attempt had been made to solve the traffic problem with the help of this IoT technology, and also, an algorithm had been developed to solve this issue.

4.1 Modules of the Current System

The current intelligent traffic light control system uses four modules which are

1. GPIO
2. User interface
3. Computer vision library
4. Bridging source (Fig. 1).

4.1.1 GPIO

This unit is mainly used to check the status of the input and output functions of the controller through which the both types of data will be processed.

4.1.2 User Interface

This unit plays the key role in the entire system model. By using this unit, the users or the operators can interact with the developed model and can observe the results, can modify the timings of the model, can modify the lights at regular interval of times, etc. In other words, all the functions being performed by the current model will be accessed by any user through this unit only.

4.1.3 Computer Vision Library

This unit is mainly used for the processing of the images or pictures received from the cameras located at various locations on roads, identifying the objects with the help of Python programming models developed, etc.

4.1.4 Bridging Source

The main source or the content processing will be implemented at this unit only. This unit plays key role in performing the processing of the images, number of vehicles on roads, at various interval of time the number of vehicles on roads all this sort of data will be processed and analyse data at this unit.

5 Working of the Current Model

The input image or the normal image captured from the cameras placed at various locations will be sent to the unit for further processing. The pictures are selected or collected from the cameras at regular intervals of time. Then, the image arrived as an input image will be processed by converting the same input image to greyscale image. Thereafter, the greyscale image will be converted to the monochrome image. Thereafter, the edges of images will be detected by identifying the edges of the input images (Fig. 2).

Build a breadboard circuit. Install and prepare the Raspberry Pi single-board computer. Integrate our Java swing applications with Pi4J library for the above circuit. Download and install the OpenCV library. Implement the designed system using the OpenCV library and Java swings to use the cohesion and the counter algorithms to approximately count the number of vehicles in the given frame, and accordingly, control the time out of green light on the board. Test the system with different images.

6 Results and Discussions

In order to understand the working of the current model, several input images at various time intervals and various scenarios had chosen, and input images were collected. These images were given as input and observed the output of the model.

The input image for the first case is as in Fig. 3.

The output of the current input is as in Fig. 4.

Expected Result: 20–30 *vehicles approximately.*

Actual Result: Same *as expected TEST PASSED.*

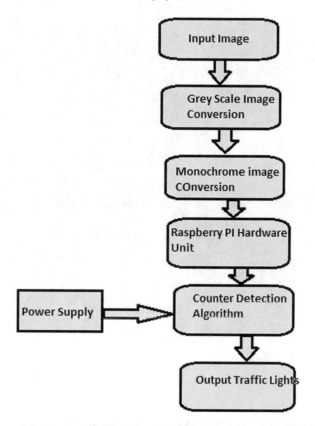

Fig. 2 Architecture for intelligent traffic lights control system

Fig. 3 Input of the first case

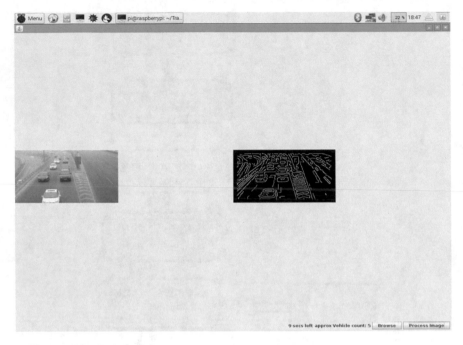

Fig. 4 Output of the first case

Test Case 2:

In the current case, the objective is to test if an image with vehicles is being detected well. The input is as in Fig. 5.

The output for the current input is as in Fig. 6.

Fig. 5 Input of the second case

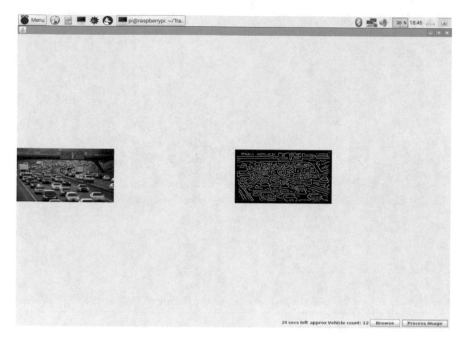

Fig. 6 Output of the second case

Expected Result: 20–30 vehicle approximately.

Actual Result: Same as expected TEST PASSED.

In the third case, the target is to identify the vehicles on the road, and the input of the model is Fig. 7.

The output of the current model is Fig. 8.

Fig. 7 Input of the third case

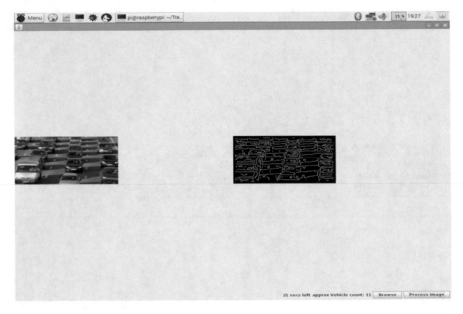

Fig. 8 Output of the third case

Expected Result: 25 vehicles approximately.

Actual Result: 21 vehicles identified approximately TEST PASSED.
  In the current test case, the vehicles on the road lane are to be identified.
  The input of the model is Fig. 9.
  The output of the model is Fig. 10.

Fig. 9 Input of the fourth case

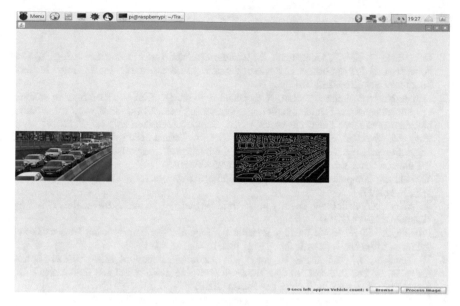

Fig. 10 Out of the fourth case

Expected Result: 20 vehicles approximately.

Actual Result: 25 vehicles identified TEST PASSED.

As a whole, the current considered model works well for various scenarios of traffic vehicles at various junctions with various numbers of vehicles. The proposed model is working correctly by identifying the number of vehicles, and based on that traffic density values and number of vehicles, the traffic lights are being operated either in green in colour or red colour.

7   Conclusion

Several works had done on the concept of identifying the traffic on roads and trying to reduce the congestion on roads. Several authors had followed different strategies to solve the same problem. In the current work, the vehicle numbers are being counted, and based on the number, the traffic lights were controlled with the combination of both hardware and software units. The Raspberry Pi unit had used for the hardware purpose, and Python-based algorithm had developed as a supplement from the software side of technical support. The data or the images being collected from various cameras had been sent as an input to the unit, thereafter, the processed images will give the number of vehicles on roads, and based on such numbers, the traffic lights will be controlled by the IoT unit.

References

1. Deol, R.K., Fiaidhi, J., Mohammed, S.: Intruder detection system using face recognition for home security IoT applications: a python Raspberry Pi 3 case study. Int. J. Secur. Technol. Smart Dev. **5**(2), 21–32 (2018)
2. Harika, N., Vamsilatha, M., Rao, N.T., Bhattacharyya, D., Kim, T.: Performance analysis and implementation of traffic monitoring system using wireless sensor network for reducing blockage in traffic on Indian city roads. J. Stat. Comput. Algorithm **1**(1), 15–26 (2017)
3. Ragu, V., Sivamani, S., Lee, M., Cho, H., Cho, Y., Park, J., Shin, C.: Analysis of traffic congestion and identification of peak hour for intelligent transportation services based on IoT environments. Int. J. Smart Home **11**(11), 25–32 (2018)
4. Y. Wiseman, Computerized traffic congestion detection system. Int. J. Transp. Logist. Manage. **1**(1), 1–8 (2017)
5. K. Vidhya, A. BazilaBanu, Density based traffic signal system. Int. J. Innov. Res. Sci., Eng. Technol. **3**, 10–18 (2013)
6. Promila, A., Sinhmar, Z.: Intelligent traffic light and density control using IR sensors and microcontroller. Int. J. Adv. Technol. Eng. Res. **2**, 11–21 (2013)
7. Viswanathan, V., Santhanam, V.: Intelligent traffic signal control using wireless sensor networks. In: 2nd International Conference on Advance in Electrical and Electronics Engineering, vol. 133, pp. 48–57. Springer, Berlin (2013)
8. J.A. Omina, An intelligent traffic light control system based on fuzzy logic algorithm. Int. Acad. J. Inf. Syst. Technol. **1**(5), 1–17 (2015)
9. J.-J. Kim, Y.-S. Lee, J.-Y. Moon, J.-M. Park, Bigdata based network traffic feature extraction. Int. J. Commun. Technol. Soc. Network. Serv. **6**(1), 1–6 (2018)
10. Anitha, J., Thirupathi Rao, N., Bhattacharyya, D., Kim, T.: An approach for summarizing Hindi text using restricted Boltzmann machine in deep learning. Int. J. Grid Distrib. Comput. **10**(11), 99–108 (2017)

A Comparative Study on Construction of 3D Objects from 2D Images

Mohan Mahanty◉, Panja Hemanth Kumar◉, Manjeti Sushma◉,
Illapu Tarun Chand◉, Kombathula Abhishek◉,
and Chilukuri Sai Revanth Chowdary◉

Abstract The 3D image construction for the 2D images is a longstanding problem, explored by the number of computer graphics, computer vision, and machine learning research organizations from decades. After the evolution of deep learning architectures, scholars have shown their interest in developing 3D images from the 2D greyscale or RGB images because this perspective shows a significant influence in the discipline of computer vision. The applications of this conversion found in various fields of medical image analysis, robotic vision, game design, lunar explorations, 3D modelling, military, geographical structuring, physics, support models, etc. Potentially, when a 2D image converted to 3D representation, the same image can be viewed from different angles and directions. The 3D structure, which in turn generated, is much more informative than 2D images as it contains information about distance from the camera to a particular object. In this paper, we discussed various exiting methods for generating 3D representations from 2D images, using 3D representational data as well as without 3D representational data and proposing a novel approach for the construction of 3D models from existing 2D images using GAN. Generative adversarial networks (GANs) have shown tremendous results in generating new fake data from existing data, where we cannot detect the false data. Various other architectures of GAN, like HOLO-GAN, IG-GAN, have also been proposed to meet the need to convert 2D to 3D representation, which produced excellent results. After analysing, we provide an extensive comparative review on methods and architectures, which can convert 2D images to 3D objects and express our thoughts on the ideas proposed. Further, the concept of GAN extended to represent 360 view images, panorama images in 3D structures, which plays a vital role in spherical view analysis and synthesis, virtual reality design, augmented reality design, 3D modelling of data, etc.

M. Mahanty (✉) · P. H. Kumar · I. T. Chand · K. Abhishek · C. S. R. Chowdary
Department of Computer Science and Engineering, Vignan's Institute of
Information Technology, Visakhapatnam, India

M. Sushma
Department of Computer Science and Engineering, Anil Neerukonda Institute of
Technology & Sciences, Visakhapatnam, India
e-mail: msushma.16.cse@anits.edu.in

© The Author(s), under exclusive license to Springer Nature Singapore Pte Ltd. 2021
S. K. Saha et al. (eds.), *Smart Technologies in Data Science and Communication*,
Lecture Notes in Networks and Systems 210,
https://doi.org/10.1007/978-981-16-1773-7_17

205

Keywords 2D images · 3D images · Deep learning · Generative adversarial network (GAN)

1   Introduction

The 3D object construction from a 2D image is attracting many researchers all around the globe because of its exciting applications. The existing computer vision in 2D drastically extended to the 3D vision, which is more informative in the sense of various views, camera angles, depth cues, light, and other factors affecting the 3D view. In this modern world, the birth of technologies like virtual reality, augmented reality, and others involved with 3D modelling shows the importance of perceiving the 2D world directly into a 3D world. The 3D view provides more information as well as gives a better experience than a 2D image. Booming technologies have emerged using this, but until now, extensive research is done on 2D images than on 3D objects. With this, we can conclude that there is a possible requirement to convert 2D images to 3D structures to fill the gap between them. Computer science techniques can help in constructing 3D structures from 2D images in various ways. Many researchers and scholars have done extensive research in this field of study. The important aspect of creating 3D structures from 2D images is depth. The presence of depth varies from 3 to 2D images. One of the methods is calculation of depth of the image by finding its depth map from the image, and finding various contours of the given images and merging these contours to construct the 3D structure is another method. In this manner, many scholars have proposed their work using different ways. The deep learning has once changed the approach to the construction of 3D from a 2D image. Many scholars started using convolutional neural networks (CNN) in this conversion process. Convolutional neural networks is an outstanding approach to deal with computer vision and this particular problem of conversion of 2D to 3D structures. Scholars even developed some techniques using convolutional neural networks (CNN) and using generative adversarial networks (GANs).

2   Literature Review

The main aim of constructing 3D structures from 2D images is to estimate the depth of each object in the 2D image. We can calculate depth in various ways and methods. Even obtaining the contours of an image can help us to calculate depth [1] Kadhim et al. have proposed their work on the visualization of a 3D object using 2D images by finding contours of a given image and merging to construct the 3D object. The 3D object is a 3D volume which has parallel slices containing useful information in the form of numerical data. Surface rendering and volume rendering are two methods described, and surface rendering is done using contours and merging them, whereas, in the volume rendering method, high-resolution data is preserved but requires much

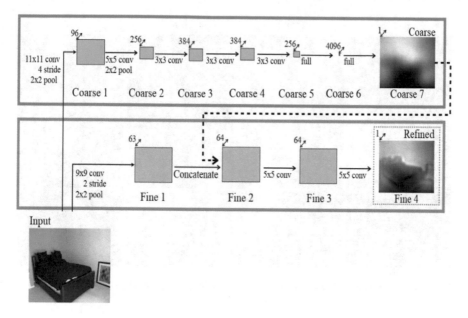

Fig. 1 Deep network architecture of both global coarse scale and local fine scale networks [2]

more information than surface rendering. Even though the 3D object constructed, it carries very little information, which may not be useful for further analysis. Calculating all the depths of objects in the 2D image gives rise to the depth map, Eigen et al. have proposed usage of the deep network for prediction of depth map [2]. As shown in Fig. 1, constructing a 3D structure from a given 2D image is an arduous job as 2D images carry very less information, which may not be sufficient. Eigen et al., in their research article, have taken both local and global information of the given image in different cues as shown in Fig. 2. From the aspects explained from the research article written by Galabov [3], the algorithms used to convert 2D to a 3D object, based on cues, are divided into two types: monocular depth cues and multi-ocular depth cues.

With the advancements in neural networks, Sinha et al. have proposed work in constructing 3D surfaces employing deep residual networks in their research paper [4]. Proposed work states that the objects classified into non-rigid and rigid objects, and based on it, different architectures proposed to generate surfaces. As shown in Fig. 3, for each input image, it also gives the view also for a rigid body as a piece of extra information. However, the method proposed by Sinha et al. is only limited to genus-0 surfaces.

The extensive work is done by constructing 3D models using convolutional neural networks from single images, which are contributed by Tatarchenko et al. [5]. As shown in Fig. 4, this architecture is resulting in an RGB image of the particular object as well as the depth map of the given image, which contains valuable information regarding the geometry for 3D structure. Because of this valuable information, the

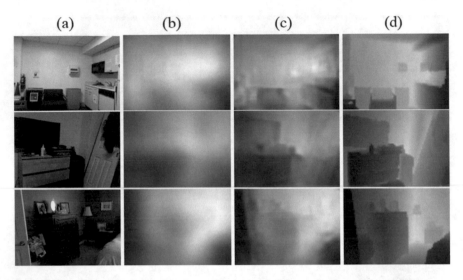

Fig. 2 **a** Input, **b** output of coarse scale, **c** refined output of fine scale, **d** ground truth [2]

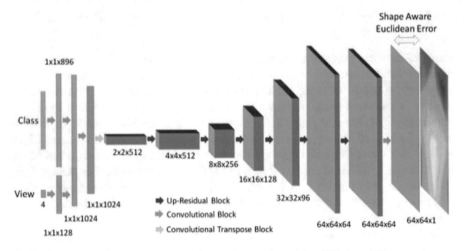

Fig. 3 Generative network for geometry-image feature channel for rigid shape [4]

voxels-based 3D model can be generated, which can further be used for analysis. By merging different views, a 3D point cloud is generated, which is useful to guess the invisible parts of the object. Tatarchenko et al. have developed the quality of generated images as shown in Fig. 5, use an elegant architecture, and the interesting thing is that it also applied for non-homogeneous backgrounds.

After this extensive research carried in convolutional neural networks (CNN), due to the generative capability of generative adversarial networks (GANs), scholars started migrating to it. Generative adversarial networks (GANs) are called the best

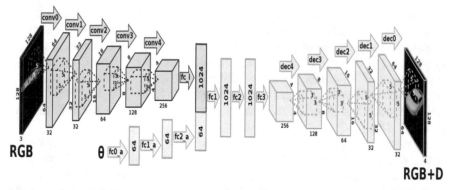

Fig. 4 Architectural design of the encoder–decoder network [5]

Fig. 5 Leftmost images show the input, network generated images in the top, and below are ground truth [5]

modern invention in the discipline of deep learning due to the capability of generating fake data from the existing original data, which are not distinguishable due to its design. Goodfellow et al. first introduce generative adversarial networks (GANs) [6]. Scholars started using this simple idea to generate data in different aspects such as in computer vision, robotic vision, natural language processing, time-series data, music, and voice. Gui et al. have shown the capabilities of GAN in his research paper describing different GANs [7] and its capabilities in generating data from existing data.

Generative adversarial networks (GANs) are the hybrid model of deep learning techniques that consist of two simultaneous and dynamic trained models. Generative adversarial nets developed by Goodfellow et al. [6] have a major impact on generating data from exiting patterns of the data which are not distinguishable from the original data. It consists of two dynamic training models: first (generator) network trained to produce fake data, and the second (discriminator) network trained to distinguish the generated data from real data. They develop some variations in the basic idea of generative adversarial networks (GANs), such as cyclic consistency loss mechanism by Zhu et al. in his research paper [8]. The modes of output even can be controlled by proposed work by Mirza et al. in their research paper [9], which is the primary development of the proposed work of Goodfellow et al. [6]. Different variants of GAN are proposed with different features in research papers [10–14]. The extensive review of GAN and its applications and various methods can be found in research papers [7, 15].

As shown in Fig. 6, the generative adversarial networks (GANs) mainly consist of two parts, namely generator and discriminator. The generator takes input as the random noise vector and generates fake data. The discriminator on the other side takes input from both generator, fake data, and the real data from the input source. The discriminator outputs the probability of similarity between the fake data and real data. Depending on the probability estimated by the discriminator network, then back propagation is done to the generator network to produce even more similar fake data to the real data. After each iteration, it revises that the discriminator network has weights and required biases in order to maximize its accuracy in the task of classification, which indeed means to produce correct predictions by giving maximum probability as possible: input data as real and generated fake data as fake. Generator network weights and biases are revised to maximize the probability that the discriminator network misclassifies fake generated data as real and continues this process until a particular state achieved, called an equilibrium state. Reaching equilibrium, when two players are involving in a game, the state in the game where neither of the players can improve their performance by implementing several methods is termed to be Nash equilibrium. We intend the dataset of real data that we make a generator to learn to

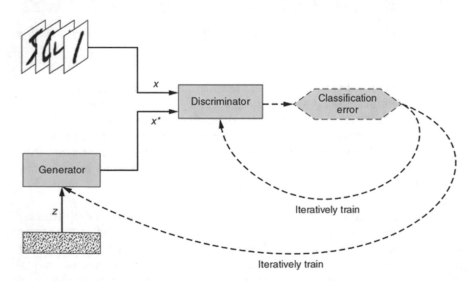

Fig. 6 Architecture of generative adversarial network

emulate with near-perfect quality fake data from it. The discriminator network is given real data as the input (x). A vector (generated from the raw input (z)) is given as input to the generator net-work, considered as initial point for synthesizing (or) generating fake data. We picture the process described above, in Fig. 6.

The generator network depicted with function $G()$, and discriminator network depicted with function $D()$. The generator network takes a random number in the form of a vector, z as the input with density p_z. In contrast, the discriminator network takes real input, x, and density generated by a discriminator network is p_g. $G()$ function returns density p_g, whereas $D()$ function returns probability. Equation (1) describes the overall error generated by both generator and discriminator.

$$E(G, D) = \frac{1}{2}\mathbb{E}_{x \sim p_t}[1 - D(x)] + \frac{1}{2}\mathbb{E}_{z \sim p_z}[D(G(z))]$$

$$E(G, D) = \frac{1}{2}\left(\mathbb{E}_{x \sim p_t}[1 - D(x)] + \mathbb{E}_{x \sim p_g}[D(x)]\right) \tag{1}$$

In the process of training of the generator, maximize the error while we try to minimize it for the discriminator. As both generator and discriminator networks compete with each other, which treated as the minimax game. The described game is described mathematically in Eq. (2).

$$\max_{G}\left(\min_{D} E(G, D)\right) \tag{2}$$

Equation (3) describes the overall equation of generative adversarial networks.

$$\min_{G} \max_{G} V(D, G) = \mathbb{E}_{x \sim p_{\text{data}}(x)}\left[\log D(x)\right]$$

$$+ \mathbb{E}_{z \sim p_z(z)}\left[\log(1 - D(G(z)))\right] \tag{3}$$

The ideal case is when the generator network produces the same density as the discriminator density, and perfect convergence has taken place. The whole method depicted in below graphical representation. The discriminator network calculates the probability by comparing the distributions. The probability distributions comparison is made between them by the method Kullback-Leibler divergence. Suppose p diverges from a second expected probability distribution q, then, Eq. (4) describes the KL divergence.

$$D_{KL}(p||q) = \sum_{i=1}^{N} p(x_i) . \log \frac{p(x_i)}{q(x_i)} \tag{4}$$

Because of asymmetric nature, the distance between any two distributions may play some role, which cannot be estimated by Kullback-Leibler divergence. Jensen-Shannon divergence is symmetric and can be used to calculate the distance between two probability distributions. Table 1 describes the functioning of various GANs.

Table 1 Brief description of generator and discriminator networks and their functioning

	Generator network	Discriminator network
Input	Vector of random numbers or, in some cases, particular numbers	The discriminator is fed input from two different sources: (1) Real data directly from the source (2) Fake data generated by the generator network
Output	Fake data that is mostly similar to the real data in the given dataset	The probability at which the fake data generated is similar to the real data
Goal	Continuously generate fake data which is indiscernible from the real data	Continuous distinguishing between the fake data from the generator network and the real data from the training dataset

Various architectures are related to GAN in the aspect of the generation of fake or duplicate images from real images which cannot be distinguished. Generative methods result in unique samples from high-dimensional distributions (like Gaussian distribution, Poisson distribution, etc.) including images, voice and some of the methods are described. GAN reaches Nash equilibrium when the following conditions are met:

- Generator network generates fake data which are indiscernible from the real data from the training dataset, so no more iteration is required.
- Discriminator network can guess randomly where a particular example is real or fake is best.

Figures 7 and 8 shows the learning procedure of the generator to generate fake data from distributions [16]. The GAN suffers from the problem as it is nearly impossible to find the Nash equilibrium for GANs because of the immense complexities involved in reaching convergence in non-convex games as instated by Farnia et al. in their research paper [17]. GAN convergence remains one of the most important open research questions in GAN-related research. The generative modelling is the reverse process of object recognition, and it constructs the image from the pixels instead of classification of the pixels. From the technique of generative modelling, we are interested in exploring the construction of 3D structures using 2D images. Zhu et al.

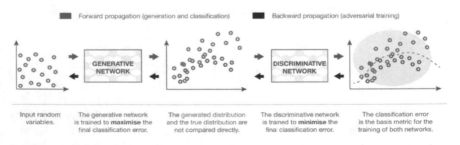

Fig. 7 Architecture of generative adversarial network (GAN) [16]

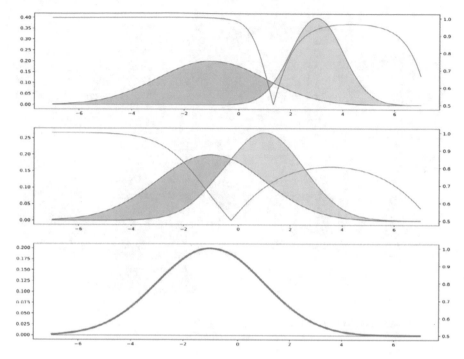

Fig. 8 Description of the idea of adversarial. The blue distribution is real, and orange distribution is generated

have come up with their work using a 2D image enhancer, and they generate A 3D model with the concept of adversarial. Zhu et al. suggested a method [18] to train two networks simultaneously. As shown in Fig. 9, it can learn from both 2D data and 3D data. 2D images are given as input to the enhancer network (deep convolutional neural network), generates the feature vectors, which are given as input to 3D Generator (GAN).

Edward et al. have proposed an adversarial system for 3D object generation and reconstruction [19], in order to capture more complicated distributions. It uses Wasserstein training objective with gradient penalty. Usage of this in their work showed improvement in instability. Improvements to the 3D-GAN [18] are shown clearly with their work. The combination of the IWGAN algorithm with a new novel generative model claimed as 3D-IWGAN. The architecture of 3D-IWGAN is depicted in Fig. 10.

Chen et al. contributed a method which is the differentiable interpolation-based renderer. Constructing 3D structures from 2D objects, not used for further analysis, is of no use. Because of the rasterization step involved, operations related are mostly differentiable, which may not be accessible for many machine learning techniques. Chen et al. proposed a differentiable method which is used not only for generating 3D object from a single image but also preserving geometry, texture, and light. As shown in Figs. 11 and 12, HOLO-GAN proposed by Phuoc et al. has shown tremendous

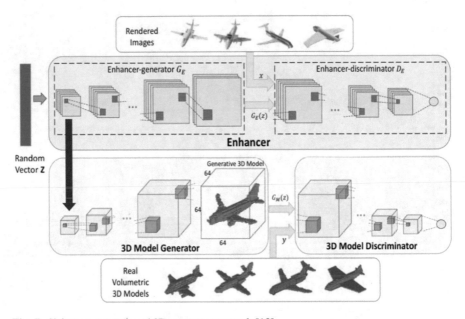

Fig. 9 Enhancer network and 3D generator network [18]

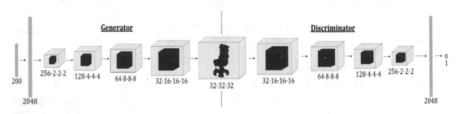

Fig. 10 3D-IWGAN architecture [19]

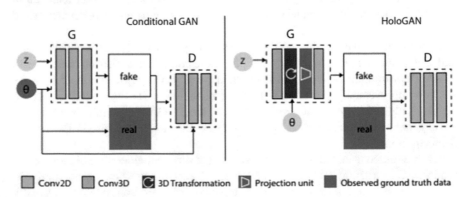

Fig. 11 Difference between conditional GAN and HOLO-GAN [20]

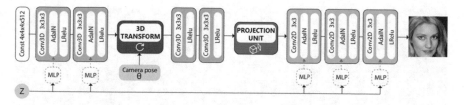

Fig. 12 HOLO-GAN generator network [20]

results in constructing 3D objects from 2D images [20]. HOLO-GAN has dynamic control with the pose and view of generated objects through rigid body transformations, which learned 3D features. In conditional GAN [9], the pose observed and information to the discriminator are given as input, whereas HOLO-GAN does not require additional pose labels in the process of training, as the pose information is not fed as input to the discriminator.

In the research article by Phuoc et al. [20], the GAN integrated with 3D transformation, randomly rotated 3D features during the training process of HOLO-GAN, makes the difference with 3D-GAN, shown in Figs. 13 and 14.

Lunz et al. proposed an extensible training procedure for 3D generative models from given 2D data, which employs a non-differentiable renderer in his research article [21]. To deal with the problem of non-differentiability, Lunz et al. have introduced proxy neural renderer, which can eliminate the problem by allowing back propagation as well as discriminator output matching. Figures 15 and 16 describe the architecture and functionality of IG-GAN.

From the above figure, the existence of the proxy neural network is depicted, which is eliminating the problem of non-differentiability.

Fig. 13 Results obtained from HOLO-GAN [20]

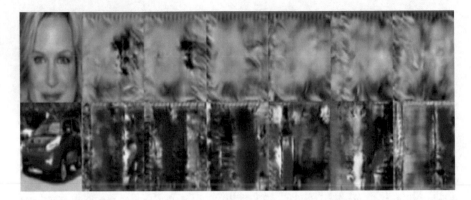

Fig. 14 Results obtained from 3D-GAN with using 3D transformations [20]

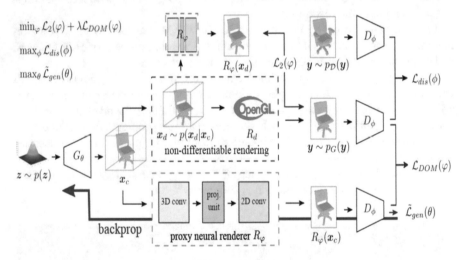

Fig. 15 Architecture and training set-up of IG-GAN [21]

3 Results and Analysis

By the extensive analysis, finding contours and merging them to find the depth of objects have a problem of low resolution, high level of noise, etc. It is suggestable to utilize other methods as finding the depth map of the image. Depth cues are also helpful as it carries more information about the image. A deep network with convolutions can convert the 2D image to a 3D structure. Convolutional neural networks (CNN) are also given good results, which are further modified to increase the performance of the network. Developments of generative adversarial networks (GAN) changed the scenario to not only construct the 3D objects but also to analyse them. On a comparative study on different methods of constructing 3D objects from 2D images, we can say that results produced by generative adversarial networks are

Fig. 16 Results obtained on IG-GAN on chanterelle mushroom dataset [21]

entirely satisfactory in various factors. The results generated by different research articles with different metrics are analysed. On analysis and study based on the comparison, we try to produce results of different methods with advantages and disadvantages. As shown in Fig. 17, generating 3D shape surfaces from 2D images using deep neural networks is only limited to genus-0 surfaces, and no feedback mechanism is defined. From Sinha et al. research article [4], both non-rigid and rigid bodies are feed, and obtained results are shown in Fig. 18.

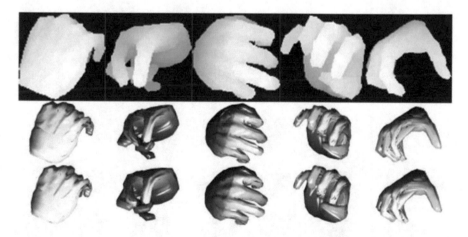

Fig. 17 Results generated by Surf-Net [4] on the non-rigid body. The topmost row is a depth maps, and the following are ground truths followed by the generated surfaces

Fig. 18 Results generated by Surf-Net on a rigid body. The topmost row is the given image, and the following row contains ground truth followed by the generated surface in the alternate view [4]

From work proposed by Tatarchenko et al. [5], different views are generated to the 3D object using convolutional neural networks from a single image used ShapeNet dataset. Tatarchenko et al. have used the nearest neighbour baseline approach, in the determination of error in colour and depth. Zhu et al. [18] considered that ModelNet (Figs. 19 and 20) dataset had produced results when compared to 3D-GAN proposed by Wu et al. [22]. ModelNet contains two sub-datasets, namely ModelNet10 and ModelNet40. Zhu et al. even produced results with the proposed model with and without the enhancer network.

From work initiated by Chen et al. [23], the experiments are done on the ShapeNet dataset on various aspects such as predicting 3D structures from single images: its

Fig. 19 3D objects generated by the proposed model by Zhu et al. [18]

Fig. 20 3D objects generated by 3D-GAN by Wu et al. [18]

geometry and colour, predicting 3D structures from single images: its geometry, texture, and light results with adversarial loss. Phuoc et al. [20] have done experiments on cars dataset and provided results for not only the proposed network but also for the other GAN networks such as DC-GAN, LS-GAN, and WGAN-GP. Lunz et al. have experimented on the ShapeNet dataset and provided results. The quality of the generated 3D models is measured by rendering them to 2D images and calculating Frchet inception distance (FID). All the results are tabulated below with the dataset used and metrics used for comparison in their research articles. Some other interesting works on GAN are image-to-image translation models [8, 24, 25]. Iizuka et al. propose the image in painting model [26]. Table 2 describes the comparison between various architectures.

Table 2 shows the comparison between various 2D to 3D conversion methods provided by scholars, and in most of the cases, the dataset is ShapeNet (CAR), and

Table 2 Comparison between various architectures

S. No.	Architecture	Dataset	Metrics		Quantitative result	
			First metric	Second metric	First metric	Second metric
1	Modified 3d-Gan with enhancer network	Modelnet (Modelnet40)	Mean average precision	Classification accuracy	44.44	87.85
2	Dib-R (geometry and colour)	Shapenet (Car)	3d-Iou	F-Score	78.8	53.6
3	DIB-R (geometry, texture, light)	Shapenet (Car)	Texture	Light	0.0217	9.709
4	Dib-R (adversarial loss)	Shapenet	2d-Iou	Key-Point	0.243	0.972
5	CNN For Multiview-3D Model	Shapenet	Error in colour	Error in depth	0.013	0.028

Table 3 Comparison of architectures of GANs proposed

S. No.	Architecture	Dataset	Metrics	Quantitative result
1	DC-GAN	Cars	Kernel inception distance	4.78 ± 0.11
2	LS-GAN	Cars	Kernel inception distance	4.99 ± 0.13
3	WGAN-GP	Cars	Kernel inception distance	15.57 ± 0.29
4	HOLO-GAN	Cars	Kernel inception distance	2.16 ± 0.09

generative adversarial networks used for the construction of 3D objects from 2D images because of its capability of generating models in a new way. DIB-R model even uses generative adversarial networks (GANs) in its architecture. DIB-R model used in the case of retrieval of information such as texture, colour, light, and other characteristics of an image (Table 3).

Frchet inception distance is the estimate of Wassertein-2 distance between these two Gaussians, primarily used to quantitative results of the quality of generated samples, which is described mathematically in Eq. (5).

$$\text{FID}(r, g) = ||\mu_r - \mu_g||_2^2 + Tr\left(\sum_r + \sum_g - 2\left(\sum_r \sum_g\right)^{\frac{1}{2}}\right) \qquad (5)$$

From Eq. (5), Frchet inception distance (FID) can be considered better in terms of computational efficiency, robustness, and discriminability. Even though it takes the first two order moments of the distribution, it is considered to be one of the best metrics for comparison of GAN, as stated in the research article [27]. On the other hand, the inception score considered as a key metric in the de-tection of intra-class mode dropping, i.e. when a model generates one and only one image per a particular class has a high inception score but not good FID. The inception score measures the diverse nature and qualitative results of generated samples, while FID measures the Wassertein-2 distance between the generated distribution and real distribution. When the difference in distribution is measured, it can evaluate the unobserved insides of the object. FID [27] is more consistent with human-based judgments and can handle noise more consistently than the inception score. As FID, considered to be better than KID, we consider that results generated by IG-GAN are considered to be more accurate in terms of robust and computationally efficient results through FID.

4   Conclusion

From the comparative study of various architectures, every research work has its advantages and disadvantages. Filling the gaps between them and utilizing the models where those are suitable show satisfactory results. Eliminating the problem of non-differentiability to the rasterization process is also useful in terms of recovering texture, colour, light, and other properties. Generative capability and other modifications to the adversarial idea are recommended for this task, which can be seen from the comparative results. At the same time, the discriminative capability is continuously increasing the capability of the generator, which provides the best results. We encourage scholars in performing research in GAN for the construction of 3D objects from 2D images, which have good scope for the future.

5 Future Scope

Moreover, GANs are used in the medical field in structuring DNA, doctor recommendations, and more interestingly, used in the detection of pandemic virus COVID-19. Khalifa et al., in their research paper [9], used GAN on the X-ray dataset of the chest in the detection of COVID-19, and Waheed et al. in his research paper [10] used GAN along with data augmentation in the detection of COVID-19. The above described shows that scholars are much interested in the idea of GAN and its development because of its tremendous capabilities. Generative methods result in unique samples from high-dimensional distributions (various distribution include Gaussian distribution, Poisson distribution, etc.) including images, voice and some of the methods are described. In the process of constructing 3D structures from given 2D images, HOLO-GAN learns through a differentiable projection unit which deals with occlusions. More precisely, the 4D tensor received by the projection unit, which has 3D features and results in a 3D tensor, which is 2D features. Furthurly, we are experimenting with generative adversarial networks on panorama images to construct 3D objects. We are trying to deduce the results in using GAN in the spherical analysis, which can be integrated into technologies like virtual reality, augmented reality, robotic vision, etc.

References

1. K.K. Al-shayeh, M.S. Al-ani, Efficient 3D object visualization via 2D images. J. Comput. Sci. **9**(11), 234–239 (2009)
2. D. Eigen, C. Puhrsch, R. Fergus, Depth map prediction from a single image using a multi-scale deep network. Adv. Neural Inf. Process. Syst. **3**(January), 2366–2374 (2014)
3. A Real Time 2D to 3D Image Conversion Techniques. Accessed 19 June 2020 (Online). Available https://www.researchgate.net/publication/272474479
4. A. Sinha, A. Unmesh, Q. Huang, K. Ramani, SurfNet: generating 3D shape surfaces using deep residual networks, in *Proceedings of 30th IEEE Conference on Computer Vision Pattern Recognition, CVPR 2017*, vol. 2017-Janua, 2017, pp. 791–800. https://doi.org/10.1109/CVPR.2017.91
5. M. Tatarchenko, A. Dosovitskiy, T. Brox, Multi-view 3D models from single images with a convolutional network, in Lecture Notes Computer Science (including Subser. Lect. Notes Artif. Intell. Lect. Notes Bioinformatics), vol. 9911, LNCS, pp. 322–337, 2016, https://doi.org/10.1007/978-3-319-46478-7_20

6. I.J. Goodfellow et al., Generative adversarial nets. Adv. Neural Inf. Process. Syst. **3**(January), 2672–2680 (2014)
7. J. Gui, Z. Sun, Y. Wen, D. Tao, J. Ye, A review on generative adversarial networks: algorithms, theory, and applications, vol. 14, no. 8, pp. 1–28, 2020. (Online). Available https://arxiv.org/abs/2001.06937
8. J.Y. Zhu, T. Park, P. Isola, A.A. Efros, Unpaired image-to-image translation using cycle-consistent adversarial networks, in *Proceedings of IEEE International Conference on Computer Vision*, vol. 2017-Octob, 2017, pp. 2242–2251. https://doi.org/10.1109/ICCV.2017.244
9. M. Mirza, S. Osindero, Conditional Generative Adversarial Nets, pp. 1–7, 2014 (Online). Available https://arxiv.org/abs/1411.1784
10. T. Karras, S. Laine, T. Aila, A style-based generator architecture for generative adversarial networks, in *Proceedings of IEEE Computer Social Conference on Computer Vision Pattern Recognition*, vol. 2019-June, 2019, pp. 4396–4405. https://doi.org/10.1109/CVPR.2019.00453
11. C. Wang, C. Xu, X. Yao, D. Tao, Evolutionary Generative Adversarial Networks, Mar 2018, Accessed 19 June 2020 (Online). Available https://arxiv.org/abs/1803.00657
12. T. Karras, T. Aila, S. Laine, J. Lehtinen, Progressive growing of GANs for improved quality, stability, and variation, in *6th International Conference on Learning Representation ICLR 2018—Conference Track Proceedings*, 2018, pp. 1–26
13. H. Zhang et al., StackGAN++: realistic image synthesis with stacked generative adversarial networks. IEEE Trans. Pattern Anal. Mach. Intell. **41**(8), 1947–1962 (2019). https://doi.org/10.1109/TPAMI.2018.2856256
14. A. Banerjee, D. Kollias, Emotion Generation and Recognition: A StarGAN Approach, 2019 (Online). Available https://arxiv.org/abs/1910.11090
15. S. Desai, A. Desai, Int. J. Tech. Innov. Mod. Eng. Sci. (IJTIMES) **3**(5), 43–48 (2017)
16. Understanding Generative Adversarial Networks (GANs). https://towardsdatascience.com/understanding-generative-adversarial-networks-gans-cd6e4651a29. Accessed 19 June 2020
17. F. Farnia, A. Ozdaglar, GANs May Have No Nash Equilibria, 2020, pp. 1–38 (Online). Available https://arxiv.org/abs/2002.09124
18. J. Zhu, J. Xie, Y. Fang, Learning adversarial 3D model generation with 2D image enhancer, in *32nd AAAI Conference on Artificial Intelligent, AAAI 2018*, 2018, pp. 7615–7622
19. E. Smith, D. Meger, Improved Adversarial Systems for 3D Object Generation and Reconstruction, no. CoRL, 2017, pp. 1–10 (Online). Available https://arxiv.org/abs/1707.09557
20. T. Nguyen-Phuoc, C. Li, L. Theis, C. Richardt, Y.L. Yang, HoloGAN: unsupervised learning of 3D representations from natural images, in *Proceedings of 2019 International Conference on Computer Vision Working ICCVW 2019*, no. Figure 1, 2019, pp. 2037–2040. https://doi.org/10.1109/ICCVW.2019.00255
21. S. Lunz, Y. Li, A. Fitzgibbon, N. Kushman, Inverse Graphics GAN: Learning to Generate 3D Shapes from Unstructured 2D Data, 2020 (Online). Available https://arxiv.org/abs/2002.12674
22. Z. Wu et al., 3D ShapeNets: A Deep Representation for Volumetric Shapes. Accessed 19 June 2020 (Online). Available https://3dshapenets.cs.princeton.edu
23. W. Chen et al., Learning to Predict 3D Objects with an Interpolation-based Differentiable Renderer, 2019, pp. 1–12 (Online). Available https://arxiv.org/abs/1908.01210
24. M.Y. Liu, T. Breuel, J. Kautz, Unsupervised image-to-image translation networks. Adv. Neural Inf. Process. Syst. **2017-Decem**(Nips), 701–709 (2017)
25. T.-C. Wang, M.-Y. Liu, J.-Y. Zhu, A. Tao, J. Kautz, B. Catanzaro, High-Resolution Image Synthesis and Semantic Manipulation with Conditional GANs (2018)
26. S. Iizuka, E. Simo-Serra, H. Ishikawa, Globally and locally consistent image completion. ACM Trans. Graph **36** (2017). https://doi.org/10.1145/3072959.3073659
27. A. Borji, Pros and Cons of GAN Evaluation Measures

A Deep Learning-Based Object Detection System for Blind People

Kalam Swathi⊙, **Bandi Vamsi**⊙, **and Nakka Thirupathi Rao**⊙

Abstract Visual impairment is one of the top disabilities among men and women across the world of all ages. Object detection is the primary task for them, and it can be implemented by deep learning techniques. Earlier implementation techniques involve in object detection with a strategy of single labeling. The proposed model uses classification techniques which reduce the recognize time of multi-objects with best time complexities and can help the visually impaired people in assisting the accurate navigation, in both indoor and outdoor circumstances. The proposed hybrid model is a combination of U-Net with base as residual network (ResNet) which improves accuracy in detection of objects in indoor and outdoor for visually impaired people.

Keywords Object detection · Deep learning · U-Net · ResNet · Visually impaired people

1 Introduction

1.1 Computer Vision

Computer vision is an area of computer science which allows computers to visualize similar to humans. Computer vision is trending in the area of segmentation, feature extraction as well as object detection [1]. It is excelling in the areas of medical, defense, traffic monitoring, robotics, automation and surveillance. Computer vision in computer technology is for analyzing the visual world [2]. In the early 1950s, initial work was started in computer vision where determination of keen edges aligns the simpler objects such as circles and squares by the techniques of first neural networks. Later in 1970s, optical character recognition came into existence, and later facial detection from images and videos was developed in 1990s. In classifying

K. Swathi (✉) · B. Vamsi · N. T. Rao
Department of Computer Science and Engineering, Vignan's Institute of Information Technology, Visakhapatnam 530049, India

© The Author(s), under exclusive license to Springer Nature Singapore Pte Ltd. 2021
S. K. Saha et al. (eds.), *Smart Technologies in Data Science and Communication*,
Lecture Notes in Networks and Systems 210,
https://doi.org/10.1007/978-981-16-1773-7_18

223

the objects' accuracy from digital images such as cameras and videos, computers use deep learning models [3].

1.2 Object Detection

Objection detection is one of the basic computer vision problems which is providing important data for understanding images and videos relating to many applications like face recognition, gender recognition, image classification, human behavior analysis, etc. [4]. To understand a complete image, we need to concentrate not just on primary points like image classification but also on classifying estimation of location of objects and concepts contained in each image. With the development of neural networks and progress in learning systems, there is a great impact on object detection techniques [5]. The object detection in an image requires focus to many factors such as the luminescent conditions, the range of the object as well as the orientation.

The object detection is the main action in the classification of the image, and also numerous researchers have proposed different ways to identify and situate the object. Upgraded object detection admits several objects in a single image [6], for instance, in certain circumstances, like the football field, an offensive player, a defensive gamer, a ball and so on [7]. To obtain this X, the Y coordinate model is implemented for the bounding box and detecting everything inside the region. The face recognition technology for object detection has developed the latest type that concedes the human face in the whole image and also finds as an individual specifically. Using pattern discovery, a duplication of the shapes, colors, and also various other visual indicators are in the picture. The image classifications are used to bring together into numerous divisions [8]. The feature similarities can be attached to pattern detection that classifies the similitude among the matched objects. The staggering advancement of the computer vision includes these advanced impacts. Around 50–90%, the accuracy in identifying the objects and dividing them right into specific categories has actually rapidly been increased in less than a decade [9].

1.3 Deep Learning

Deep learning is an artificial intelligence (AI) feature that imitates the operations of the human brain in processing data and creating patterns for usage in decision making [10]. Deep learning is a subset of artificial intelligence in expert system that has networks with the ability of learning unsupervised from information that is unstructured as shown in Fig. 1.

The data, which typically is unstructured, is so huge that it could take decades for people to comprehend it and also extract relevant details. Firms recognize the unbelievable possibility that can result from unraveling this wealth of details and are significantly adapting to AI systems for automated assistance [11]. The individual

Fig. 1 Relation of deep
learning with artificial
intelligence [13]

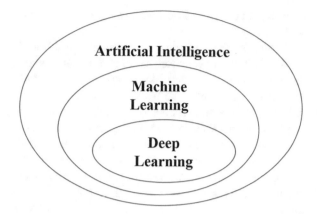

layers of neural networks can also be thought of as a sort of filter that works from gross to refined, increasing the possibility of discovering and outputting a correct outcome [12].

2 Related Work

Mouna et al. have proposed a research work titled "Recognizing signs and doors for Indoor Wayfinding for Blind and Visually Impaired Persons" [14]. In this work, the author has proposed an indoor and outdoor signaling model for blind people used for system detection for accessing environments. The indoor signaling system is used for recognizing objects in indoor environment with safety measures. This model has trained to recognize exit, doors, etc., to reduce dangers. The trained data is given for algorithm classification,, and it is given as input to knowledge phase. The tested data is used for decision making to check weight layers to produce the output. The residual architecture has some skip connections. These add with previous layers output to the ahead layers that are used for improving the convolutions performance. The dataset is having 800 subjects in which 80% used for training and 20% used for testing. This model has attained a decent accuracy for indoor environment with 99.8% and for outdoor environments.

Zhongen et al. have proposed a research work titled "A Wearable Device for Indoor Imminent Danger Detection and Avoidance with Region-Based Ground Segmentation" [15]. In this work, the author has proposed an inertial measurement unit (IMU) for helping blind people to detect the surroundings with safety awareness. The segmented image is act as a path for detecting 3D objects. The RGB images from 3D camera are obtained to accelerate the velocity by using IMU. This model lifts images and depths to 3D and reconstructs it based on height and orientation. By using region of interest, the ground objects are detected to increase the orientation.

The dataset contains 91 subjects in which 80% used for training and 20% used for testing.

Mouna et al. have proposed a research work titled "Deep Learning-Based Application for Indoor Scene Recognition" [16]. In this work, the author proposed a deep neural network used for recognizing objects for blind people. This model works for indoor environments. The feature extraction is used to train the model using class labels. The trained classifier is given for testing to predict the indoor objects. The convolution network is used for recognizing and classifying indoor environments. This model has attained decent accuracy with 95.06% for indoor objects detection.

Hiroki. K et al. have proposed a research work titled "Two-mode Mapless Visual Navigation of Indoor Autonomous Mobile Robot using Deep Convolutional Neural Network" [17]. In this work, the author proposed an autonomous model using convolutions for navigating environments. The model navigates landmark images for navigating detection of objects by calculating orientation and position of the target image. The dataset is having 712 subjects in which 80% used for training and 20% used for testing. The model uses Cartesian coordinates to move left and right direction of the targeted position.

3   System Architecture and Methodology

In Fig. 2, the proposed model takes input from available images. Apply data preprocessing techniques to remove the noise. The preprocessed data is divided into two sets for training and testing purpose. The trained data is given to U-Net, ResNet and hybrid architectures for performance evaluation. Out of available models, the best accurate model is given for the detection of objects available in indoor and outdoor environments.

3.1   U-Net Architecture

Figure 3 explains about the U-Net architecture which is symmetric and consists of two major parts: One is called contracting path where general convolution process is carried out and generally called as down sampling. In the right most path, expansion process is carried out and generally called as up sampling. The size of the input image is $160 \times 160 \times 3$. The encoder is applied in the form of two convolutions with a size of $80 \times 80 \times 32, 64$. Then operation of max pooling is applied with a size of $40 \times 40 \times 64$ with a stride of 3. Similarly, the same procedure is applied to remaining convolutions, and it is called as down sampling. The input for the decoder is with a size of convolution 10×10 with 256 filters. The operation of up sampling is done with the same size to generate the two convolutions with $20 \times 20 \times 256, 128$.

Fig. 2 Proposed model

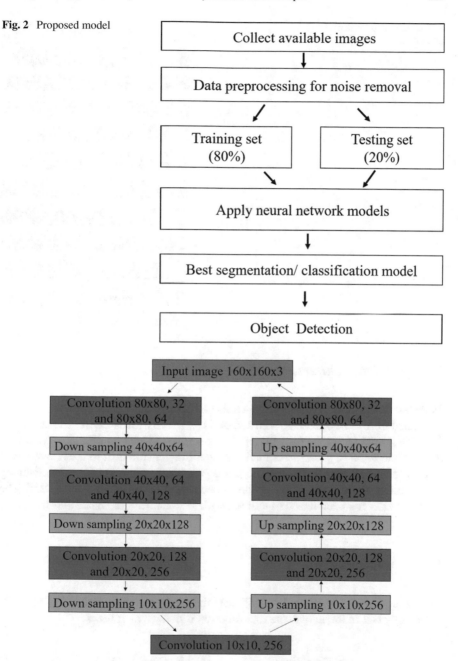

Fig. 3 U-shaped network architecture

Fig. 4 Residual network
architecture

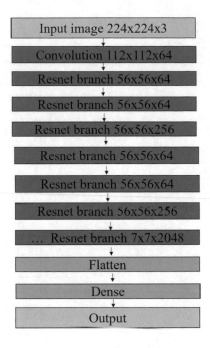

3.2 ResNet Architecture

In Fig. 4, the input image is taken with a size of $224 \times 224 \times 3$. The convolution-1 is generated from the input image with a size of 112×112 with 64 filters. The activation function is applied for the convolution-1 to generate the max pooling operation with a size of $56 \times 56 \times 64$. The residual branches are generated from the above layers with 64 filters of size 56×56. The same procedure is applied for the remaining convolutions and residual branches up to adder with a size of $56 \times 56 \times 256$. The dense layer is considered as output layer with shape 2 from the flatten layer.

3.3 Loss Function

Let the input image be labeled with 'X' and the predicted image be labeled with 'Y'. For every pixel in the image, the loss function is defined as follows:

$$\text{Loss Function}(X, Y) = \frac{1}{n} \sum_{i=1}^{n} \text{loss function}(X, Y) \tag{1}$$

Table 1 Performance evaluation of U-Net

Train loss	Train accuracy	Validation loss	Validation accuracy
2.965	94.82	2.952	92.91

Table 2 Performance evaluation of ResNet

Train loss	Train accuracy	Validation loss	Validation accuracy
1.225	84.38	1.3452	83.51

Table 3 Performance evaluation of U-Net with base ResNet

Train loss	Train accuracy	Validation loss	Validation accuracy
3.952	95.36	4.139	93.36

4 Results and Discussion

In Table 1, we can observe that the performance evaluation of U-Net architecture has a minimum validation loss with respective trained loss with 2.965%. This model has received a decent accuracy for training and validation with 94.82 and 92.91%.

Table 2 shows about the performance evaluation of ResNet architecture. By comparing the ResNet and U-Net models, it is observed that ResNet model is giving minimum loss values under training and validation phase, whereas U-Net model is giving decent accuracy when compared to ResNet with 92.9% than 83.51%.

Table 3 shows about the performance evaluation of U-Net architecture with base ResNet. By comparing the existing models with proposed hybrid model, it is observed that hybrid model is giving decent high accuracy with 95.36 by comparing with 94.82 and 95.36%.

Figure 5 shows about all the available objects with multi-labeling for visually impaired people in traffic environment. In this, the model has predicted objects like cars, trucks, bicycles and other objects.

5 Conclusion and Future Work

The proposed hybrid model is tested with existing deep neural networks like U-Net and ResNet for detecting objects used for visually impaired people in indoor and outdoor surroundings. The proposed model is producing decent accuracies for training and validation when coming to the loss parameter, and there is an increase in value. It is also observed that hybrid architectures even improve the performance of the model but at the same time increase the loss factor. So, in future, there is a scope to develop a model where loss rate is minimized.

Fig. 5 Model output for visually impaired people in traffic environment

References

1. T. Nabil, MultiResUNet: rethinking the U-Net architecture for multimodal biomedical image segmentation. Neural Netw. **121**, 74–87 (2020)
2. S. Asim, Convolution neural network based object detection: a review. J. Crit. Rev. **7**(11), 786–792 (2020)
3. Z. Zhong, Object detection with deep learning: a review. IEEE Trans. Neural Netw. Learn. Syst. **99**, 1–21 (2019)
4. B. Ando, A smart multi sensor approach to assist blind people in specific urban navigation tasks. IEEE Trans. Neural Syst. Rehabil. Eng. **16**(6), 592–594 (2008)
5. C. Phanikrishna, Contour tracking based knowledge extraction and object recognition using deep learning neural networks, in *2nd International Conference on Next Generation Computing Technologies (NGCT)*, vol. 14 (2016), pp. 352–354
6. F.M. Hasanuzzaman, Robust and effective component-based banknote recognition for the blind. IEEE Trans. Syst., Man, Cybern., Part C (Appl. Rev.) **42**(6), 1021–1030 (2012)
7. N. Sasikala, Unifying boundary, region, shape into level sets for touching object segmentation in train rolling stock high-speed video. IEEE Access **23**(6), 70368–70377 (2018)
8. N. Venkatesh, Object detection using machine learning for visually impaired people. Int. J. Curr. Res. Rev. **12**(20), 157–167 (2020)
9. M. Mekhalfi, A compressive sensing approach to describe indoor scenes for blind people. IEEE Trans. Circuits Syst. Video Technol. **25**(7), 1246–1257 (2014)
10. L. Zhang, Using multi-label classification for acoustic pattern detection and assisting bird species surveys. Appl. Acoust. **110**, 91–98 (2016)
11. C. Cadena, A fast, modular scene understanding system using context-aware object detection, in *IEEE International Conference on Robotics and Automation (ICRA)*, vol. 26 (2015), pp. 4859–4866
12. M. Anila, Study of prediction algorithms for selecting appropriate classifiers in machine learning. J. Adv. Res. Dyn. Control Syst. **9**(18), 257–268 (2017)
13. M. Adel, Advanced methods for photovoltaic output power forecasting: a review. Appl. Sci. **10**(2), 487 (2020)
14. A. Mouna, Recognizing signs and doors for indoor way finding for blind and visually impaired persons, in *5th International Conference on Advanced Technologies for Signal and Image Processing* (2020)

15. L. Zhongen, A wearable device for indoor imminent danger detection and avoidance with region-based ground segmentation. IEEE Access **8**, 184808–184821 (2020)
16. A. Mouna, Deep learning based application for indoor scene recognition. Neural Process. Lett. **51**, 1–2 (2020)
17. K. Hiroki, Two-mode mapless visual navigation of indoor autonomous mobile robot using deep convolutional neural network, in *IEEE/SICE International Symposium on System Integration Honolulu* (2020), pp. 536–541

Pothole Detection Using Deep Learning

Pathipati Bhavya⊚, **Ch. Sharmila**⊚, **Y. Sai Sadhvi**⊚,
Ch. M. L. Prasanna⊚, **and Vithya Ganesan**⊚

Abstract In the present-day scenario, the government needs accurate information for effective road maintenance at regular intervals but road inspection requires enormous amounts of manpower every year and this obviously slows down the process due to the distance involved. So, detection of potholes on roads is noticeably required by the government for maintaining road which can be done by the techniques of deep learning. The main purpose of the project is to classify the images of roads based on condition/status, that is either it is a plain road or road with potholes. This model initially takes the pictures of the roads which is our dataset as input. These inputs are into the deep learning classification algorithms to classify the images of roads, and this classification can be helpful to assess road condition. This project replaces external manpower for road maintenance. This model is useful for the government for better road maintenance with less manpower in a small period of time.

Keywords Pothole · Pre-trained CNN model · ResNet50 · Pothole roads · Deep learning

1 Introduction

In the field of effective computing, many techniques and tools to detect images with pothole roads were proposed. Pothole is the kind of damage in roads which is caused as a result of cracks and by water and traffic and its minimum plan dimension is 150 mm [1]. Any works based on this pothole detection have been carried out using

P. Bhavya (✉) · Ch. Sharmila · Y. Sai Sadhvi · Ch. M. L. Prasanna · V. Ganesan
Department of Computer Science and Engineering, Koneru Lakshmaiah Education Foundation, Vaddeswaram, AP, India
e-mail: 170030996@kluniversity.in

Department of Electronics and Communication Engineering, Koneru Lakshmaiah Education Foundation, Vaddeswaram, AP, India

© The Author(s), under exclusive license to Springer Nature Singapore Pte Ltd. 2021   233
S. K. Saha et al. (eds.), *Smart Technologies in Data Science and Communication*,
Lecture Notes in Networks and Systems 210,
https://doi.org/10.1007/978-981-16-1773-7_19

different algorithms and various image processing techniques [2, 3]. By using pre-trained CNN model ResNet50 we build a model which can classify the images of plain road and pothole road [4, 5].

1.1 Objective

- Analyse the condition of the road based on the potholes.
- Assess the road condition.
- Provide a classification of roads which are plain and which are not plain (road with pothole).

1.2 About the Project

Our project is to utilize deep learning tools for detecting whether it has potholes or it is plain road.

The dataset is organized into two folders (train, test) and contains subfolders for each folder. There are hundreds of images of roads (JPEG, JPG) and two categories of roads such as

- Pothole road
- Plain road.

1.3 Purpose

- Government needs accurate information for effective road maintenance at regular intervals
- But, road inspection requires an enormous amount of manpower every year. This obviously slows down the process due to the distance involved. Technique for road damage detection is highly effective in road management.

1.4 Scope

Pictures of roads with potholes and pictures of plain roads will be sent to our deep learning model and then it will give the output as whether it is plain road or pothole road.

2 Methodology

The total scenario has to be divided into various classes as shown in the below figure. Erasis basically a back end for the image classification when we want to employ spyder IDE (Python) which comes with any packages [6, 7]. Data has to be divided into training, validation and testing sets and data preprocessing is one. Networks have to be built which consist of many layers keConv2D layer, Flatten layer, Dense layer, Max Pooling layer [8, 9]. The model has to be tested with random images picked on a dataset and if the accuracy is not good, then some traditional algorithms like augmentation, feature extraction have to be employed. In the below activity diagram, the dataset from the user is passed on to the machine for preprocessing [10]. The machine efficiently interacts with the server in order to perform the operations at every stage. Once the layers are built, the training can be started in search of good accuracies [11]. Improving accuracies can be done by using traditional methods like augmentation, feature extraction and fine tuning [12]. Once good accuracy is seen, we can terminate the training process.

RS may be a document created by the system analyst after ants are collected from various customers [13]. It defines how software will interact across various platforms. The requirements received from clients were written in a natural language [14, 15]. It is a responsibility of the system analyst to prepare document of the wants in technical language [16, 17]. The production of the software requirement specification states the goals and objectives of a software, describing it as the context of the computER20 base system. The SRS includes an formation description, functional description, behavioural description and validation criteria.

RS should implement with the following features: the requirements of user are expressed in natural language.

Technical requirements will be expressed in a structured language [18]. Description of design must be written in the form of pseudocode. Format of forms and GUI screen-prints will be displayed. Conditional and mathematical notations for DFDs, etc. will be displayed [19].

The purpose of the document is to present the software requirements in a precise manner. This document provides the functional, performance and the verification requirements of the software to be developed in general. Users might invite illegal, impractical solutions or experts may implement the wants incorrectly [20]. This leads to an enormous raise in the price if not nipped within the bud.

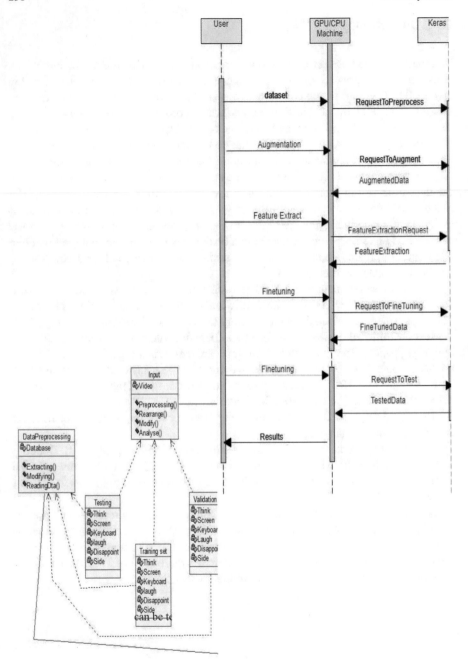

Requirements can be tested against following conditions:

1. If they can practically implement the system.
2. If they are valid as per the functionality and domain of the software.
3. If there are any ambiguities.
4. If they are complete.
5. If they can be demonstrated.

A requirement may be a statement about what the proposed system will do this all customers agree must be made true so as for the customer's problem to be adequately solved.

Requirements will be divided into functional and non-functional requirements.

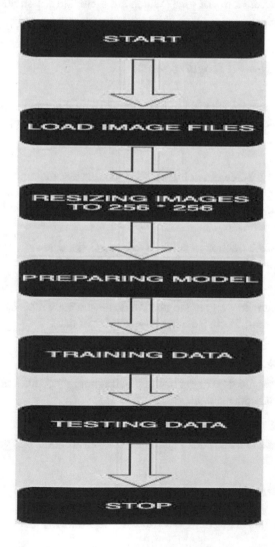

2.1 Functional Requirements

A functional requirement defines a function of a system or its component. A function can be described as a set of inputs, the behaviour and outputs. It also depends upon the software, expected users and the type of system where the software is used. Functional user requirements may be statements of what the system should do but functional system requirements should also describe clearly about the system services in detail.

The functional requirements in our project are:

1. Analyse the condition of the road based on the potholes.
2. Assess the road condition.
3. Provide a classification of roads which are plain and which are not plain(road with pothole).

2.2 Non-functional Requirements

A non-functional requirement is a requirement criteria that can be used to judge the operation of a system, rather than specific behaviours. Non-functional requirements are called qualities of a system, there are as follows:

1. Performance
2. Accuracy (The exactness in the trained model is learning and also classifying.)
3. Our model is trained successfully with an accuracy of 96% against custom datasets Processing time (How long is acceptable to perform key functions is export/import data?).

Our model takes a fraction of seconds to process the data.

1. Capacity and scalability
2. Storage

(How much data are we going to need to achieve what we need)?
Our model is trained with a couple of gigabytes of data.

3. Recovery
4. Restore time

(How long it should take to get back up and running?) It takes around 2 h to get backup and run the entire model.

5. Backuptime

(How long does it take to back your data?) It takes around 2 h to get back the data.

2.2.1 System Requirements

System requirements are nothing but software requirements and Hardware requirements.

Software Requirements

Operating system: WINDOWS/LINUX (PREFERABLY) Programming languages:
Python
   Backend: Keras, TensorFlow
   Editor: Pycharm or jupyter notebook
   Packages: NumPy, pandas, matplotlib, cv2, TensorFlow

Hardware Requirements

Laptop with the following requirements: Processor: Intel i5 or higher
   RAM: 8 GB or higher Disk space: 500 GB HDD

3 Module Description

3.1 Flask Framework

Flask is the micro-web framework written in Python. This has no database abstraction
layer, form validation where pre-existing of third-party libraries which provide the
common functions in it. The Flask which extent can add application features as if
they are implemented in the Flask. Extensions are updated far more frequently than
the core of the Flask program.
   Extensions exist for the object-relational mappers and many common frameworks
which are related tools.
   Many applications use the Flask framework which includes Pinterest and
LinkedIn.

3.1.1 Transfer Learning

Transfer learning usually focuses on storing the knowledge gained while solving one
problem.
   Applying it to different but related problems.

1. For example, the knowledge gained while trying to recognize cars could apply
    when trying to recognize trucks.
2. By using transfer learning helps us to increase our model accuracy to 96%.
3. However, there are some measures taken by us to use transfer learning.
4. ResNet50 is the pre-trained model that we are using to implement our model.
5. Measures are taken while using TL.

Step 1: ResNet50 is trained to classify 1000 different images (Classes).

But in our model we required to classify two classes (pothole or plain). To overcome this problem we removed the last output layer.

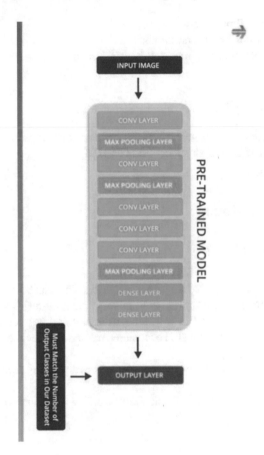

Step 2: After preparing the Base_model we added up to the stack(sequential()).

1. After that, we added up three more dense layers with ReLU as our activation function.
2. At last, we added our output layer with 2 nodes (2 nodes are required for 2 classes to classify) with softmax as our activation function which calculates the probability of 2 classes (pothole or not).
3. How computer treats images.

Greyscale Image 6 x 6

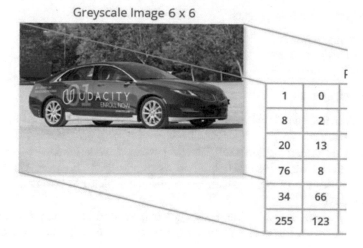

1	0
8	2
20	13
76	8
34	66
255	123

3.1.2 CNN (Convolutional Neural Networks)

In neural networks, convolutional neural networks are one of the main categories to do image recognition, image classifications. Object detections, recognition of faces, etc., are some of the areas where convolutional neural networks are used mostly.

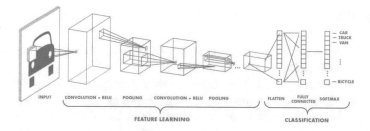

4 Algorithm Analysis

Existing system:

1. Existing systems acquire road status information using gyroscope and accelerate sensors which are called logging.
2. This existing model will divide the road status into three types (normal, pothole and bump) were considered on a road
3. This model requires manpower, dedicated vehicles and time to know the status of each road.

Drawbacks of existing system:

1. More manpower
2. Complexity of work
3. Time consuming

Proposed system:

1. To overcome the drawbacks of the existing system, the proposed system has been evolved.
2. Using deep learning tools for detecting whether a road is having potholes or it is plain road, which is helpful in assessing road condition and provides a classification of roads which are plain and which are not plain(road with potholes).
3. Deep learning classification techniques can be used to classify the roads into two categories. Based on potholes, we can predict the road condition using deep learning techniques.

Advantages of proposed system:

1. Easy to analyse
2. Less time consuming and easy to control.
3. External evaluation can be reduced.
4. Easy to identify the plain roads and roads with potholes.

5 Results

Pothole detection using deep learning is trained successfully using python with an accuracy of 96% against custom dataset.

This can further be applied to various real-time image classification problems on detection of potholes on roads by giving images as data. In the case of tools for recognizing potholes on roads, attention should be paid to the appearance of potholes. Apart from potholes or plain roads, other objects such as vehicles may also be sometimes considered as potholes by some deep learning model. So, in order to reduce those problems, we used the ResNet model which is a pre-trained deep learning model. Using this ResNet model, we build a CNN(convolutional neural network) model. Once the model gets trained successfully, a lot of human work in classifying images based on scans can be reduced to minimum. This model can be used by the government for better road maintenance with less manpower in a small period of time.

References

1. H. Song, K. Baek, Yungcheolbyun, Pothole Detection Using Machine learning, in *2018 Advanced Science and Technology Research Gate* (2018)
2. A. Kulkarni, N. Mhalgi, S. Gurnani, N. Giri, Pothole Detection System using Machine Learning on Android. In 2014 Semantic Scholar
3. A. Rasyid, M.R.U. Albaab, M.F. Falah, Y.Y.F. Panduman, A.A. Yusuf, D.K. Basuki, A. Tjahjono, R.P.N. Budiarti, S. Sukaridhoto, F. Yudianto, H. Wicaksono, Pothole visual detection using machine learning method integrated with internet of thing video streaming platform, in *2019 International Electronics Symposium (IEC)*, 2019
4. A.A. Angulo, J.A. Vega-Fernández, L.M. Aguilar-Lobo, S. Natraj, G. Ochoa-Ruiz, Road damage detection acquisition system based on deep neural networks for physical asset management, in *2019 ITESM Campus Guadalajara*
5. B. Varona, A. Monteserin, A. Teyseyre, A deep learning approach to automatic road surface monitoring and pothole detection. Pers. Ubiquitous Comput. **24** (2019)
6. A. Karmel, M. Adhithiyan, P. Senthil Kumar, Machine learning based approach for pothole detection, in *2018 International Journal of Civil Engineering and Technology (IJCIET)* (2018)
7. S. Arjapure, D.R. Kalbande, Road pothole detection using deep learning classifiers. Int. J. Recent Technol. Eng. (IJRTE)
8. A. Dhiman, R. Klette, Pothole detection using computer vision and learning. IEEE Trans. Intell. Transp. Syst. (2019)
9. E.N. Ukhwah, E.M. Yuniarno, Y.K. Suprapto, Asphalt pavement pothole detection using deep learning method based on YOLO neural network. Int. Seminar Intell. Technol. Appl. (ISITIA) (2019)
10. V. Pereira, S. Tamura, S. Hayamizu, H. Fukai, A deep learning-based approach for road pothole detection in timor leste, in *2018 IEEE International Conference on Service Operations and Logistics, and Informatics (SOLI)* (2018)
11. H. Chen, M. Yao, Q. Gu, Pothole detection using location-aware convolutional neural networks. Int. J. Mach. Learn. Cybern. **11** (2020)
12. S. Gupta, P. Sharma, D. Sharma, V. Gupta, N. Sambyal, Detection and localization of potholes in thermal images using deep neural networks, in *Multimedia Tools and Applications*, vol. 79 (Springer, 2020)
13. P.V. Rama Raju, G. Bharga Manjari, G. Nagaraju, Brain tumour detection using convolutional neural network. Int. J. Recent Technol. Eng. (IJRTE) (2019)
14. Aparna, Y. Bhatia, R. Rai, V. Gupta, N. Aggarwal, A. Akula, Convolutional neural networks based potholes detection using thermal imaging. J. King Saud Univ. Comput. Inf. Sci. (2019)
15. H. Maeda, Y. Sekimoto, T. Seto, T. Kashiyama, H. Omata, Road damage detection using deep neural networks with images captured through a smartphone. Comput. Vis. Pattern Recogn. (2018)
16. A. Kumar, Chakrapani, D. JyotiKalita, V.P. Singh, A modern pothole detection technique using deep learning, in *2020 2nd International Conference on Data, Engineering and Applications (IDEA)*
17. C. Chun, S. Shim, S.-K. Ryu, Development and evaluation of automatic pothole detection using fully convolutional neural networks. J. Korea Inst. Intell. Transp. Syst. (2018)
18. C. Koch, I. Brilakis, Pothole detection in asphalt pavement images. Adv. Eng. Inform. Sci. Direct (2011)
19. K.E. An, S.W. Lee, S.-K. Ryu, D. Seo, Detecting a pothole using deep convolutional neural network models for an adaptive shock observing in a vehicle driving, in *2018 IEEE International Conference on Consumer Electronics (ICCE)*
20. H. Maeda, Y. Sekimoto, T. S.T. Kashiyama, H. Omata, Road damage detection and classification using deep neural networks with smartphone images. Natl. Inst. Inf. Commun. Technol. (NICT) (2018)

Automatic Determination of Harassment in Social Network Using Machine Learning

Bhanu Prakash Doppala⬤, S. NagaMallik Raj⬤,
Eali Stephen Neal Joshua⬤, and N. Thirupathi Rao⬤

Abstract Globally the number of Internet users are very high, and the majority of the users are youngsters. They will be participating in many activities in social networks like twitter, Facebook, etc. With lightning speed of Internet, everyone can explore the information of unknown. As a result, much more cyber-based crimes and harassments are raising day by day. Artificial intelligence can bring out solution for such issues. Lot of research has been taken place for the identification of online harassments through comments and messages posted over the platforms. Sometimes context of the statement matters for judging the comment. We propose a mechanism for identifying online harassment based on context by using one of the familiar online platform called Twitter. For this research work, we have used few machine learning algorithms.

Keywords Cyber bullying · Social networking · Cyber stalking · Cyber harassment · Machine learning

1 Introduction

Internet became natural part of everyday life impacting children and adolescents. Social media and networking provide vast amount of information with a button click. There is a scope for every individual to explore meet the unknown and collaborate. There can be many advantages like acquiring knowledge, exploring new things, but on the downside, it has many disadvantages like online harassment. Because of the age factor and experience, study on cyber bullying has been started 16 years ago and a report about online harassment has been published by Mitchell et al. which was noted as the first article in this area. Since that time many articles have been published throughout the globe. Social bullying is nothing but the negative or indecent comment made over known or unknown individual using abusive language.

B. P. Doppala (✉) · S. NagaMallik Raj · E. Stephen Neal Joshua · N. Thirupathi Rao
Vignan's Institute of Information Technology (A), Visakhapatnam, India

© The Author(s), under exclusive license to Springer Nature Singapore Pte Ltd. 2021   245
S. K. Saha et al. (eds.), *Smart Technologies in Data Science and Communication*,
Lecture Notes in Networks and Systems 210,
https://doi.org/10.1007/978-981-16-1773-7_20

Bullying is common in school, but how often online harassment is caused by incidents at school is unknown. As mentioned earlier, online contacts that young people do not know personally are often harassed online [1]. Online harassment excludes and may be less threatening, as harassers cannot use physical attributes such as height and tone of voice to intimidate. [2] Studies using convenience samples of adolescents have significant overlap between victims of bullying in school and online—harassment detected. It is also not clear how often online harassment would qualify as bullying according to the definition used in the research for bullying in school with three elements: (1) aggressive acts, including verbal acts associated with harmful intent were executed, (2) repetition and (3) an imbalance of power between the perpetrator and target. Although we had no information on the intent of molesters, 62% of teens molested online were not desperate, which means that they did not consider the experience serious or harmful [3].

Most common reasons for cyber harassment are appearance, academic achievement, race, financial status, religion, etc. Figure 1 presents those factors upon which people will be bullied along with the percentage.

Technology became natural need these days where cyber bullying needs to be addressed on urgent basis. Children may believe that they can post anything on the social media platform without any consequences. Number of social media applications and platforms are increasing day by day where it is a clear opportunity for bullies to target people and can terrorize them. Major problem is these issues will not be reported on time. Statistics says that every one child in ten is becoming victims for this bullying. Parents also need to have exposure on online platforms where cyber bullying occurs.

Many of us have misconception that emails will be used for professional communication but in fact every individual acquired a separate email id someone even with

Fig. 1 Reasons for cyber harassment [12]

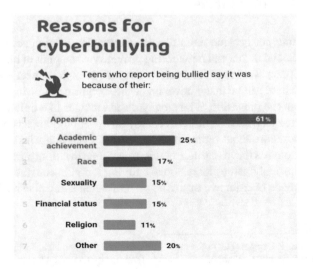

more than one account. Research says that more than 60% of teenagers and children uses emails to communicate with friends. Here the bigger threat is stackers may take advantage of this by sending inappropriate images and content to large group of emails at single click. Sharing personal passwords may also be sometimes misused. Computerized context and phrase monitoring system may help to avoid these circumstances.

Nowadays all the social media platforms have made their forums available to access on both mobile and desktop versions. Most of the teenagers and children having their own device to communicate. This facility can be used for positive and negative ways. Sometimes they were more addicted to access the device even in schools and colleges.

Many websites are built to send text messages to mobile phones for free. Therefore, these platforms became a means to harass individuals with different types of text messages. People may think that these messages are anonymous and cannot be tranced out.

More than 70% of teenagers have profiles on different and multiple social media websites and forums. Every site is provided with special settings which helps the individual to hide their personal information which is not to be shared to every unknown individual. Extremity of knowledge on these things may also create an impact on individual to engage in harmful behaviours. Figure 2 shows various social networking plats where cyber bullying takes place.

For creating this specific model, we explored two studies: one is model represented in code and the other is quantity of positives that the model returned. For understanding this, we have taken help of decision matrix in order to identify true positives and true negatives as well as false positives and false negatives.

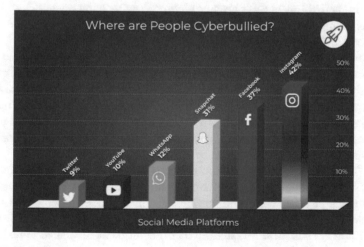

Fig. 2 Social media platforms

The consecutive section of the paper concentrates on related work, problem statement, proposed system, as well as data collection pre- and post-processing, etc.

2 Related Work

This section discusses the research that has been done on the prevalence and seriousness of cyber bullying over past decade. Additionally, it discusses bullying of women over men and the recommended remedies to be taken against to bullying.

Belsey et al. in 2004 [4] identified that bullying will happen through different communication modes involving personal information of the victim. According to the National Crime Prevention Council, they identified that an average of 40% students are being bullied every year.

Willard et al. in 2006 made a research on bullying and came up with different types of bullying that can create problem to an individual [5], which involves blazing, harassment, impersonation, derogation, cyber stalking, etc. Blazing is a kind of fighting electronically using messages with indecent language. Harassment involves sending messages which are continuously insulting the victim. Impersonation is a form of ruining someone life by pretending like others. Derogation is sending or making gossip over the Internet about a person and damaging one's reputation. Cyber stalking is giving intense pressure on the victim in the form of threatening messages.

Hinduja et al. in 2006 [6] defined in their study that bullying is a repeated harm inflicted through electronic media in the form of text messages.

William et al. in 2007 specified in their article mentioning that over 13 Million children in the USA aged 6–17 became victims in cyber harassment [7]. According to the poll conducted by First Crime: Invest in kids group declared that one-third of teens group and one-sixth of primary school aged got bullied.

Walrave et al. in 2008 mentioned about private networking where people having their own personal email and account on different social networking platforms [8]. Mostly children and teenagers are meeting several people over internet in the form of friends from other places or as a part of dating. This process of friendships may lead the victim to share their personal information to the unknown during the process. Sometimes parents also do not know what actually their kid id doing online or bullied uncles they share with their parents.

Shariff et al. in 2009 expressed few concerns related to bullying like anonymity where people are able to hide their names behind the screen to protect their identity and posts negative comments towards victim [9]. Sometimes supporting the perpetrators than victim also matters. One more concern is bullying based on gender.

Yilu Zhou et al. in 2012 investigated on different methods related to text mining towards the detection of content which is offensive in protecting teenagers and children [10]. They have proposed a lexical syntactical feature to catch the abusive content on social platform.

3 Problem Statement

Due to these prevailing situations any individual may be a child or a teenager who involved in this bullying were not able to sleep properly. Some of them were addicted to some dangerous habits to get rid of this problem. So, we need a system to be built as a solution to such kind of problems. At times messages or comments posted on the social platform may be of short in size but carries different meanings. So, researchers must work on a mechanism to build a solution against to social harassment based on the context of messages.

4 Proposed System

We proposed context-based cyber bullying where we have collected data records from one of the major repositories of social networking like Facebook and Twitter. With the proposed system, we can identify cyber harassment easily based on additional pattern matching approach.

4.1 Data Extraction

We have extracted data (tweets) from Twitter using user profile. Simply login to Twitter account and navigate to developers profile. Click on Produce New Application and fill necessary information for the purpose of the work. Produce your own Twitter application for API key and get access to the token. Install Tweepy on your machine and access the tokens obtained from the portal. This allows you to access the comments made for several posts available in Twitter (Fig. 3).

4.2 Feature Selection

Feature selection plays prominent role in machine learning approaches. Reducing or removing unwanted features from the dataset always helps the algorithm to perform well. As if we make meaningful input for best outputs. Feature selection is vital in classifier performance. Features generally of kind lexical and syntactical, where lexical features are words and phrase where the appearance and frequencies matter. Syntactical fails to understand the whole sentence and the order of words if scrambled.

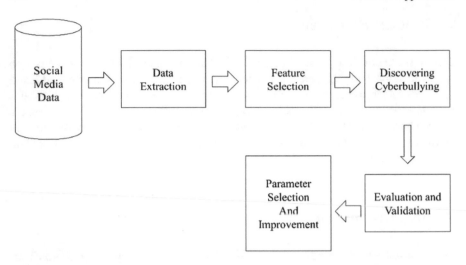

Fig. 3 Proposed system architecture

5 Pre-processing

Data available in the social networking platforms need to be processes as it contains more noise over time. Pre-processing mechanism involves in transforming the raw data into required form. Based on the need of algorithm to be processed, information needs to be fed by removing unwanted data.

Here in this research work tweets are pre-processed by importing nltk, re packages in python. This mechanism eliminates terms and conditions, removes hashtags and the removal of unwanted information incorporated with plain text in a tweet.

Complete procedure of pre-processing has been represented in Fig. 4.

6 Results and Discussion

For the better understanding of context-based identification, we have taken a very famous movie series "Harry Potter" term regularity (TF)—inverse record regularity (TFIDF) is used to identify the usage of a particular word in a file. We are processing a compilation of few documents. These are the textual information available from very famous movie series Harry Potter I to VII contained with all the special lines and captions in the movie series. Term regularity can be defined as how often a word occurs in the document. For an instance consider a word "Expelliarmus" has been used in the series used 36 times in Movie VII.

Later to that Regularity (TF) is actually: TF $(t, d) = 36$

Where the term t = "Expelliarmus" and the record d = Movie VII—Harry Potter as well as the Deathly hallows.

Fig. 4 Pre-processing

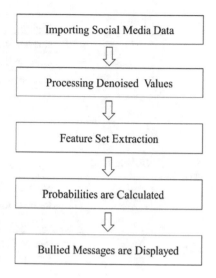

The Record Frequency (DF) is a frequency determines the number of documents that hold a word. If the phrase "Expelliarmus" appears in series I, II III as well as VII, then the DF for $t =$ " Expelliarmus" is actually DF $(t) = 4$, DF (t) represents Documentation Frequency "Expelliarmus".

It is actually labelled as: IDF $(t) = \log$ DF

Where IDF (t) stands for inverse document regularity of t where t is "Expelliarmus" listed here. The IDF for the phrase $t =$ " Expelliarmus" is IDF $(t) =$ record 4 ≈ 1.098.

With the help of IDF while browsing for a word, we can easily identify this terminology for finding which documents are suitable. As a specific example, we intend to see the Harry Potter series if it has just "Expelliarmus".

While hunting for the word in the total series of Harry Potter where Expelliarmus is obtained with highest ranking to the film VII, Table 1 indicates different words or phrases which have been used in different series. So, the hunt for the word "Expelliarmus" in the event of seven movies of Harry Potter will certainly offer the highest

Table 1 Term frequency values

Term	M-I	M-II	M-III	M-IV	M-V	M-VI	M-VII
Harry	1120	1560	1820	2910	3820	2790	2690
Ron	370	610	790	990	1212	870	1010
Hermione	251	260	600	790	1230	620	1410
Professor	190	490	370	305	510	250	0
Snape	130	0	240	0	0	390	0
Moody	0	0	0	260	0	0	0
Wand	0	0	0	0	0	0	610

Table 2 Statistics on data

Parameter	Instances
Number of conversations	1506
Cyberbullied	785
Number of distinct words	5628
Max conversation happened	800 characters
Min conversation happened	60 characters

Table 3 Performance of classifiers

Classifier	2-Gram	3-Gram	4-Gram
Support vector machine	89.46	89.93	**90.05**
Random forest	85.65	**88.96**	87.65
Logistic regression	73.76	**86.06**	82.36

Source: Bold showcases the increase in the percentage acquired from specific model

rank to film I as much more appropriate. Observe Table 1 to understand the made phrase frequency for the phrases Harry, Ron, Hermione, Professor, Snape, Moody and Wand in the 7 Harry Potter movies.

From Table 1, we can see different words that appear in several series of Harry Potter movies. Few words we can see appearing in all the movies, and few words are confined to specific movie. Implementation on data obtained kaggle which was collected for this experiment [11].

This dataset contains questions and answers either cyberbullied or not. Table 2 shows the statistics of the data used for experimental results. We applied support vector machine algorithm, random forest and logistic regression.

After pre-processing and feature extraction is done, we will split the data into testing and training. Several experiments on different n-gram language models, Table 3 summarizes the accuracies of SVM, random forest and logistic regression, and Fig. 5 displays the comparison among the classifiers.

7   Conclusion

In this research work, our proposed mechanism on context-based identification of online harassment online yielded good result. Main purpose of this paper is providing data on automated acknowledgement of online harassment which may be created, and we have used few classifiers for the work where SVM produced best result of 90.05% for 4-Gram and random forest and logistic regression produced best result for 3-Gram with 88.96 and 86.06%.

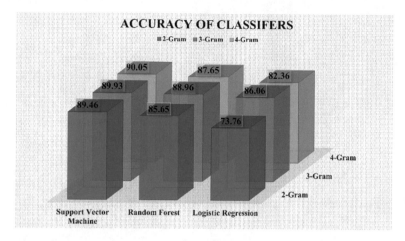

Fig. 5 Comparison between classifiers

References

1. T.R. Nansel, M. Overpeck, R.S. Pilla, et al., bullying behaviours among US youth: prevalence and association with psychosocial adjustment. Accessed 5 Nov 2020
2. D.L. Espelage, S.M. Swearer, Research on school bullying and victimization: what have we learned and where do we go from here? School Psychol Rev **32**(3), 365–383 (2003)
3. J. Wolak, K. Mitchell, D. Finkelhor, *Online victimization: 5 years later* (National Center for Missing & Exploited Children, Alexandria, 2006)
4. B. Belsey, What is cyberbullying? (2004). Retrieved 4 Apr 2009, from www.bullying.org/ext ernal/documents/ACF6F8.pdf
5. N.E. Willard, Educators guide to cyberbullying: addressing the harm caused by online social cruelty (2006). Retrieved 4 Apr 2009
6. J.W. Patchin, S. Hinduja, Bullies move beyond the schoolyard; a preliminary look at cyberbullying. Youth Violence Juv. Justice **4**(2), 148–169 (2006)
7. C.R. Cook, K.R. Williams, N.G. Guerra, L. Tuthill, Cyberbullying: what it is and what we can do about it. NASP Commun. **36**(1) (2007)
8. W. Heirman, M. Walrave, Assessing concerns and issues about the mediation of technology in cyberbullying. Cyberpsychol.: J. Psychosoc. Res. Cybersp. **2**(2), Article 1 (2008)
9. S. Shariff, *Confronting Cyber ~ Bullying: What Schools Need to Know to Control Misconduct and Avoid Legal Consequences* (Cambridge University Press, New York, 2009)
10. Y. Chen, Y. Zhou, S. Zhu, H. Xu, Detecting offensive language in social media to protect adolescent online safety, in *Privacy, Security, Risk and Trust (PASSAT), 2012 International Conference on and 2012 International Conference on Social Computing (SocialCom)*. IEEE, 2012, pp. 71–80
11. K. Reynolds, A. Kontostathis, L. Edwards, Using machine learning to detect cyberbullying, in *2011 10th International Conference on Machine Learning and Applications and Workshops*, vol. 2. IEEE, 2011, pp. 241–244
12. https://firstsiteguide.com/cyberbullying-stats/. Accessed 5 Nov 2020

Enhanced Performance of ssFFT-Based GPS Acquisition

Harshali Mane, Matcha Venu Gopala Rao, V. Rajesh Chowdhary, and S. M. M. Naidu

Abstract This paper presents Improved Sub-Sampled Fast Fourier transform (IssFFT) that lowers computation steps, simplifies the hardware implementation and correspondingly minimizes the correlation time in comparison with ssFFT GPS acquisition. The software GPS receiver generates an *In-phase* signal *I* and a *Quadrature* signal *Q* but most of the times signal lies in In-phase instead of Quadrature. This research decreases computational load in acquisition and obtain a accurate code phase efficiently in less time. So to reduce the computation time, *only In-phase processing* has taken into consideration instead of both processing.

Keywords GPS receiver · GPS acquisition · ssFFT · Subsampling · In-phase component

1 Introduction

Global Navigation Satellite Systems (GNSS) support enhanced precision and reliability for determining the location of the user in different applications, consisting of accurate aviation for safety and efficiency of flights, launching rockets or guiding missiles in defense, autonomous ground vehicles or self-driving cars, and other modernized civilian services [1]. When receiver is first powered on, it is unaware

H. Mane (✉) · M. V. G. Rao
Department of ECE, KLEF, Guntur, India
e-mail: harshamane@kluniversity.in; harshalim@isquareit.edu.in

M. V. G. Rao
e-mail: mvgr03@kluniversity.in

H. Mane · V. Rajesh Chowdhary · S. M. M. Naidu
Department of E&Tc, International Institute of Information Technology (I2IT), Hinjawadi, Pune, India
e-mail: vrajeshc@isquareit.edu.in

S. M. M. Naidu
e-mail: mohans@isquareit.edu.in

© The Author(s), under exclusive license to Springer Nature Singapore Pte Ltd. 2021
S. K. Saha et al. (eds.), *Smart Technologies in Data Science and Communication*,
Lecture Notes in Networks and Systems 210,
https://doi.org/10.1007/978-981-16-1773-7_21

of which satellites are in space, the acquisition must be performed on all satellites
[1–3]. The receiver must search for current position of satellites, all possible PRNs,
checking all possible Doppler and code shifts, until satellites are found [4]. The
primary objective of GPS acquisition is to find out these parameters depending upon
availability of satellites and signal strength. The acquisition involves "2-D search
operation" for available satellite signal, which includes "Frequency search" and
"code-phase search" [5]. Acquisition needs rough estimation of these two important parameters of GPS signal-carrier frequency (fc) and code phase (ph) because,
only after removing both from the modulated signal, tracking can be obtained [6, 7].
The acquisition performance defines the area of the search space to a great extent.
This area is the multiplication of the PRN length (1023) times the carrier frequency
as shown below.

$$\underbrace{1023}_{\text{code phases}} \times \underbrace{\left(2\frac{10000}{500}+1\right)}_{\text{frequencies}} = 41943 \text{ combinations.}$$

This research is based on ssFFT-based GPS acquisition algorithm [8]. It explores
the basic concepts required for ssFFT-based GPS acquisition algorithm and further
discusses the proposed IssFFT algorithm. Then ssFFT-based GPS acquisition algorithm is extensively discussed with simulation results and observed with an improvement of subsampling factor up to 12 [9]. Further the ssFFT algorithm is investigated
and improved by considering only *In-phase components* and interesting results are
obtained up to the subsampling by 4.

2 ssFFT-Based Acquisition

The ssFFT-based GPS acquisition algorithm is depicted in Fig. 1. It is consisting the
compressive sensing-based subsampling (CS-SS) which sub-samples the incoming
GPS data as well as locally generated carrier replica and PRN code [10]. This is the
fastest computational algorithm for GPS acquisition [11].

The flowchart of the ssFFT algorithm can be obtained by modification of parallel
code phase search acquisition algorithm flowchart [12, 13]. The modifications
occurred are shown with blue boxes, refer Fig. 2. Thus subsampling are added at
three places—(i) subsampling incoming GPS data, (ii) subsampling locally generated data, and (iii) subsampling of PRN code. The flowchart of ssFFT algorithm is
shown in Fig. 2.

While computing ssFFT algorithm, FFT of a decimated GPS signal is considered
as illustrated in Fig. 2. The decimation is done by d^{th} factor, where 'd' is subsampling
factor. Prior to down sampling, a low pass filter is used in practice to eliminate the
aliasing frequency components at the output as shown in Fig. 3a but by eliminating
LPF, aliased decimator saves processing time, refer Fig. 3b.

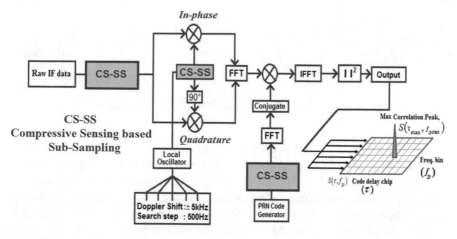

Fig. 1 Block diagram of sub-sampled-based GPS acquisition

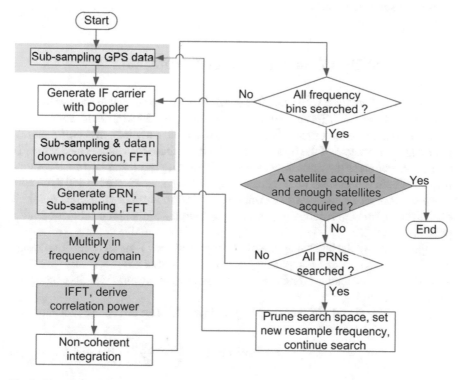

Fig. 2 Flowchart of the proposed acquisition algorithm

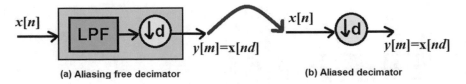

(a) Aliasing free decimator   (b) Aliased decimator

Fig. 3 Adapted LPF process, **a** Aliasing free decimator, **b** Aliased decimator

It is observed that the computation complexity has increased whereas in reverse computation time decreases exponentially, with increase in subsampling factor as shown in Fig. 4a, b. For $d = 12$, the computation time for GPS acquisition is 0.032310 s. Whereas previous computation time is 0.276479 s, thus the computation time becomes 8.5570 faster than original. The code phase variations are negligible, refer Fig. 4c. Figure 5 shows ssFFT acquisition results where subsampling by 'd' factor has been done, all the results show correct correlation peak even with less number of samples processed.

3 Proposed Technique IssFFT-Based Acquisition

The GPS signal is mixed with a locally generated oscillator signal to remove the high frequencies in signal. This generates an *In-phase* signal, I and a *Quadrature* signal. As the I signal contain of modulated C/A, one should only take into consideration the I signal. The most of the times signal lies in In-phase instead of Quadrature. To reduce the computation time, *only In-phase processing* has taken into consideration instead of both processing. When considering both signals, we need to add them after down-conversion To take FFT of incoming signal which further multiplies with FFT of CA code. This approach is to achieve the desired acquisition or autocorrelation peak in less time.

Figure 6 shows the block diagram of the improved ssFFT-based acquisition where only 12,000 samples of I component has been considered for processing. The GPS acquisition results obtained for *In-phase* processing gives accurate results as compare to *ssFFT* processing.

The received signal after subsampling by factor 'd' can be represented as:

$$s'(t) = \sqrt{P'_C}\, D(t - \tau)\, x(t - \tau) \cos\left(2\pi \left(\frac{f_{IF} + f_D}{d}\right)t\right) + n'(t) \tag{1}$$

where $s'(t)$ is a received GPS signal after subsampling, P'_C is power of the received GPS signal after subsampling, and $n'(t)$ is the noise added after subsampling. The frequency gets divided by 'd.' Thus the new sampling frequency can be given as, $f_{sn} = \frac{f_{IF} + f_D}{d}$. The purpose of subsampling is to reduce the sampling rate of a signal while simultaneously reducing the bandwidth proportionally causes the power loss

Fig. 4 ssFFT algorithm
performance parameters

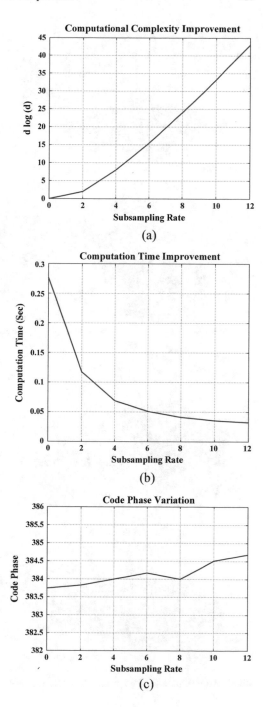

(a)

(b)

(c)

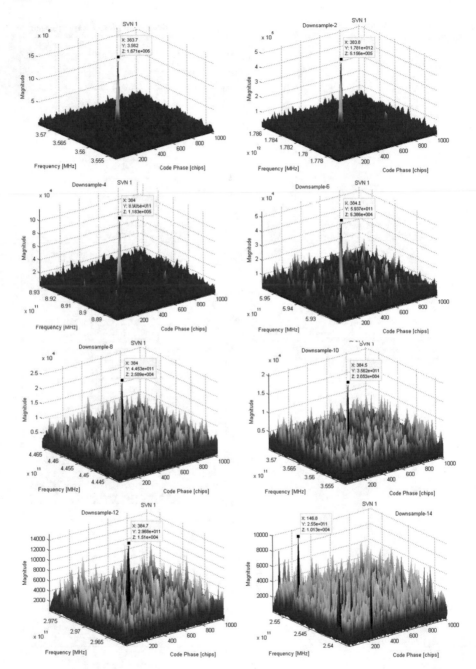

Fig. 5 ssFFT-based GPS acquisition results (different 'd')

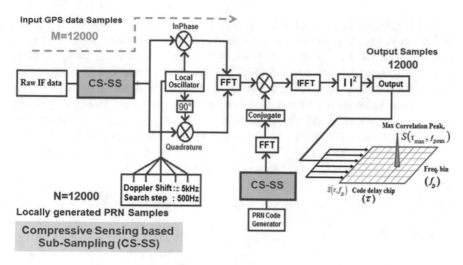

Fig. 6 Block diagram of the improved ssFFT based acquisition

of $P = \frac{1}{\pi} \int_0^{2\pi B} S(\omega)\delta\omega$, where B is single sided spectral bandwidth and $S(\omega)$ is the power spectral density which has given as:

$S(\omega) = \frac{A^2 T_c}{2} \frac{\sin^2(\omega T_c/2)}{(\omega T_c/2)^2}$ where A is signal amplitude and T_c represent code chip rate.

Here, the decimated signal of length is $(N/2)/d$, the computation complexity of IssFFT is $(N/d) \log(N/d)$. Therefore, the proposed algorithm is $[(N/2) \log(N/2)]/[((N/2)/d) \log((N/2)/d)] = (d/2) \log(d/2)$ times faster than ssFFT-based algorithm.

4 Result and Discussion

To detect the correlation peak, its needed to find out two unknown components: (a) the actual Doppler frequency offset and (b) the phase offset of the local oscillator at the receiver. The performance of the proposed ssFFT and improved ssFFT algorithm is evaluated by conducting several experiments with incremental subsampling factor 0, 2 and 4. The 20 ms duration of GPS L1 band signals (1575.42 MHz). Here, carrier frequency of 3.562 MHz and sampling frequency 11.999 MHz. The performance parameters considered are as follows: computational time improvement, code phase variation versus subsampling rate, percentage of efficiency, computational complexity factor, correlation Peak versus code phase, detection of carrier frequency versus subsampling rate, acquisition loss. The MATLAB ver.9.11 (R2019b) under Windows 10, 64 bit operating system is used for simulation. The IssFFT-based GPS simulation results are discussed in detail below.

From Fig. 7, it is observed, for fs = 11.999 MHz, carrier frequency and code phase are unaltered, but the peak value of spike has been reduced from $p = 1.671 \times 10^6$ to $p = 9.0064 \times 10^5$ and found the improvement in the computation time by 36.22%, i.e., from $t = 0.4433$ s to $t = 0.2827$ s. The reason for reduction in computation time is illuminating the *Quadrature* processing steps. Similarly, fs = 5.999 MHz there is an improvement in the computation time by 87.48%, i.e., from $t = 0.4433$ s to $t = 0.0555$ s with correlation peak reduction from $p = 1.1826 \times 10^5$ to $p = 5.7600 \times 10^4$, refer Fig. 8. The improvement in the computation time by 91.08%, i.e., from $t = 0.4433$ s to $t = 0.03958$ s for fs = 2.999 MHz. Thus, the significant change in computation time is observed from Fig. 9.

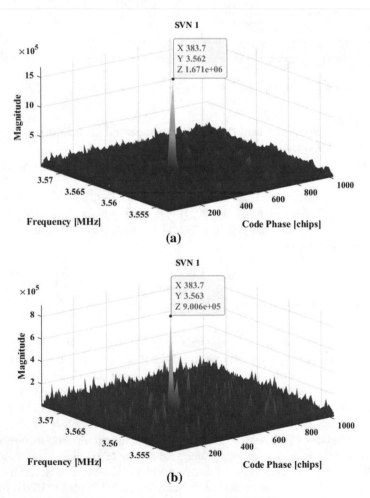

Fig. 7 Acquisition with ssFFT and only In-Phase at fs = 11.999 MHz

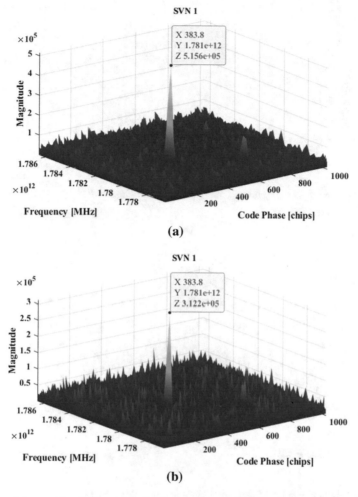

Fig. 8 Acquisition with ssFFT and only In-Phase at fs = 5.999 MHz

5 Conclusion

The papers analyze of ssFFT-based GPS acquisition and improved sub-sampled FFT (IssFFT): *In-phase components only* for software defined GPS receiver. The performance of IssFFT is evaluated by conducting several experiments. IssFFT exploits the properties sparse GPS signal, that lowers computation steps, cut down the computing time and hence hardware complexity for finding correlation peak. The proposed IssFFT algorithm computational complexity improved by a factor $(d/2)\log(d/2)$. The simulation results demonstrate that the computation time is improved by 91.08% for subsampling rate $d = 4$. The computational time improvement for $d = 4$ observed is 89.62% for ssFFT and 91.08% for *In-phase components*.

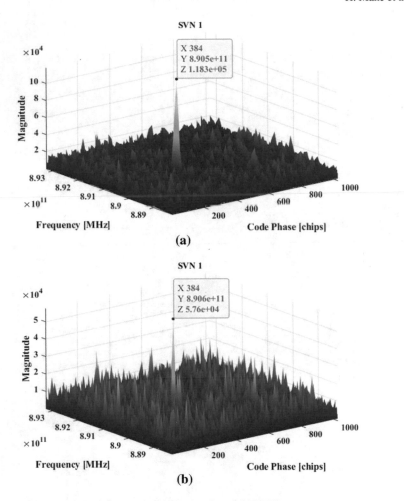

Fig. 9 Acquisition with ssFFT and only In-Phase at fs = 2.999 MHz

Acknowledgments The authors would like to express their gratitude to KLEF, Guntur as well as to the International Multilateral Regional Cooperation Division, Department of Science and Technology, Government of India for grant (IMRC/AISTDF/CRD/2018/000037) toward this project on ionospheric studies over Indian subcontinent. The development of IssFFT-based GPS acquisition is done in collaboration with KLEF, Guntur and I2IT, Pune.

References

1. E.D. Kaplan, C. Hegarty, *Understanding GPS: Principles and Applications* (Artech House, Norwood, 2005)
2. Y. Zhang, Research on Acquisition-tracking Algorithm and Anti-jamming Performance for BD-II Signal. Master Thesis, Xidian University, Xi'an, China (2014)
3. J. Yuan, S. Ou, Study on parallel acquisition algorithm for GPS receiver. J. Chongqing Univ. Posts Telecommun. (Nat. Sci. Ed.) **25**(4), 470–474 (2013)
4. Wu Zhiqiang, Acquisition algorithm research and FPGA implementation of GNSS receiver. Master Thesis, Beijing Telecommunication University, Beijing, China (2011)
5. M.M. Chansarkar, L. Garin, Acquisition of GPS signals at very low signal to noise ratio, in *Proceedings of the 2000 National Technical Meeting of The Institute of Navigation*, Anaheim, CA (2000), pp. 731–737
6. S.K. Shanmugam, R. Watson, J. Nielsen, G. Lachapelle, Differential signal processing schemes for enhanced GPS acquisition (2005)
7. A.J. Jahromi, B. Ali, N. John, L. Gérard, GPS spoofer countermeasure effectiveness based on signal strength, noise power, and C/N0 measurements. Int. J. Satellite Commun. Network. 181–191 (2012)
8. M.V.G. Rao, D.V. Ratnam, Faster GPS/IRNSS acquisition via sub-sampled fast Fourier transform (ssFFT) and thresholding, in *IEEE International Conference*, Indicon-13, IIT, Bombay, India (2013)
9. V.G.R. Matcha, D.V.R., Faster acquisition technique for software defined GPS receivers. Def. Sci. J. **65**(1), 5–11 (2015). https://doi.org/10.14429/dsj.65.5579
10. E.J. Candes, M. Wakin, An introduction to compressive sampling. IEEE Signal Process. Mag. **25**(2), 21–30 (2008)
11. M. Harshali, M. Venu Gopala Rao, Sub sampling based software GPS receiver. Int. J. Eng. Technol. **7**(2.7), 722–724 (2018)
12. H. Mane, V.G.R. Matcha, Evaluation of performance of software GPS receiver using sub-sampling and thresholding. Int. J. Simul. Syst., Sci. Technol. **19**(6), 22.1–22.7 (2018)
13. H. Mane, M.V.G.R. Matcha, Performance evaluation of ssFFT based GPS acquisition in the harsh environmental conditions. J. Green Eng. **9**(3), 411–426 (2019)

Optimized Deep Neural Model for Cancer Detection and Classification Over ResNet

Pavan Nageswar Reddy Bodavarapu⬤, P. V. V. S. Srinivas⬤,
Pragnyaban Mishra⬤, Venkata Naresh Mandhala⬤, and Hye-jin Kim

Abstract Deep learning models can assist dermatologist in classification of cancer into benign or malignant with high success rate. So, we proposed a new convolutional neural network model for classification of skin cancer, brain tumor, breast cancer, and colon cancer into benign or malignant. The proposed model and ResNet-50 model are trained on training set and compared by evaluating them on testing set. The insights of the results show that proposed model performs better than ResNet-50 in every category. The proposed model achieved 95% accuracy for brain tumor classification; the ResNet-50 model obtained 82% accuracy for brain tumor classification. The proposed model clearly outperformed ResNet-50 model. The average time taken per each epoch for proposed model is 4 s and for ResNet-50 model is 9 s.

Keywords Breast cancer · Skin cancer · Brain tumor · Colon cancer · Deep learning · Convolutional neural network

P. N. R. Bodavarapu · P. V. V. S. Srinivas (✉) · P. Mishra · V. N. Mandhala
Department of Computer Science and Engineering, Koneru Lakshmaiah Education Foundation,
Vaddeswaram, Guntur, India
e-mail: cnu.pvvs@kluniversity.in

H. Kim
Kookmin University, Seoul 02707, Republic of Korea

1 Introduction

Finding out the cancer in first stage can provide a higher chance of success rate for curing a cancer patient. A cancer detecting software can help the doctor in finding the cancer at early stages. Machine learning models and image processing play main role in cancer detection. Convolutional neural networks can assist in classifying the cancer. The convolutional neural network contains three layers, namely convolution layer, polling layer and a fully connected layer [1]. For classifying skin legions convolutional neural networks perform efficiently by training them end-to-end. In skin cancer, classification deep learning performs better than dermatologist and malignant dermal tumors are difficult to classify [2].

Melanocytic tumors can be classified by an ensemble method, combination of back propagation and fuzzy neural networks, outperforming the traditional methods, by achieving 91.11% accuracy on Caucasians race dataset [3]. For real-time mobile applications, MobileNet model is more efficient, because of their simple architecture. This model can achieve up to 83.1% accuracy for skin cancer classification [4]. The features that rely on exact estimation of lesion perimeter are thinness, roundness, etc. The advantage with feature selection is, decreases feature storage, classifier complexity is reduced, training and testing time decreases, prediction accuracy also increases. ReliefF, mutual information-based feature selection and correlation-based feature selection are the three filtering methods that have better performance on many datasets [5]. For detecting melanoma skin cancer, asymmetry, border, color, diameter, etc., are various parameters for checking melanoma. These parameters help us in detection of melanoma skin cancer [6].

For classifying colon cancer in Indian population, the images are extracted based on texture of the images. Haralick texture features, local binary pattern, histogram of oriented gradient, Gabor features, histogram features, gray level run length matrix features are the texture features extracted [7]. Data augmentation and transfer learning have ability to obtain better results for deep learning models. The limitation with deep learning methods is they should be feed with more samples of data. This can be handled by using transfer learning or data augmentation techniques [8]. To achieve improved detection rate and increase the efficiency of medical diagnosis in colonoscopy, deep learning approach can be used. VGGNets and ResNets can achieve more than 98% accuracy using global average pooling in colonoscopy medical diagnosis [9]. Various magnifications like $10\times$, $20\times$ and $40\times$ magnifications can be used to detect the colon cancer effectively [10]. The three basic steps involved in detection of cancer are preprocessing, feature extraction and classification.

2 Related Work

Sean Mobilia et al. [11] proposed a system that uses the spectral data to classify the cancerous and non-cancerous cells, image segmentation and data augmentation techniques help to achieve better performance with traditional classifiers. After comparing the performance of hyperspectral images and panchromatic grayscale images, hyperspectral images are very useful in classification of cancerous and non-cancerous cells with 74.1% accuracy on testing samples. Tina Babu et al. [12] proposed a system which consists of three stages, namely preprocessing, feature extraction and classification. In preprocessing stage the techniques used are normalization and K-Means Clustering. A novel hybrid feature was designed by combining the texture features like histogram, Gabor, Grey level run length matrix, histogram of oriented gradient. The morphological features that determine the malignancy of tissue are eccentricity, area, perimeter, convex area, orientation and Euler number. Rathore et al. [13] proposed a new image segmentation and classification technique for colon biopsy images. The phases involved in image segmentation are preprocessing, object definition, feature extraction and region demarcation. This proposed technique achieved 87.59% 92.33% accuracies for segmentation and classification, respectively.

Noreen et al. [14] proposed a method, to early diagnosis of brain tumor, designed a multi-level feature extraction to resolve the problem of feature extraction from bottom layers of pre-trained models. This proposed method achieved 99.51% accuracy on testing samples. The Inception-v3 model and DenseNet201 achieved an accuracies of 99.34% 99.51%, respectively, while using features concatenation approach. Sultan et al. [15] proposed a deep learning model to classify various brain tumors. To increase the performance of model in less time the images are resized to 128 × 128 × 1 pixels. The proposed model contains 16 layers from input layer to classification layer, with a softmax layer just before the classification layer to predict the outcome. To avoid overfitting, techniques like data augmentation, dropout layers and L2 regularization are used. The optimizer used in this model is stochastic gradient descent with momentum. This model achieved 97.54%, 95.81% and 96.89% accuracies on classification of classifying meningioma, glioma and pituitary tumor types, respectively.

Deepak et al. [16] proposed a system for classification of glioma, meningioma and pituitary tumors. The limitation in deep learning models is limited data access. This paper provides an observation that leads to, transfer learning technique can overcome the limited availability of data. Data preprocessing stage in this system consists of three phases, namely (1) collecting MRI images, (2) min-max normalization and (3) resizing images to 224 × 224. Amin et al. [17] proposed a unique approach to overcome problems of automated brain tumor classification. The model is build on long short-term memory (LSTM) to detect stroke and gliomas in MR images. For classification, a fully connected and softmax layer is used. This method obtained an 987% accuracy.

Parveen et al. [18] proposed a new hybrid technique for classifying brain tumor. This hybrid technique is created by combining the support vector machines and fuzzy c-means clustering, where fuzzy c-means clustering is applied for image segmentation and support vector machines for classification. The enhancement techniques used for improving the quantity of MRI images are contrast improvement and mid-range stretch. The classification performance of the SVM classifier with linear kernel function achieved 91.66% accuracy, while the classification performance of the SVM classifier with polynomial kernel function achieved 87.50% accuracy.

Tan et al. [19] proposed a convolutional neural network, which is able to detect the abnormalities in mammogram images and classify them into malignant, benign and normal. The images are resized to 48×48 pixels, since the raw data size is large and can upset the accuracy and computation time of the model. This model can classify mammogram images into malignant, benign and normal with more than 80% accuracy. Sara Alghunaim et al. [20] addressed the problem, prediction of breast cancer in big data context and used apache spark as platform to increase the classification efficiency of traditional machine learning models for breast cancer. After analyzing the results, SVM classifier shows highest accuracy with gene expression dataset than other classifier on gene expression dataset. Deniz et al. [21] used transfer learning and feature extraction methods for classification of breast cancer. For extracting the features AlexNet and VGG16 models are used. After comparing the results, transfer learning method achieved best results, when compared to deep feature extraction.

Bazazeh et al. [22] compared three techniques, namely support vector machine, random forest and Bayesian networks for detecting breast cancer, the various characteristics that can help in detecting breast cancer are clump thickness, cell size uniformity, cell shape uniformity, single epithelial cell size, Bare nuclei, Bland chromatin, etc., whose values range between 1 and 10. After analyzing the results, SVM classifier shows better performance in detecting breast cancer. Al-hadidi et al. [23] proposed a new approach for breast cancer detection. The back propagation neural network used in this paper consists of one hidden layer, 240 neurals in first layer and triangular function. The regression value for this neural network is greater than 93%. Zheng et al. [24] proposed a deep learning assisted efficient AdaBoost algorithm for breast cancer detection. The survival rate for age group 30–40 is 80%. The segmentation accuracy of proposed method for train dataset is 96% at 20 epochs and 100% at 100 epochs, similarly for test dataset the segmentation accuracy is 99% at 100 epochs. The proposed method displays better performance in breast cancer detection, when compared to traditional methods. Titoriya et al. [25] proposed an approach for classification of breast cancer images, where the architecture is trained at various magnifications. The proposed approach consists of three steps, namely (1) image acquisition and processing, (2) transfer learning and (3) classification. The average accuracy of proposed approach is 95.28%, 95.6%, 95.5%, 93.08% for $40\times$, $100\times$, $200\times$, $400\times$ magnifications, respectively.

3 Proposed Work

3.1 Model Description

In this proposed work, the datasets used for skin cancer classification, brain cancer classification, breast cancer classification and colon cancer classification are publicly available datasets. All the images are converted to JPG Format and then resized all the resulting images to 224 × 224 pixels. Then transfer all the JPG images to an array image, then Normalization is applied on all the images and the labels are assigned to all the images (i.e., benign, malignant). The datasets are split in ratio 80:20 for training and testing sets. The model is trained using the training set and evaluated by using testing set and the performance metrics are displayed.

Input image size of proposed convolutional neural network model is 224 × 224 × 3. The proposed model contains six convolution layers, six pooling layers and one dense layer. In pooling layer, max pooling is used, which helps to find the maximum value. The activation function used in output layer is sigmoid, since it is most useful for binary classification, the result is either 0 or 1. For classification result to be 1, the sigmoid neuron's output must be greater than or equal to 0.5, else the classification result is 0.

Step 1: Collect all the images that are publicly available and store them in their respective datasets.

Step 2: Convert all the images into JPG Format.

Step 3: The resulting images are resized to 224 × 224 pixels.

Step 4: Apply normalization to all images and store in their respective directories.

Step 5: Assign labels to all the images (i.e., benign, malignant).

Step 6: Divide the datasets to training set and testing set in the ratio 80:20.

Step 7: Train the convolutional neural network model on training set

Step 8: Evaluate the model using testing set and display the performance metrics (Fig. 1).

Fig. 1 Architecture of above proposed convolutional neural network model

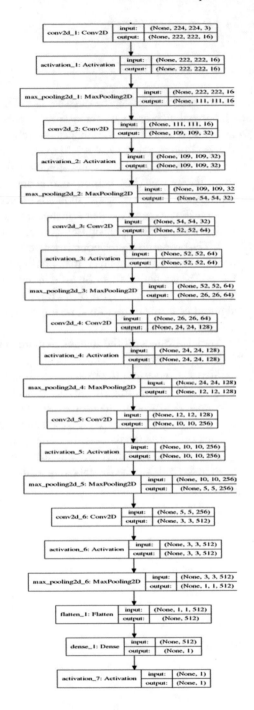

3.2 Algorithm

$Input: ImgDatasetcontainingimagesofcancer.$

$Output: ClassifythemasBenignorMalignant.$

$Begin$

$ifsize(ImgDataset[\,] \neq \emptyset)then//$

$foralltheimagesinImgDataset$

$JPG(ImageDataset[\quad],JPG)//$Converting to JPG

$Resizing(ImgDataset[\quad],224,224)/$
/Resizing Images $Transfer(ImgDataset[\quad],arrayimg)$

$Normalization(ImgDataset[\quad])//$Apply Normalization

$Label\ (ImgDataset[\quad],Benign,Malignant)//$Assign Labels

$Endloop$

$Endif$

CANCER_CNN(ImgDataset [ ]) // Cancer Classification Model

TrainingSt, TestingSt ← Split(ImgDataset [ ], 80, 20)

Shuffle(TrainingSt, TestingSt) //Shuffling

CANCER_CNN MODEL ← CANCER_CNN MODEL(TrainingSt) //Training

Evaluation ← CANCER_CNN MODEL(TestingSt) //Testing

Return Benign/Malignant

End

3.3 Performance Metrics

Performance metrics used to evaluate model are: accuracy, precision, recall and F1-score. The mathematical notations used for calculating these performance metrics are shown below,

$$Accuracy = \frac{TP + TN}{TP + FN + FP + TN} \tag{1}$$

$$Precision = \frac{TP}{TP + FP} \qquad (2)$$

$$Recall = \frac{TP}{TP + FN} \qquad (3)$$

$$F1 \text{-} score = 2 \times \frac{Precision \times recall}{Precision + Recall} \qquad (4)$$

where 'TP' stands for True Positive, 'TN' denotes True Negative, 'FP' represents False Positive and 'FN' indicates False Negative.

4 Experiment and Results

In this work, we proposed a new convolutional neural network model, which helps in classifying the benign and malignant classes of different cancer images. The system contains mainly four stages, namely (1) preprocessing, (2) assign labels, (3) implementing convolutional neural network model and (4) classification. There are many cancers, namely (1) skin cancer, (2) breast cancer, (3) colon cancer, etc. The proposed model is designed in such a way that it can perform efficiently on skin cancer, breast cancer, colon cancer and brain tumor. All the images used in this work are publicly available datasets. During preprocessing stage all the images are converted to JPG format and the images are resized to 224×224 pixels. Then all the images are transferred to an array image and normalization is applied. Later labels are assigned to them (i.e., benign, malignant). For dividing datasets into training and testing sets, we used 80:20 ratio. The proposed model and ResNet-50 are then trained with training set, then proposed model and ResNet-50 are compared by evaluating them on testing set.

The performance metrics, accuracy, precision, recall and F1-score, are displayed. After analyzing the results, the proposed model is more efficient and performs better than ResNet-50 in skin cancer, brain tumor, colon cancer and breast cancer datasets. The proposed model achieved 95% accuracy for brain tumor dataset, whereas ResNet-50 obtained only 82% accuracy. Here proposed model clearly outperformed the ResNet-50, the train loss for proposed model is only 0.006 on brain tumor dataset, but for ResNet-50, the train loss is 0.38 on brain tumor dataset. The proposed model is able to classify both benign and malignant classes, the F1-score of both benign and malignant classes of brain tumor is 0.95. Similarly, the proposed model achieved 82%, 96% and 72% accuracies for skin cancer, colon cancer and breast cancer classification, respectively. ResNet-50 obtained 80%, 94% and 71% accuracies for skin cancer, colon cancer and breast cancer, respectively. After analyzing all the performance metrics, the proposed model is more efficient than ResNet-50 in every category.

4.1 Dataset and Execution

The proposed model and ResNet-50 model are implemented in Kaggle platform. Two CPU cores, 13 GB Gigabytes of RAM and 16 GB GPU, are used for execution of these models. The datasets that are used in this work are publicly available cancer datasets. The datasets are split into training and testing sets in the ratio 80:20. The average time taken per each epoch for proposed model is 4 s and average time taken per each epoch for ResNet-50 model is 9 s. Below are the some sample images.

4.2 Graphical Representation

See Figs. 2, 3, 4, 5, 6 and 7.

4.3 Comparison Table

See Tables 1 and 2.

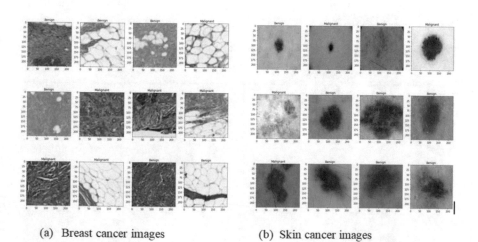

(a) Breast cancer images   (b) Skin cancer images

Fig. 2 Images of breast cancer and skin cancer

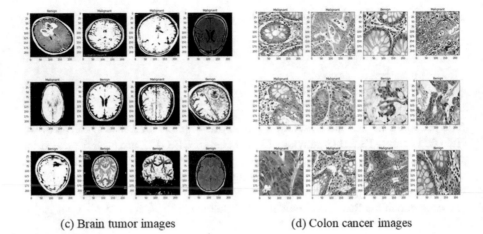

(c) Brain tumor images   (d) Colon cancer images

Fig. 3 Images of brain tumor and colon cancer

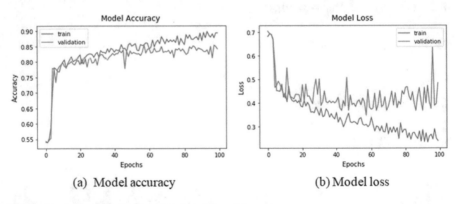

(a) Model accuracy   (b) Model loss

Fig. 4 Accuracy and loss of proposed model on skin cancer

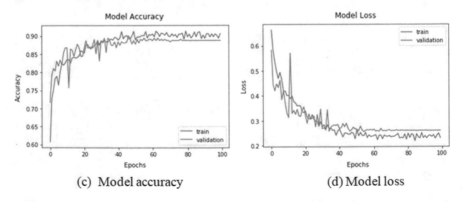

(c) Model accuracy   (d) Model loss

Fig. 5 Accuracy and loss of proposed model on breast cancer

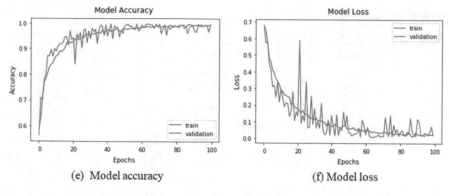

(e) Model accuracy   (f) Model loss

Fig. 6 Accuracy and loss of proposed model on brain tumor

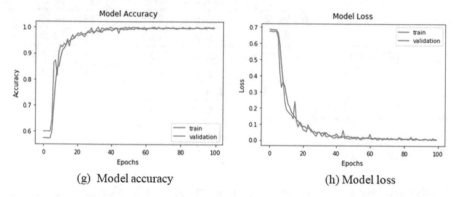

(g) Model accuracy   (h) Model loss

Fig. 7 Accuracy and loss of proposed model on colon cancer

Table 1 Comparison table of accuracy and loss

S. No.	Type of cancer	Model	Train accuracy	Test accuracy	Train loss	Test loss
1	Skin cancer	proposed model	0.89	0.82	0.22	0.52
		ResNet-50	0.82	0.80	0.40	0.42
2	Brain tumor	proposed model	0.99	0.95	0.006	0.49
		ResNet-50	0.92	0.82	0.38	0.99
3	Breast cancer	proposed model	0.89	0.72	0.26	0.64
		ResNet-50	0.93	0.71	0.18	0.66
4	Colon cancer	proposed model	0.99	0.96	0.003	0.40
		ResNet-50	0.97	0.94	0.25	0.78

Table 2 Comparison table of performance metrics

S. No.	Type of cancer	Model	Class	Precision	Recall	F1-score
1	Skin cancer	proposed model	Benign	0.94	0.72	0.82
			Malignant	0.74	0.95	0.83
		ResNet-50	Benign	0.94	0.69	0.80
			Malignant	0.72	0.94	0.82
2	Brain tumor	proposed model	Benign	0.98	0.93	0.95
			Malignant	0.93	0.98	0.95
		ResNet-50	Benign	0.80	0.87	0.84
			Malignant	0.85	0.77	0.81
3	Breast cancer	proposed model	Benign	0.78	0.63	0.70
			Malignant	0.69	0.82	0.75
		ResNet-50	Benign	0.71	0.84	0.77
			Malignant	0.72	0.55	0.63
4	Colon cancer	proposed model	Benign	0.94	0.98	0.96
			Malignant	0.98	0.94	0.96
		ResNet-50	Benign	0.94	0.94	0.94
			Malignant	0.94	0.94	0.94

5 Conclusion

In this paper, we proposed a new convolutional neural network model for classification of skin cancer, brain tumor, breast cancer and colon cancer into benign or malignant. The proposed model contains six convolution layers, six pooling layers and a dense layer. Since it is a binary classification, the results being either benign(0) or malignant(1), the activation function used for output layer is sigmoid. All the datasets used in this paper are publicly available cancer datasets. In preprocessing stage, all the images are converted to JPG format and resized to 224 × 224 pixels and transferred to an array image, then normalization is applied. After analyzing the results, it is clear that the proposed model outperformed the ResNet-50 model on skin cancer, brain tumor, breast cancer and colon cancer. The proposed model achieved 82%, 96% and 72% accuracies for classification of skin cancer, colon cancer and breast cancer into benign or malignant, respectively. The performance metrics show that proposed model performs better than ResNet-50 in every category.

References

1. M. Hasan, S.D. Barman, S. Islam, A.W. Reza, Skin cancer detection using convolutional neural network, in *Proceedings of the 2019 5th International Conference on Computing and Artificial Intelligence* (2019), pp. 254–258

2. A. Esteva, B. Kuprel, R.A. Novoa, J. Ko, S.M. Swetter, H.M. Blau, S. Thrun, Dermatologist-level classification of skin cancer with deep neural networks. Nature **542**(7639), 115–118 (2017)
3. F. Xie, H. Fan, Y. Li, Z. Jiang, R. Meng, A. Bovik, Melanoma classification on dermoscopy images using a neural network ensemble model. IEEE Trans. Med. Imaging **36**(3), 849–858 (2016)
4. S.S. Chaturvedi, K. Gupta, P.S. Prasad, Skin lesion analyser: an efficient seven-way multi-class skin cancer classification using MobileNet, in *International Conference on Advanced Machine Learning Technologies and Applications* (Springer, Singapore, 2020), pp. 165–176
5. M.E. Celebi, H.A. Kingravi, B. Uddin, H. Iyatomi, Y. Alp Aslandogan, W.V. Stoecker, R.H. Moss, A methodological approach to the classification of dermoscopy images. Comput. Med. Imaging Graphics **31**(6), 362–373 (2007)
6. Shivangi Jain, Nitin Pise, Computer aided melanoma skin cancer detection using image processing. Procedia Comput. Sci. **48**, 735–740 (2015)
7. T. Babu, T. Singh, D. Gupta, S. Hameed, Colon cancer detection in biopsy images for Indian population at different magnification factors using texture features, in *2017 Ninth International Conference on Advanced Computing (ICoAC)* (IEEE, 2017), pp. 192–197
8. A. Tsirikoglou, K. Stacke, G. Eilertsen, M. Lindvall, J. Unger, A study of deep learning colon cancer detection in limited data access scenarios. *arXiv preprint* arXiv:2005.10326 (2020)
9. W. Wang, J. Tian, C. Zhang, Y. Luo, X. Wang, J. Li, An improved deep learning approach and its applications on colonic polyp images detection. BMC Med. Imaging **20**(1), 1–14 (2020)
10. T. Babu, D. Gupta, T. Singh, S. Hameed, Colon cancer prediction on different magnified colon biopsy images, in *2018 Tenth International Conference on Advanced Computing (ICoAC)* (IEEE, 2018), pp. 277–280
11. S.J. Mobilia, Classification of hyperspectral colon cancer images using convolutional neural networks (2019)
12. T. Babu, D. Gupta, T. Singh, S. Hameed, R. Nayar, R. Veena, Cancer screening on indian colon biopsy images using texture and morphological features, in *2018 International Conference on Communication and Signal Processing (ICCSP)* (IEEE, 2018), pp. 0175–0181
13. S. Rathore, M.A. Iftikhar, CBISC: a novel approach for colon biopsy image segmentation and classification. Arab. J. Sci. Eng. **41**(12), 5061–5076 (2016)
14. N. Noreen, S. Palaniappan, A. Qayyum, I. Ahmad, M. Imran, M. Shoaib, A deep learning model based on concatenation approach for the diagnosis of brain tumor. IEEE Access **8**, 55135–55144 (2020)
15. H.H. Sultan, N.M. Salem, W. Al-Atabany, Multi-classification of brain tumor images using deep neural network. IEEE Access **7**, 69215–69225 (2019)
16. S. Deepak, P.M. Ameer, Brain tumor classification using deep CNN features via transfer learning. Comput. Biol. Med. **111**, (2019)
17. J. Amin, M. Sharif, M. Raza, T. Saba, R. Sial, S.A. Shad, Brain tumor detection: a long short-term memory (LSTM)-based learning model. Neural Comput. Appl. **32**(20), 15965–15973 (2020)
18. A. Singh, Detection of brain tumor in MRI images, using combination of fuzzy c-means and SVM, in *2015 2nd International Conference on Signal Processing and Integrated Networks (SPIN)* (IEEE, 2015), pp. 98–102
19. Y.J. Tan, K.S. Sim, F.F. Ting, Breast cancer detection using convolutional neural networks for mammogram imaging system, in *2017 International Conference on Robotics, Automation and Sciences (ICORAS)* (IEEE, 2017), pp. 1–5
20. S. Alghunaim, H.H. Al-Baity, On the scalability of machine-learning algorithms for breast cancer prediction in big data context. IEEE Access **7**, 91535–91546 (2019)
21. E. Deniz, A. Şengür, Z. Kadiroğlu, Y. Guo, V. Bajaj, Ü. Budak, Transfer learning based histopathologic image classification for breast cancer detection. Health Inf. Sci. Syst. **6**(1), 18 (2018)

22. D. Bazazeh, R. Shubair, Comparative study of machine learning algorithms for breast cancer detection and diagnosis, in *2016 5th International Conference on Electronic Devices, Systems and Applications (ICEDSA)* (IEEE, 2016), pp. 1–4
23. A. Alarabeyyat, M. Alhanahnah, Breast cancer detection using k-nearest neighbor machine learning algorithm, in *2016 9th International Conference on Developments in eSystems Engineering (DeSE)* (IEEE, 2016), pp. 35–39
24. J. Zheng, D. Lin, Z. Gao, S. Wang, M. He, J. Fan, Deep learning assisted efficient AdaBoost algorithm for breast cancer detection and early diagnosis. IEEE Access (2020)
25. A. Titoriya, S. Sachdeva, Breast cancer histopathology image classification using AlexNet, in *2019 4th International Conference on Information Systems and Computer Networks (ISCON)* (IEEE, 2019), pp. 708–712

A Steady-State Analysis of a Forked Communication Network Model

Sk. Meeravali, N. Thirupathi Rao, Debnath Bhattacharyya, and Tai-hoon Kim

Abstract Communication between the people around the world is increasing a lot. As a result, several types of network models are being designed to meet the requirements of the users for data transmission with respect to both voice and video communication. There is always a scope for the development of new network models for better and faster communication. As a result, the number of researchers is getting interested toward the utilization of these sensor networks and also working with the wireless sensor devices. In the current article, an attempt has been made to develop a forked communication network model with queuing model-based equations are developed for better understanding of the model to implement and verify its performance at various situations. The results are calculated by the help of MATLAB and MathCAD software. The results can be observed in detail in the results section for a better understanding of the working of the network model.

Keywords Communication networks · Forked model · Utilization · Delay · Throughput · Uniform batch size · Probability generating function (PGF)

Sk. Meeravali
Department of Computer Science and Engineering, Malla Reddy University, Hyderabad, Telangana, India

N. Thirupathi Rao (✉)
Department of Computer Science and Engineering, Vignan's Institute of Information Technology, Visakhapatnam, AP, India

D. Bhattacharyya
Department of Computer Science and Engineering, K L Deemed to be University, KLEF, Guntur 522502, India

T. Kim
School of Economics and Management, Beijing Jiaotong University, Chungwon-daero, Beijing, China

1 Introduction

The demand for improving the quality of service in communication systems has opened to investigate and efficient devices for data/voice transmission. A forked communication network model with non-homogenous bulk arrivals is designed and analyzed. Here, it is assumed that the data/voice packets received in the first buffer must be transmitted to the second or third buffers without having any loss of packets at the first node [1–3]. This network model is suitable for some communication systems like satellite communications, radio communications, etc.

However, in some other tele and satellite communication systems, the packets arrive to the first buffer. After getting transmitted from the first node, the packet may be forwarded to the second or third buffer connected to the second, third transmitter in tandem to the first node or gets terminated with certain probabilities. This scenario is visible at telecommunications where the domestic calls get terminated at the local exchange and the outstation calls are forwarded to the nodal exchange, and this mode of transmission is treated as phase mode of transmission. The realistic scenario of converting messages into a number of packets known as bulk arrivals with random batch size is considered such that to understand and analyze the précised performance of the considered model.

Hence, for efficient performance evaluation of this type of communication networks, we developed and analyzed a mathematical model. It is assumed that the transmission time in each node follows Poisson processes with dynamic bandwidth allocation [4–6].

2 Literature Review

In [1, 2], the authors had discussed in detail about the design of the communication network model with two and three nodes based on the queuing model having bulk arrivals at each node. Here, the arrival of packets to each node in the network model follows the bulk arrival model and also the non-homogeneous Poisson arrivals. The design had analyzed and results were presented in detail in the result section.

In [3], the authors had discussed in detail about the design of a queuing model-based communication network with the mode of random phases of transmissions. Here, the arrival of packets to the network and the leaving of packets from the network model considered will follow the random mode of transmissions for both arrivals and for transmissions.

In [4], the authors had discussed in detail by considering the load dependent queuing models such that to control the congestion in communication networks. The results are giving better understanding for further models to be developed and encouraged.

In [5], the authors had considered the new conditions and new constraints for the design and analysis of the queuing model-based communication networks. Here, the authors had considered the interdependent queuing models such that the working of one node or the processing of data packets at each node will be dependent on the data or the packets at the other node.

In [6], the authors had considered a new model of networks which will be working on the basis of queuing models. The new technique or the method that they had considered was time shared systems having the bulk arrival of packets to the nodes in the network model.

In [7], the authors had considered the new model of application having the phase type transmission being provided to the output. The performance or the output of the current model can be produced and can be utilized by using two processes or the other phenomenon can be considered are like the phase type transmission. This phase type transmission of packets from the network can be processed at two option levels. The first option of the case was the time dependent. The second option that was considered for the better communication expectation from the current model was the state dependent model of phase type transmission. The state of each node in the network model will be considered for the better understanding and for the better processing of the network model considered here in the current case of the phase type of transmission.

3 Communication Network Model

In the current section, a communication network model which is a forked one is designed and developed under both arrival and transmission phases. Here, the node 1 is connected to the second node and third node which were in tandem mode or in the series mode. The data packets will arrive to the first node at first and then those packets are converted into packets and stored at the buffer placed at node 1 [7–9]. Once the packets were reached to node one, then the packets may go to either the second node or the third node [10]. The assumption made of the current network model was that the arrival of packets to both nodes and buffers will follow the bulk in size and also will follow the batch sizes in various sizes [11].

The network model considered in the current work can be observed at Fig. 1 as,

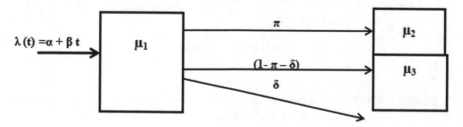

Fig. 1 Communication network model considered

In the current model, the number of packets at each node can be taken in the probability values and assumed that the n1 number of packets at first node, the n2 number of packets at second node, and the n3 number of packets at node 3. The PGF of the current model is,

$$
\begin{aligned}
P^{1}_{0,n2,n3}(t) = &-(\alpha + \beta(t) + n_2\mu_2 + n_3\mu_3)P_{0,n2,n3}(t) \\
&+ \Pi\mu_1 P_{1,n2-1,n3}(t) + (1 - \pi - \delta)\mu_1 P_{1,n2,n3-1}(t) \\
&+ \delta\mu_1 P_{1,n2,n3}(t) + (n_2 + 1)\mu_2 P_{0,n2+1,n3}(t) \\
&+ (n_3 + 1)\mu_3 P_{0,n2,n3+1}(t)
\end{aligned}
\tag{1}
$$

$$
\begin{aligned}
\mathbf{P}^{1}_{0,n2,n3}(t) = &-(\alpha + \beta(t) + n_2\mu_2 + n_3\mu_3)\mathbf{P}_{0,n2,n3}(t) \\
&+ \Pi\mu_1\mathbf{P}_{1,n2-1,n3}(t) + (1 - \pi - \delta)\mu_1\mathbf{P}_{1,n2,n3-1}(t) + \delta\mu_1\mathbf{P}_{1,n2,n3}(t) \\
&+ (n_2 + 1)\mu_2\mathbf{P}_{0,n2+1,n3}(t) + (n_3 + 1)\mu_3\mathbf{P}_{0,n2,n3+1}(t) \\
\mathbf{P}^{1}_{n_1,0,n_3}(t) = &-(\lambda(t) + n_1\mu_1 + n_3\mu_3)\mathbf{P}_{n_1,0,n_3}(t) + \lambda(t)\left[\sum_{k=1}^{n_1} c_k\,\mathbf{P}_{n_1-k,0,n_3}(t)\right] \\
&+ (n_1 + 1)\delta\mu_1\mathbf{P}_{n_1+1,0,n_3}(t) + (n_1 + 1)(1 - \pi - \delta) \\
&\mu_1\mathbf{P}_{n_1+1,0,n_3-1}(t) + \mu_2\mathbf{P}_{n_1,1,n_3}(t) + (n_3 + 1)\mu_3\mathbf{P}_{n_1,0,n_3+1}(t)
\end{aligned}
\tag{2}
$$

By solving above equation models with various parameters, the final PGF of the model considered was,

Using the initial conditions,

$$
\begin{aligned}
P = \exp\Bigg\{ &\sum_{n=1}^{\infty}\sum_{r=1}^{n}\sum_{i=0}^{r}\sum_{j=0}^{i} c_n \binom{n}{r}\binom{r}{i}\binom{i}{j}(-1)^i \\
&\left[(z_1 - 1) + \left(\frac{(z_2 - 1)\mu_1\pi}{\mu_2 - \mu_1}\right) + \left(\frac{(z_3 - 1)\mu_1(1 - \pi - \delta)}{\mu_3 - \mu_1}\right)\right]^{r-i} \\
&\left(\frac{(z_2 - 1)\mu_1\pi}{\mu_2 - \mu_1}\right)^{i-j}\left(\frac{(z_3 - 1)\mu_1(1 - \pi - \delta)}{\mu_3 - \mu_1}\right)^{j} \\
&\left[\frac{\begin{aligned}[\alpha + \beta t] - \alpha e^{-[\mu_1(r-i)+\mu_2(i-j)+\mu_3 j]t}][\mu_1(r-i)+\mu_2(i-j)+\mu_3 j] \\ +\beta[e^{-[\mu_1(r-i)+\mu_2(i-j)+\mu_3 j]t} - 1]\end{aligned}}{[\mu_1(r-i) + \mu_2(i-j) + \mu_3 j]^2}\right]\Bigg\}
\end{aligned}
\tag{3}
$$

In order to analyze the performance of the considered model, several performance metrics like the throughput of the network model and end to end delay are calculated and represented in the results section.

4 Performance Analysis of the Model Under Equilibrium State

In order to understand the performance of the currently considered model, two states can be considered for the better performance network model. The first state is the equilibrium state and the second state is the transient state. In the current model, we had considered the equilibrium state to analyze and understand the performance of the currently considered model. The performance measures considered here are a and b.

The number of packets and their probability of function is,

$$C_k = 1/((b - a + 1)) \text{ for } k = a, a + 1, \ldots, b$$

The mean number of packets in a message is $((a + b))/2$ and its variance is $1/12$ $[b - a + 1]^2$.

Substituting the values of C_k in PGF, the actual joint PGF of the buffers is,

$$P_{0,0,0}(t) = \exp\left\{\sum_{n=1}^{\infty}\sum_{r=1}^{n}\sum_{i=0}^{r}\sum_{j=0}^{i}\left[\frac{1}{b-a+1}\right]\binom{n}{r}\binom{r}{i}\binom{i}{j}(-1)^{i+r}\right.$$

$$\left[1 + \left(\frac{\mu_1\pi}{\mu_2 - \mu_1}\right) + \left(\frac{\mu_1(1 - \pi - \delta)}{\mu_3 - \mu_1}\right)\right]^{r-i}\left(\frac{\mu_1\pi}{\mu_2 - \mu_1}\right)^{i-j}\left(\frac{\mu_1(1 - \pi - \delta)}{\mu_3 - \mu_1}\right)^{j}$$

$$\left.\left[\frac{\begin{array}{c}[(\alpha + \beta t) - \alpha e^{-[\mu_1(r-i)+\mu_2(i-j)+\mu_3 j]t}][\mu_1(r-i) + \mu_2(i-j) + \mu_3 j]\\+\beta[e^{-[\mu_1(r-i)+\mu_2(i-j)+\mu_3 j]t} - 1]\end{array}}{[\mu_1(r-i) + \mu_2(i-j) + \mu_3 j]^2}\right]\right\}$$

Intensifying the equation $P(Z_1, t)$, the PGF of the second buffer in the network model is,

$$P_{0,\ldots}(t) = \exp\left\{\sum_{n=1}^{\infty}\sum_{r=1}^{n}\left[\frac{1}{b-a+1}\right]\binom{n}{r}(-1)^r\right.$$

$$\left.\left[\frac{[(\alpha + \beta t) - \alpha e^{-r\mu_1 t}](r\mu_1) + \beta[e^{-r\mu_1 t} - 1]}{(r\mu_1)^2}\right]\right\} \tag{4}$$

At the first buffer, the packets mean number can be represented as,

$$L_1(t) = \left[\frac{[(\alpha + \beta t) - \alpha e^{-\mu_1 t}](\mu_1) + \beta[e^{-\mu_1 t} - 1]}{(\mu_1)^2}\right]\left[\frac{a + b}{2}\right] \tag{5}$$

At the first node, the utilization is,

$$U_1(t) = 1 - \exp\left\{\sum_{n=1}^{\infty}\sum_{r=1}^{n}\left[\frac{1}{b-a+1}\right]\binom{n}{r}(-1)^r\right.$$
$$\left.\left[\frac{\left[(\alpha+\beta t)-\alpha e^{-r\mu_1 t}\right](r\mu_1)+\beta\left[e^{-r\mu_1 t}-1\right]}{(r\mu_1)^2}\right]\right\}$$

(6)

At the first node, the throughput is,

$$Thp_1(t) = \mu_1\left[1 - \exp\left\{\sum_{n=1}^{\infty}\sum_{r=1}^{n}\left[\frac{1}{b-a+1}\right]\binom{n}{r}(-1)^r\right.\right.$$
$$\left.\left.\left[\frac{\left[(\alpha+\beta t)-\alpha e^{-r\mu_1 t}\right](r\mu_1)+\beta\left[e^{-r\mu_1 t}-1\right]}{(r\mu_1)^2}\right]\right\}\right]$$

(7)

At the first buffer, the average delay is,

$$W_1(t) = \frac{\left[\left[\frac{\left[(\alpha+\beta t)-\alpha e^{-\mu_1 t}\right](\mu_1)+\beta\left[e^{-\mu_1 t}-1\right]}{(\mu_1)^2}\right]\left[\frac{a+b}{2}\right]\right]}{\left[\mu_1\left[1-\exp\left\{\sum_{n=1}^{\infty}\sum_{r=1}^{n}\left[\frac{1}{b-a+1}\right]\binom{n}{r}(-1)^r\left[\frac{\left[(\alpha+\beta t)-\alpha e^{-r\mu_1 t}\right](r\mu_1)+\beta\left[e^{-r\mu_1 t}-1\right]}{(r\mu_1)^2}\right]\right\}\right]\right]}$$

(8)

At the second buffer, the mean numbers of packets are,

$$L_2(t) = \left(\frac{\mu_1\pi}{\mu_2-\mu_1}\right)\left[\left[\frac{\left[(\alpha+\beta t)-\alpha e^{-\mu_1 t}\right](\mu_1)+\beta\left[e^{-\mu_1 t}-1\right]}{(\mu_1)^2}\right]\right.$$
$$\left.-\left[\frac{\left[(\alpha+\beta t)-\alpha e^{-\mu_2 t}\right](\mu_2)+\beta\left[e^{-\mu_2 t}-1\right]}{(\mu_2)^2}\right]\right]\left[\frac{a+b}{2}\right]$$

(9)

At the second node, the utilization is,

$$U_2(t) = 1 - \exp\left\{\sum_{n=1}^{\infty}\sum_{r=1}^{n}\sum_{i=0}^{r}\left[\frac{1}{b-a+1}\right]\binom{n}{r}\binom{r}{i}(-1)^{i+r}\left[\frac{\mu_1\pi}{\mu_2-\mu_1}\right]^r\right.$$
$$\left.\left[\frac{\left[(\alpha+\beta t)-\alpha e^{-[\mu_1(r-i)+\mu_2 i]t}\right][\mu_1(r-i)+\mu_2 i]+\beta\left[e^{-[\mu_1(r-i)+\mu_2 i]t}-1\right]}{[\mu_1(r-i)+\mu_2 i]^2}\right]\right\}$$

(10)

At the second node, the throughput is,

$$Thp_2(t) = \mu_2\left[1 - \exp\left\{\sum_{n=1}^{\infty}\sum_{r=1}^{n}\sum_{i=0}^{r}\left[\frac{1}{b-a+1}\right]\binom{n}{r}\binom{r}{i}(-1)^{i+r}\left[\frac{\mu_1\pi}{\mu_2-\mu_1}\right]^r\right.\right.$$

$$
\left[\frac{\left[(\alpha + \beta t) - \alpha e^{-[\mu_1(r-i)+\mu_2 i]t}\right][\mu_1(r-i)+\mu_2 i] + \beta\left[e^{-[\mu_1(r-i)+\mu_2 i]t} - 1\right]}{[\mu_1(r-i)+\mu_2 i]^2}\right]\Biggr\}\Biggr]\Biggr]
$$

(11)

At the third buffer, the mean number of packets can be represented as,

$$
L_3(t) = \left(\frac{\mu_1(1 - \pi - \delta)}{\mu_3 - \mu_1}\right)\left[\left[\frac{\left[(\alpha + \beta t) - \alpha e^{-\mu_1 t}\right](\mu_1) + \beta\left[e^{-\mu_1 t} - 1\right]}{(\mu_1)^2}\right]\right.
$$
$$
\left. - \left[\frac{\left[(\alpha + \beta t) - \alpha e^{-\mu_3 t}\right](\mu_3) + \beta\left[e^{-\mu_3 t} - 1\right]}{(\mu_3)^2}\right]\right]\left[\frac{a+b}{2}\right]
$$

(12)

At the third node, the utilization is,

$$
U_3(t) = 1 - \exp\left\{\sum_{n=1}^{\infty}\sum_{r=1}^{n}\sum_{j=0}^{r}\left[\frac{1}{b-a+1}\right]\binom{n}{r}\binom{r}{j}(-1)^j\left[\frac{\mu_1(1-\pi-\delta)}{\mu_3-\mu_1}\right]^r\right.
$$
$$
\left.\left[\frac{\left[(\alpha+\beta t)-\alpha e^{-[\mu_1(r-j)+\mu_3 j]t}\right][\mu_1(r-j)+\mu_3 j]+\beta\left[e^{-[\mu_1(r-j)+\mu_3 j]t}-1\right]}{[\mu_1(r-j)+\mu_3 j]^2}\right]\right\}
$$

(13)

At the third node, the throughput is,

$$
Thp_3(t) = \mu_3\left[1 - \exp\left\{\sum_{n=1}^{\infty}\sum_{r=1}^{n}\sum_{j=0}^{r}\left[\frac{1}{b-a+1}\right]\binom{n}{r}\binom{r}{j}(-1)^j\left[\frac{\mu_1(1-\pi-\delta)}{\mu_3-\mu_1}\right]^r\right.\right.
$$
$$
\left.\left.\left[\frac{\left[(\alpha+\beta t)-\alpha e^{-[\mu_1(r-j)+\mu_3 j]t}\right][\mu_1(r-j)+\mu_3 j]+\beta\left[e^{-[\mu_1(r-j)+\mu_3 j]t}-1\right]}{[\mu_1(r-j)+\mu_3 j]^2}\right]\right\}\right]
$$

(14)

At the third buffer, the average delay is,

$$
W_3(t) = \frac{\left[\left(\frac{\mu_1(1-\pi-\delta)}{\mu_3-\mu_1}\right)\left[\left[\frac{[(\alpha+\beta t)-\alpha e^{-\mu_1 t}](\mu_1)+\beta[e^{-\mu_1 t}-1]}{(\mu_1)^2}\right] - \left[\frac{[(\alpha+\beta t)-\alpha e^{-\mu_3 t}](\mu_3)+\beta[e^{-\mu_3 t}-1]}{(\mu_3)^2}\right]\right]\left[\frac{a+b}{2}\right]\right]}{\left[\mu_3\left[1-\exp\left\{\sum_{n=1}^{\infty}\sum_{r=1}^{n}\sum_{j=0}^{r}\left[\frac{1}{b-a+1}\right]\binom{n}{r}\binom{r}{j}(-1)^j\left[\frac{\mu_1(1-\pi-\delta)}{\mu_3-\mu_1}\right]^r\left[\frac{[(\alpha+\beta t)-\alpha e^{-[\mu_1(r-j)+\mu_3 j]t}][\mu_1(r-j)+\mu_3 j]+\beta[e^{-[\mu_1(r-j)+\mu_3 j]t}-1]}{[\mu_1(r-j)+\mu_3 j]^2}\right]\right\}\right]\right]}
$$

(15)

5 Evaluation Metrics of the Current Model Under Equilibrium State

In order to analyze the performance of the current model, several input parameters are considered and given as input to the model and tried to analyze the behavior of the currently considered model with steady-state mode of operations. The performance of the model is analyzed and presented in the tabular formats and presented in the graphical representations are shown as follows,

$t = 0.2, 0.5, 0.8, 1.2, 2.0$ s
$a = 1, 2, 3, 4, 5$
$b = 10, 15, 20, 25, 30$
$\alpha = 3.0, 3.5, 4.0, 4.5, 5.0$ (with multiplication of 104 packets/second)
$\beta = 1.5, 2.0, 2.5, 3.5, 4.0$ (with multiplication of 104 packets/second)
$\mu 1 = 3, 4, 5, 6$ (with multiplication of 104 packets/second)
$\mu 2 = 8, 9, 10, 11, 12$ (with multiplication of 104 packets/second)
$\mu 3 = 8, 12, 16, 20, 25$ (with multiplication of 104 packets/second)
$\delta = 0.1, 0.2, 0.3, 0.4, 0.5$.

From the above graphical representations and the tabular representations, it is observed that the consumption of the two nodes amplify while the consumption of the other node reduces which could be observed in detail at Tables 1, 2 and Fig. 2.

Throughput is another important performance measure to assess the performance of any communication network and an individual node in the network. In general, high throughput indicates effective utilization of channel bandwidths which reduces waiting time for the packets in the buffer avoiding burstness in the buffer. As a result, less number of time-out events will occur for the packets avoiding retransmission of the packets which saves the network resource second time for the same packets which were time out. The overall impact is less chance for occurrence of congestion in the network. Hence, high throughput is desirable in the design of any communication network model which can be observed at Figs. 3 and 4.

Throughput is another important performance measure to assess the performance of any communication network and an individual node in the network. In general, high throughput indicates effective utilization of channel bandwidths which reduces waiting time for the packets in the buffer avoiding burstness in the buffer. As a result, less number of time-out events will occur for the packets avoiding retransmission of the packets which saves the network resource second time for the same packets which were time out. The overall impact is less chance for occurrence of congestion in the network. Hence, high throughput is desirable in the design of any communication network model which could be observed at Table 3 and Fig. 5.

From the above graphical representations and the tabular representations, it is clearly observe that the impact of input or the inputs is having the great impact on all the nodes and their performance in the considered network model. When the input parametric value of each node changes, the mean number of packets had some impact and changing their values either by increasing or by decreasing having the impact and the same impact is going on the entire network. Similarly, as the input parametric values changes for all the three nodes, the performance of the network changes in utilization mode, throughput of the model and also the mean number of packets in

Table 1 Performance of the network under buffer emptiness

t^*	a	b	$\alpha^\#$	$\beta^\#$	$\mu1^\$$	$\mu2^\$$	$\mu3^\$$	π	$1-\pi-\delta$	δ	$P_{000}(t)$	$P_{0,,,}(t)$	$P_{,0,,}(t)$	$P_{,,,0}(t)$
0.2	5	25	4	2	4	10	20	0.3	0.3	0.4	0.4318	0.4321	0.657	0.7285
0.5	5	25	4	2	4	10	20	0.3	0.3	0.4	0.1100	0.1136	0.3391	0.5052
0.8	5	25	4	2	4	10	20	0.3	0.3	0.4	0.0321	0.0367	0.2259	0.4139
1.2	5	25	4	2	4	10	20	0.3	0.3	0.4	0.0105	0.0136	0.1600	0.3456
2	5	25	4	2	4	10	20	0.3	0.3	0.4	0.0023	0.0045	0.0926	0.2547
0.5	1	25	4	2	4	10	20	0.3	0.3	0.4	0.1208	0.1270	0.3872	0.5512
0.5	2	25	4	2	4	10	20	0.3	0.3	0.4	0.1161	0.1215	0.3739	0.5389
0.5	3	25	4	2	4	10	20	0.3	0.3	0.4	0.1133	0.1188	0.3614	0.5282
0.5	4	25	4	2	4	10	20	0.3	0.3	0.4	0.1114	0.1155	0.3419	0.5660
0.5	5	10	4	2	4	10	20	0.3	0.3	0.4	0.1189	0.1278	0.5131	0.6786
0.5	5	15	4	2	4	10	20	0.3	0.3	0.4	0.1139	0.1201	0.4369	0.6094
0.5	5	20	4	2	4	10	20	0.3	0.3	0.4	0.1115	0.116	0.3811	0.5525
0.5	5	25	4	2	4	10	20	0.3	0.3	0.4	0.1100	0.1136	0.3391	0.5052
0.5	5	25	3	2	4	10	20	0.3	0.3	0.4	0.1092	0.1121	0.3066	0.4655
0.5	5	25	3.5	2	4	10	20	0.3	0.3	0.4	0.1796	0.1840	0.4308	0.5866
0.5	5	25	4.5	2	4	10	20	0.3	0.3	0.4	0.1406	0.1446	0.3822	0.5444
0.5	5	25	5	2	4	10	20	0.3	0.3	0.4	0.0862	0.0893	0.3008	0.4688
0.5	5	25	4	1.5	4	10	20	0.3	0.3	0.4	0.0674	0.0702	0.2669	0.4351
0.5	5	25	4	2.5	4	10	20	0.3	0.3	0.4	0.1171	0.1209	0.3497	0.5161
0.5	5	25	4	3	4	10	20	0.3	0.3	0.4	0.1034	0.1068	0.3287	0.4946
0.5	5	25	4	3.5	4	10	20	0.3	0.3	0.4	0.0972	0.1004	0.3187	0.4842
0.5	5	25	4	2	3	10	20	0.3	0.3	0.4	0.0913	0.0944	0.309	0.474
0.5	5	25	4	2	4	10	20	0.3	0.3	0.4	0.107	0.1082	0.3707	0.4364
0.5	5	25	4	2	5	10	20	0.3	0.3	0.4	0.1100	0.1160	0.0.3391	0.5052
0.5	5	25	4	2	6	10	20	0.3	0.3	0.4	0.1154	0.1235	0.3224	0.4891
0.5	5	25	4	2	4	8	20	0.3	0.3	0.4	0.1234	0.1385	0.3138	0.4816
0.5	5	25	4	2	4	9	20	0.3	0.3	0.4	0.1095	0.1136	0.2972	0.5052
0.5	5	25	4	2	4	11	20	0.3	0.3	0.4	0.1098	0.1136	0.3184	0.5052

(continued)

Table 1 (continued)

t^*	a	b	$\alpha^\#$	$\beta^\#$	$\mu1^\$$	$\mu2^\$$	$\mu3^\$$	π	$1-\pi-\delta$	δ	$P_{000}(t)$	$P_{0,.,.}(t)$	$P_{.,0,.}(t)$	$P_{.,.,0}(t)$
0.5	5	25	4	2	4	**12**	20	0.3	0.3	0.4	0.1103	0.1136	0.3592	0.5052
0.5	5	25	4	2	4	10	**8**	0.3	0.3	0.4	0.1104	0.1136	0.3785	0.5052
0.5	5	25	4	2	4	10	**12**	0.3	0.3	0.4	0.1087	0.1136	0.3391	0.2972
0.5	5	25	4	2	4	10	**16**	0.3	0.3	0.4	0.1094	0.1136	0.3391	0.3785
0.5	5	25	4	2	4	10	**22**	0.3	0.3	0.4	0.1098	0.1136	0.3391	0.4480
0.5	5	25	4	2	4	10	20	0.3	0.3	0.4	0.1101	0.1136	0.3391	0.5299
0.5	5	25	4	2	4	10	20	0.3	0.3	0.4	0.1102	0.1136	0.3391	0.5630
0.5	5	25	4	2	4	10	20	0.3	0.3	0.4	0.1094	0.1136	0.3391	0.3267
0.5	5	25	4	2	4	10	20	0.3	0.6	0.1	0.1096	0.1136	0.3391	0.3703
0.5	5	25	4	2	4	10	20	0.3	0.5	0.2	0.1098	0.1136	0.3391	0.4277
0.5	5	25	4	2	4	10	20	0.3	0.4	0.3	0.110	0.1136	0.3391	0.5052
0.5	5	25	4	2	4	10	20	0.3	0.3	0.4	0.1103	0.1136	0.3391	0.6138
0.5	5	25	4	2	4	10	20	0.3	0.2	0.5	0.1103	0.1136	0.3391	0.6138

* = Seconds, # = Multiples of 10,000 messages/seconds, \$ = Multiples of 10,000 packets/second

the network being flowing from each node to the other single node or other double nodes like the second and third nodes in the network.

6 Conclusions

In the current article, a forked-based communication network model had been developed and the performance of such network model had been analyzed by using MathCAD and MATLAB software. The performance of the model had been analyzed under various conditions and with various performance metrics, and the results were displayed and discussed in detail in the results section.

Table 2 Performance of mean number of packets and utilization

t*	a	b	α#	β#	μ1$	μ2$	μ3$	π	1-π-δ	δ	L1(T)	L2(T)	L3(t)	LN(t)	U1(T)	U2(T)	U3(T)
0.2	5	20	1	0.5	10	20	35	0.3	0.3	0.4	8.7276	0.6398	0.4167	9.7841	0.5679	0.343	0.2715
0.5	5	20	1	0.5	10	20	35	0.3	0.3	0.4	15.0987	1.5874	0.8568	17.5429	0.8864	0.6609	0.4948
0.8	5	20	1	0.5	10	20	35	0.3	0.3	0.4	18.59	2.0984	1.0849	21.7733	0.9633	0.7741	0.5861
1.2	5	20	1	0.5	10	20	35	0.3	0.3	0.4	22.017	2.5434	1.2969	25.8573	0.9864	0.8400	0.6544
2	5	20	1	0.5	10	20	35	0.3	0.3	0.4	28.1206	3.2841	1.6647	33.0694	0.9966	0.9074	0.7433
0.5	3	20	1	0.5	10	20	35	0.3	0.3	0.4	13.0856	1.3758	0.7426	15.2039	0.873	0.6128	0.4488
0.2	4	20	1	0.5	10	20	35	0.3	0.3	0.4	13.5889	1.4287	0.7711	15.7886	0.8785	0.6261	0.4611
0.2	6	20	1	0.5	10	20	35	0.3	0.3	0.4	14.0921	1.4816	0.7997	16.3734	0.882	0.6386	0.4728
0.2	7	20	1	0.5	10	20	35	0.3	0.3	0.4	14.5954	1.5345	0.8282	16.9581	0.8845	0.6501	0.4840
0.2	5	18	1	0.5	10	20	35	0.3	0.3	0.4	7.5495	0.7937	0.4284	8.077	0.8722	0.4869	0.3214
0.2	5	19	1	0.5	10	20	35	0.3	0.3	0.4	10.0658	1.0583	0.5712	11.6953	0.8799	0.5631	0.3906
0.2	5	21	1	0.5	10	20	35	0.3	0.3	0.4	12.5823	1.3229	0.714	14.6191	0.884	0.6139	0.4475
0.2	5	22	1	0.5	10	20	35	0.3	0.3	0.4	15.0987	1.5874	0.8565	17.5429	0.8864	0.6609	0.4908
0.2	5	20	1.5	0.5	10	20	35	0.3	0.3	0.4	17.6152	1.852	0.9996	20.4668	0.8879	0.6934	0.5345
0.2	5	20	2.5	0.5	10	20	35	0.3	0.3	0.4	11.8562	1.2369	0.6698	13.763	0.816	0.5692	0.4134
0.2	5	20	3.5	0.5	10	20	35	0.3	0.3	0.4	13.4775	1.4122	0.7633	15.653	0.8554	0.618	0.4556
0.2	5	20	4	0.5	10	20	35	0.3	0.3	0.4	16.72	1.7627	0.9503	19.4320	0.9107	0.6992	0.5312
0.2	5	20	1	1.5	10	20	35	0.3	0.3	0.4	18.3412	1.9379	1.0437	21.3229	0.9298	0.7331	0.5619
0.2	5	20	1	2	10	20	35	0.3	0.3	0.4	14.5665	1.5411	0.8295	16.9372	0.8751	0.6503	0.4839
0.2	5	20	1	2.5	10	20	35	0.3	0.3	0.4	15.6309	1.338	0.884	18.1487	0.8932	0.6713	0.5054
0.2	5	20	1	3	10	20	35	0.3	0.3	0.4	16.1631	1.6801	0.9113	18.7545	0.8916	0.6813	0.5158

(continued)

Table 2 (continued)

t*	a	b	α#	β#	μ1$	μ2$	μ3$	π	1-π-δ	δ	L1(T)	L2(T)	L3(t)	LN(t)	U1(T)	U2(T)	U3(T)
0.2	5	20	1	0.5	10	20	35	0.3	0.3	0.4	16.6913	1.7264	0.9386	19.3603	0.9056	0.691	0.526
0.2	5	20	1	0.5	10	20	35	0.3	0.3	0.4	17.9478	1.3868	0.7556	20.0903	0.8918	0.6293	0.4636
0.2	5	20	1	0.5	10	20	35	0.3	0.3	0.4	15.0987	1.874	0.8565	17.5429	0.8864	0.6609	0.4948
0.2	5	20	1	0.5	10	20	35	0.3	0.3	0.4	12.9138	1.7256	0.9239	15.6253	0.8765	0.6776	0.5109
0.2	5	20	1	0.5	10	20	35	0.3	0.3	0.4	11.2130	1.8219	0.9688	14.001	0.8615	0.6862	0.5184
0.2	5	20	1	0.5	10	20	35	0.3	0.3	0.4	15.0987	1.8964	0.8568	17.8519	0.8864	0.7028	0.4948
0.2	5	20	1	0.5	10	20	35	0.3	0.3	0.4	15.0987	1.7294	0.8568	17.6849	0.8864	0.6816	0.4948
0.2	5	20	1	0.5	10	20	35	0.3	0.3	0.4	15.0987	1.4657	0.8568	17.4212	0.8864	0.6408	0.4948
0.2	5	20	1	0.5	10	20	35	0.3	0.3	0.4	15.0987	1.3603	0.8568	17.3519	0.8864	0.6215	0.4948
0.2	5	20	1	0.5	**8**	20	35	0.3	0.3	0.4	15.0987	1.5874	1.8964	18.0258	0.8864	0.6609	0.7028
0.2	5	20	1	0.5	**9**	20	35	0.3	0.3	0.4	15.0987	1.5874	1.3603	18.0465	0.8864	0.6609	0.6215
0.2	5	20	1	0.5	**11**	20	35	0.3	0.3	0.4	15.0987	1.5874	1.0563	17.7391	0.8864	0.6609	0.552
0.2	5	20	1	0.5	**12**	20	35	0.3	0.3	0.4	15.0987	1.5874	0.7834	17.4691	0.8864	0.6609	0.4701
0.2	5	20	1	0.5	10	**18**	35	0.3	0.3	0.4	15.0987	1.5874	0.6941	17.3802	0.8864	0.6609	0.437
0.2	5	20	1	0.5	10	**19**	35	0.3	0.3	0.4	15.0987	1.5874	1.7316	18.3997	0.8864	0.6609	0.6773
0.2	5	20	1	0.5	10	**21**	35	0.3	0.3	0.4	15.0987	1.5874	1.428	18.1141	0.8864	0.6609	0.6297
0.2	5	20	1	0.5	10	**22**	35	0.3	0.6	0.1	15.0987	1.5874	1.1424	17.8285	0.8864	0.6609	0.5723
0.2	5	20	1	0.5	10	20	**33**	0.3	0.5	0.2	15.0987	1.5874	0.8568	17.5429	0.8864	0.6609	0.4948
0.2	5	20	1	0.5	10	20	**34**	0.3	0.4	0.3	15.0987	1.5874	0.5712	17.2573	0.8864	0.6609	0.3870
0.2	5	20	1	0.5	10	20	**36**	0.3	0.3	0.4	15.0987	1.5874	0.5712	17.2573	0.8864	0.6609	0.3870

* = Seconds, # = Multiples of 10,000 messages/seconds, $ = Multiples of 10,000 packets/second

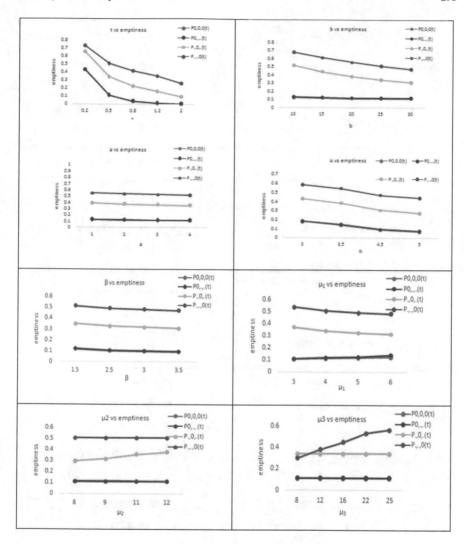

Fig. 2 Association among various parameters with the emptiness of buffers

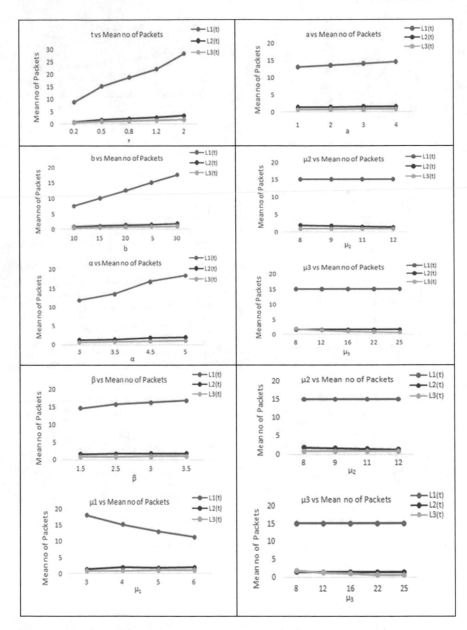

Fig. 3 Association among various parameters with the total number of packets

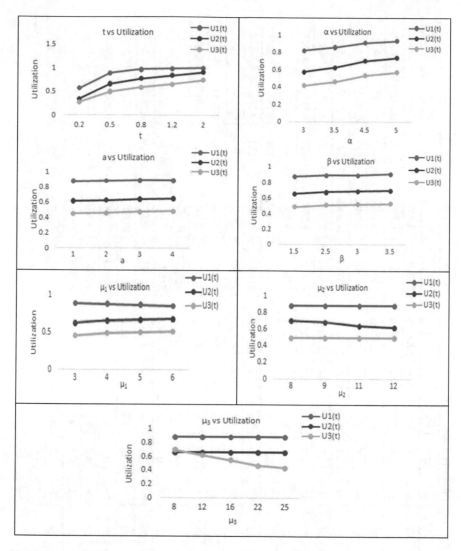

Fig. 4 Association between various parameters and the usage of nodes

Table 3 Performance of throughput and delay of the network model

t^*	a	b	$\alpha^{\#}$	$\beta^{\#}$	$\mu_1^{\$}$	$\mu_2^{\$}$	$\mu_3^{\$}$	π	$1-\pi-\delta$	δ	$Thp1(t)$	$Thp2(t)$	$Thp3(t)$	$W_1(t)$	$W_2(t)$	$W_3(t)$
0.2	**5**	**20**	**1**	**0.5**	**10**	**20**	**35**	**0.3**	**0.3**	0.4	2.2717	3.4298	5.4301	3.8419	1.865	0.0767
0.5	5	20	1	0.5	10	20	35	0.3	0.3	0.4	3.5455	6.6093	9.8956	4.2586	0.2402	0.0866
0.8	5	20	1	0.5	10	20	35	0.3	0.3	0.4	3.8533	7.7408	11.7229	4.8244	0.2711	0.0925
1.2	5	20	1	0.5	10	20	35	0.3	0.3	0.4	3.9457	8.3998	13.0884	5.580	0.3028	0.0299
2	5	20	1	0.5	10	20	35	0.3	0.3	0.4	3.9865	9.0743	14.9068	7.054	0.3619	0.1117
0.2	**3**	20	1	0.5	10	20	35	0.3	0.3	0.4	3.492	6.1276	8.9765	3.7472	0.2245	0.0827
0.2	**4**	20	1	0.5	10	20	35	0.3	0.3	0.4	3.5139	3.2612	9.2212	3.8671	0.2282	0.0836
0.2	**6**	20	1	0.5	10	20	35	0.3	0.3	0.4	3.5281	6.3855	9.455	3.9943	0.232	0.0846
0.2	**7**	20	1	0.5	10	20	35	0.3	0.3	0.4	3.5381	6.5013	9.6802	4.1251	0.236	0.0856
0.2	5	**18**	1	0.5	10	20	35	0.3	0.3	0.4	3.4888	4.8693	6.428	2.1639	0.163	0.0666
0.2	5	**19**	1	0.5	10	20	35	0.3	0.3	0.4	3.5197	5.6380	7.8126	2.8958	0.1879	0.0731
0.2	5	**21**	1	0.5	10	20	35	0.3	0.3	0.4	3.5359	6.1888	8.9502	3.5585	0.2137	0.0798
0.2	5	**22**	1	0.5	10	20	35	0.3	0.3	0.4	3.4555	6.6093	9.8956	4.256	0.02402	0.0866
0.2	5	20	**1.5**	0.5	10	20	35	0.3	0.3	0.4	3.5517	6.9338	10.6895	4.9596	0.2671	0.0935
0.2	5	20	**2.5**	0.5	10	20	35	0.3	0.3	0.4	3.264	5.692	8.267	3.6324	0.21732	0.0810
0.2	5	20	**3.5**	0.5	10	20	35	0.3	0.3	0.4	3.4216	6.1781	9.117	3.9389	0.2286	0.0838
0.2	5	20	**4**	0.5	10	20	35	0.3	0.3	0.4	3.6428	6.9919	10.6231	4.5898	0.2521	0.0895
0.2	5	20	1	**1.5**	10	20	35	0.3	0.3	0.4	3.7193	7.3313	11.2982	4.9313	0.2643	0.0934
0.2	5	20	1	**2**	10	20	35	0.3	0.3	0.4	3.5166	6.5028	9.6785	4.1423	0.237	0.0857
0.2	5	20	1	**2.5**	10	20	35	0.3	0.3	0.4	3.5727	6.7126	10.1081	4.3751	0.2434	0.0875
0.2	5	20	1	**3**	10	20	35	0.3	0.3	0.4	3.5983	6.8127	10.3162	4.491	0.2466	0.0883

(continued)

Table 3 (continued)

t*	a	b	α#	β#	μ₁$	μ₂$	μ₃$	π	1−π−δ	δ	Thp1 (t)	Thp2 (t)	Thp3 (t)	W₁ (t)	W₂(t)	W₃ (t)
0.2	5	20	1	0.5	10	20	35	0.3	0.3	0.4	3.6223	6.9098	10.5198	4.609	0.2499	0.0892
0.2	5	20	1	0.5	10	20	35	0.3	0.3	0.4	2.6754	6.2927	9.2728	6.7085	0.2204	0.0815
0.2	5	20	1	0.5	10	20	35	0.3	0.3	0.4	3.5455	6.6093	9.98956	4.2586	0.2402	0.0866
0.2	5	20	1	0.5	10	20	35	0.3	0.3	0.4	4.3823	6.7765	10.2172	2.9467	0.2546	0.0904
0.2	5	20	1	0.5	10	20	35	0.3	0.3	0.4	5.0169	6.8623	10.36745	2.1687	0.2655	0.0935
0.2	5	20	1	0.5	10	20	35	0.3	0.3	0.4	3.5455	5.6224	9.8956	4.2586	0.3373	0.0866
0.2	5	20	1	0.5	10	20	35	0.3	0.3	0.4	3.5455	6.1348	9.8956	4.2586	0.2819	0.0866
0.2	5	20	1	0.5	10	20	35	0.3	0.3	0.4	3.5455	7.0493	9.8956	4.2586	0.2079	0.0866
0.2	5	20	1	0.5	**8**	20	35	0.3	0.3	0.4	3.5455	7.4578	9.8956	4.2586	0.1824	0.0866
0.2	5	20	1	0.5	**9**	20	35	0.3	0.3	0.4	3.5455	6.6093	5.06224	4.2586	0.2402	0.3373
0.2	5	20	1	0.5	**11**	20	35	0.3	0.3	0.4	3.5455	6.6093	7.04578	4.2586	0.2402	0.1824
0.2	5	20	1	0.5	**12**	20	35	0.3	0.3	0.4	3.5455	6.6093	8.08328	4.2586	0.2402	0.1192
0.2	5	20	1	0.5	10	**18**	35	0.3	0.3	0.4	3.5455	6.6093	10.3411	4.2586	0.2402	0.0758
0.2	5	20	1	0.5	10	**19**	35	0.3	0.3	0.4	3.5455	6.6093	10.926	4.2586	0.2402	0.6351
0.2	5	20	1	0.5	10	**21**	35	0.3	0.3	0.4	3.5455	6.6093	13.4668	4.2586	0.2402	0.1272
0.2	5	20	1	0.5	10	**22**	35	0.3	0.3	0.4	3.5455	6.6093	12.5949	4.2586	0.2402	0.1134
0.2	5	20	1	0.5	10	20	**33**	0.3	0.6	0.1	3.5455	6.6093	11.4465	4.2586	0.2402	0.0998
0.2	5	20	1	0.5	10	20	**34**	0.3	0.5	0.2	3.5455	6.6093	9.8956	4.2586	0.2402	0.0866
0.2	5	20	1	0.5	10	20	**36**	0.3	0.4	0.3	3.5455	6.6093	7.7735	4.2586	0.2402	0.0738

* = Seconds, # = Multiples of 10,000 messages/seconds, $ = Multiples of 10,000 packets/second

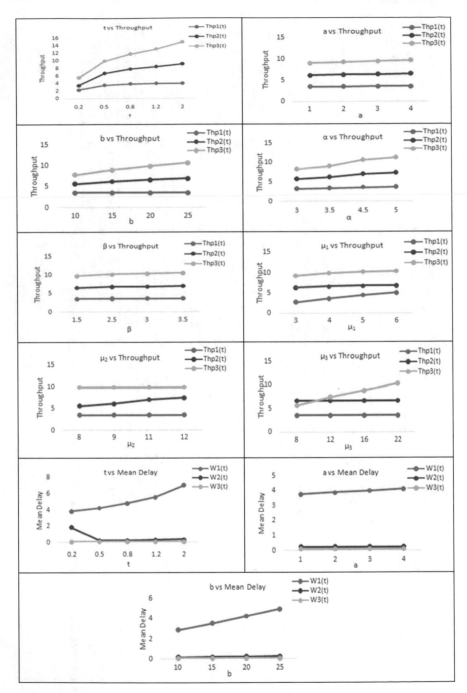

Fig. 5 Association among various parameters with throughput of the nodes

References

1. K. Srinivasa Rao et al., Stochastic control and analysis of two-node tandem communication network model with DBA and binomial bulk arrivals with phase type transmission. Int. J. Comput. Appl. **87**(10), 33–47 (2014)
2. N. Thirupathi Rao et al., Transient analysis of a two node tandem communication network with two stage compound poisson binomial bulk arrivals and DBA. Int. J. Comput. Appl. **96**(25), 19–31 (2014)
3. K. Srinivasa Rao, P.S. Varma, P.V.G.D.P. Reddy, A communication network with random phases of transmission. Indian J. Math. Math. Sci. **3**(1), 39–46 (2014)
4. K.K. Leung, Load dependent service queues with application to congestion control in broadband networks. Perform. Evaluat. **50**(1–4), 27–40 (2007)
5. N. Thirupathi Rao, D. Bhattacharyya, S. Naga Mallik Raj, Queuing model based data centers: a review. Int. J. Adv. Sci. Technol. **123**, 11–20 (2019)
6. K. Srinivasa Rao, M. GovindaRao, P. Chandra Sekhar, Studies on interdependent Tandem Queuing models with modified phase type service and state dependent service rates. Int. J. Comput. Appl. **3**(4), 319–330 (2016)
7. L. Kleinrock, R.R. Muntz, E. Rodemich, The processor-sharing queuing model for time-shared systems with bulk arrivals. Int. J. Netw. **1**(1), 13–18 (2006)
8. J. Durga Aparajitha and K. Srinivasa Rao: Two Node Tandem Queuing Model with Direct Arrivals to Both the Service Stations Having State and Time Dependent Phase Type Service. Global Journal of Pure and Applied Mathematics, 13(11), 7999–8023 (2017).
9. D. Bhattacharyya, N. Thirupathi Rao, P. Srinivas, J. Levy, Analysis of queuing applications performance using MatLab. Int. J. Adv. Sci. Technol. **27**, 1–12 (2019)
10. D.-U. Kim, M.-S. Jie, W.-H. Choi, Airport simulation based on queuing model using ARENA. Int. J. Adv. Sci. Technol. **115**, 125–134 (2018)
11. N. Miglani, G. Sharma, An adaptive load balancing algorithm using categorization of tasks on virtual machine based upon queuing policy in cloud environment. Int. J. Grid Distrib. Comput. **11**(11), 1–12 (2018)

A Comparative Analysis of With and Without Clustering to Prolonging the Life of Wireless Sensor Network

S. NagaMallik Raj⊙, Debnath Bhattacharyya⊙, and Divya Midhunchakkaravarthy ⊙

Abstract In future studies like IoT, Wireless Sensor Networks (WSN) plays a important role, such that it will connect millions of devices to exchange the sensed data from one place to another place. So, WSN comprises countless minimal effort wireless sensor nodes. Based on the application, we can deploy the sensor nodes into the environment, these nodes will sense the data and perform computations, and it sends back to the user through the base station (BS). Here, the constraint is that the sensor node will run on a battery, where life is very less. So, the challenging task is how to improve the life span of the sensor node. If the sensor node involves in unnecessary communication, utilization of the battery was also increasing, then the lifespan of the sensor node will decrease, and finally, it leads to death. So, the life of the entire WSN depends on the lifetime of sensor nodes. Today, we have so much researches on how to improve life span of sensor node. In this research work, we are addressing a clear comparison of clustering and without clustering. And also, in some papers for clustering, they prefer a static cluster head (CH). So, we are Analyzing and given a comparison of the static cluster head with our proposed Dynamic Cluster head.

Keywords Clustering · Wireless sensor networks

S. NagaMallik Raj (✉) · D. Midhunchakkaravarthy
Department of Computer Science and Multimedia, Lincoln University College, Kuala Lumpur, Malaysia
e-mail: nagamallik.Sappa@lincoln.edu.my

D. Midhunchakkaravarthy
e-mail: divya@lincoln.edu.my

D. Bhattacharyya
Department of Computer Science and Engineering, K L Deemed To Be University, KLEF, Guntur 522502, India
e-mail: debnathb@kluniversity.in

S. K. Saha et al. (eds.), *Smart Technologies in Data Science and Communication*,
Lecture Notes in Networks and Systems 210,
https://doi.org/10.1007/978-981-16-1773-7_24

1 Introduction

To monitor the environment continuously for getting a desire result, a wireless sensor network (WSN) plays a vital role. A WSN consists of sensor nodes, sink node and base station. Based on the environment, we need to select how many sensor nodes we want. After deploying sensor nodes in the given environment, it will sense the given area, and it will forward to base station through sink node. And then, that sensed data will be forwarded to user. Like that if user wants any data from that environment, the user will send query to base station, then from that base station to sink node and finally, from sink node to sensor nodes. Then, sensor nodes will work according to that query sent by user. The sensor node will send the result of that query back to user through sink node and base station. Like that WSN can be used in a wide variety of applications, and some of them are patients monitoring, surveillance, monitoring of gases, etc. [1].

In future studies like IoT, Wireless Sensor Networks (WSN) plays a important role, such that it will connect millions of devices to exchange the sensed data from one place to another place. So, WSN comprises countless minimal effort wireless sensor nodes. Based on the application, we can deploy the sensor nodes into the environment, these nodes will sense the data and perform computations, and it sends back to the user through the base station (BS). Here, the constraint is that the sensor node will run on a battery, where life is very less. So, the challenging task is how to improve the life span of the sensor node. If the sensor node involves in unnecessary communication, utilization of the battery was also increasing, then the lifespan of the sensor node will decrease, and finally, it leads to death.

So, the lifetime of the entire WSN depends on the lifetime of sensor nodes. Today, we have so much researches on how to improve life sap nob sensor node. Apart from the previous works, multi-hop in clustering with mobility (MCM) model, from that model we are considering clustering module [2].

To prolong the lifetime of WSN, we are implementing the MCM protocol [2]. In this protocol instead of a static sink node, we prefer a sink node on move, i.e., mobile sink node, which is shown in above Fig. 1. To collect the data from clusters, we have to define the path, and in that specified path, only sink node will move [3]. While coming to clusters, to find out how many clusters we need based on the distribution of sensor nodes, so many methods are there, and among that, we are preferring the average silhouette method to find out K value. Here, K value is number of clusters. Then, multi-hop routing will be applied [4, 5]. By combining these techniques, we can minimize the battery consumption of a sensor node. How it happens means, if data was sensed by one cluster, then that cluster only will send the data. Then, the remaining cluster was not involved in the communication, those energy levels will not be wasted, which means that sensor lifetime will be increased, then network lifetime will also be improved. This entire scenario is shown in above Fig. 1.

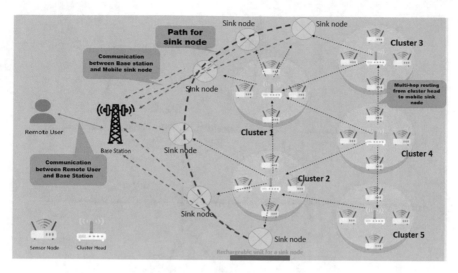

Fig. 1 MCM model

2 Comparative Analysis for Wireless Sensor Networks with Clustering and Without Clustering

In this section, we are comparing the network with and without clustering. After deploying the network with the sensor nodes, without forming any clusters, what happens means maximum all nodes will involve in the communication, so there will be more involvement of sensor node in communication, and more utilization of battery will happen, which is shown in Fig. 2. In Fig. 2, if we observe, all nodes were involved in the communication, and the node which was near to the base station will consume more energy. So, compared to all nodes, the nodes near to base station will die first. This problem is known as the black hole. If this situation happens, the networks will stop their functionality.

2.1 Plotting of Sensor Nodes in the X and Y axes

In Fig. 3, we plotted some sensor nodes in an X–Y plane. The coordinate points of sensor nodes in the X and Y axes are as follows [2].

$X = np$.array([[49, 10], [51, 16], [53, 13], [55, 12], [91, 15], [84, 17], [92, 32], [97, 11], [63, 85], [66, 85], [68, 92], [70, 86], [62, 76], [55, 78], [103, 119], [104, 110], [100, 108], [104, 105], [78, 56], [76, 58], [79, 57], [71, 46]]).

So, based on the above latitude and longitude values, the sensor nodes will be plotted in the x and y plane. For example [49, 10], one sensor node will be placed on the values latitude 49 and longitude 10. Like that remaining sensor nodes will be placed [3].

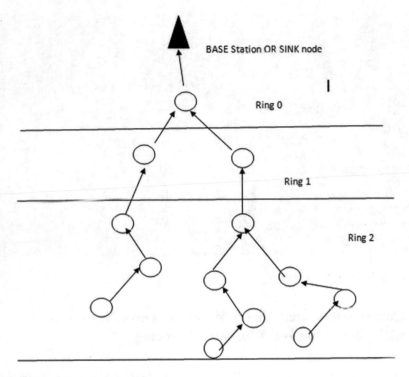

Fig. 2 WSN without clustering [1]

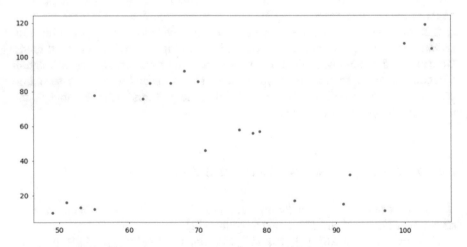

Fig. 3 Plotting of sensor nodes in the X and Y axes

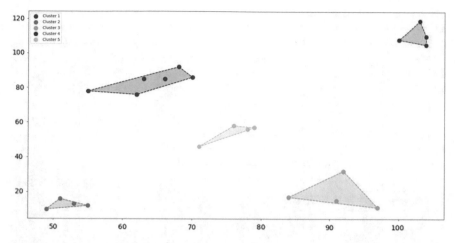

Fig. 4 Based on the *K* value, five clusters were formed [3]

2.2 Formation of Clusters

After the distribution of sensor nodes on *XY* plane based on the latitude and longitude values, we have to apply clustering techniques, to find out how many clusters we need based on the distribution of sensor nodes, so many methods are there, and among that, we are preferring the average silhouette method to find out K value [3].

Here, *K* value is number of clusters [6]. After formation of clusters, we have to elect cluster heads. Luster heads will be selected based on the energy and transmission levels [2, 4–6].

After applying the silhouette method, we get the *K* value as 5 (five). So, five clusters will be formed for the given data set, which is shown in Fig. 4. For ease of identification, each cluster was denoted in different colors [2].

KM.fit(X).

lables = KM.labels_.

sil.append(silhouette_score(X, lables, metric = 'euclidean')).

The above function returns number of clusters [3].

With the help of the above sample code for the given sensor node distribution set as input and by applying the silhouette method [7], we get the K value as 5. The total number of the cluster was 5.

3 Comparison Graph for Wireless Sensor Networks with and Without Clustering

In Fig. 5, a clear comparison is shown for ten nodes. Initially, all nodes were assigned to 1.6 J energy levels.

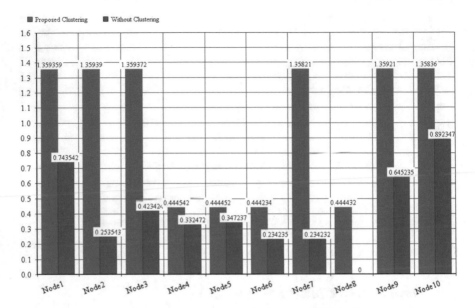

Fig. 5 Comparison graph for wireless sensor networks with and without clustering

Without clustering in a network, all nodes will be involved in communication, so all energy levels will be decreased, if we observe node 8, the energy level is 0, because node 8 is near to the base station, so the entire load was on node 8. For that reason, more power utilization was happened. If we compare iteration, for 1800th iteration, the node 8 was dead that is energy level of node 8 was zero. Then, the black hole was occurred [8, 9].

Whereas in clustering, the nodes present in that cluster only involve in communication. In Fig. 5, we can observe the energy bars of ten nodes in the proposed clustering technique. In the proposed clustering technique, no single dead node will be found even after 3lakh iterations.

This shows how we improve the network lifetime, by minimizing the power consumption of a sensor node. For that reason, we are using the clustering technique to improve the network lifetime.

4 Conclusion

WSN comprises countless minimal effort wireless sensor nodes. Based on the application, we can deploy the sensor nodes into the environment, these sensor nodes will sense the data and perform computations, and it sends back to the user through the base station (BS). Here, the constraint is that the sensor node will run on a battery, where life is very less. So, the challenging task is how to improve the life span of a sensor node. If the sensor node involves in unnecessary communication, utilization of the battery was also increasing, then the lifespan of the sensor node will decrease, and finally, it leads to death. So, the lifetime of the entire WSN depends on the lifetime of sensor nodes. Today, we have so many researches on how to improve the lifespan of the sensor node. In this research work, we are addressing a clear comparison of clustering and without clustering. And also, in some papers for clustering, they prefer a static cluster head (CH). So, we are given a comparison of the static cluster head with our proposed Dynamic Cluster head.

References

1. V. Ponnusamy, Energy analysis in wireless sensor network: a comparison. Int. J. Comput. Netw. Commun. Secur. **2**(9), 328–338 (2014)
2. S.N. Raj, Multi-hop in clustering with mobility protocol to save the energy utilization in wireless sensor networks. Wireless Pers. Commun. (2021). https://doi.org/10.1007/s11277-021-08078-y
3. D. Bhattacharyya, Low energy utilization with dynamic cluster head (LEU-DCH)—For reducing the energy consumption in wireless sensor networks, in *Machine Intelligence and Soft Computing*. Advances in Intelligent Systems and Computing (Springer, Singapore, 2021), p. 1280. https://doi.org/10.1007/978-981-15-9516 5_30
4. J.N. Al-Karaki, Routing techniques in wireless sensor networks: a survey. IEEE Wireless Commun. **11**(6), 6–28 (2004)
5. J. Gomez, PARO: supporting dynamic power controlled routing in wireless Ad Hoc networks. Wireless Netw. **9**(5), 443–460 (2003)
6. Q. Dong, Minimum energy reliable paths using unreliable wireless links, in *Proceedings of ACM MobiHoc* (2005), pp. 449–459
7. X.-Y. Li, Reliable and energy-efficient routing for static wireless ad hoc networks with unreliable links. IEEE Trans. Parallel Distrib. Syst. **20**(10), 1408–1421 (2009)
8. Y. Zhao, On maximizing the lifetime of wireless sensor networks using virtual backbone scheduling. IEEE Trans. Parallel Distrib. Syst. **23**(8), 1528–1535 (2012)
9. A. Ghabri, New fault-tolerant strategy of wireless sensor network, in *International Conference on Computer and Information Science* (2016). https://doi.org/10.1109/icis.2016.7550733

A Study on Techniques of Soft Computing for Handling Traditional Failure in Banks

T. Archana Acharya◉ and P. Veda Upasan◉

Abstract Financial turmoil (crisis) is a condition that arises due sudden decline in the nominal value of the financial assets which results in banking panics. Predicting alarming signals of crisis which is financial in nature is a tough assignment as the total economy is based on it for all industries in general and banks in particular. During the panic situation, there is coincides with the recession. The present conceptual paper gives a review of soft computing applications for predicting the crisis condition or bankruptcy which further help in promoting future empirical research to prevent bank failures and financial crises.

Keywords Financial crises · Banking failures · Financial assets · Early warning signals · Panic situations · Soft computing techniques

1 Introduction

From times immemorial, the history is telling stories of downfall of financial stability through financial crisis [1]. From credit crisis of 1772 to current pandemic crisis of 2020, the history is repeating again and again where its impact is experienced from the smallest person to the richest person in the society. Predicting warning signals [2] is a challenging task. New developing technologies their applications in this regard help not only one economy but the whole economy. One of the new techniques is soft computing techniques which is gaining progressively its presence in the world of financial.

T. Archana Acharya (✉)
Department of Management Studies, Vignan's Institute of Information Technology (A),
Beside VSEZ, Duvvada, Visakhapatnam 530049, India

P. Veda Upasan
Department of Computer Science and Systems Engineering, College of Engineering (A),
Andhra University, Visakhapatnam 530003, India

© The Author(s), under exclusive license to Springer Nature Singapore Pte Ltd. 2021 309
S. K. Saha et al. (eds.), *Smart Technologies in Data Science and Communication*,
Lecture Notes in Networks and Systems 210,
https://doi.org/10.1007/978-981-16-1773-7_25

2   Research Gap and Research Problem

Banking sector is the backbone of the whole economy. The core activity of banks is financial intermediation. During intermediation, the banks pool funds from small savers and disburse to the investors for the economic growth. In the process of intermediation, the intermediaries take the risk of the small depositors. In the sudden declining position of the nominal values of the financial assets of the banks, then its impact directly falls on the small depositors, and then the total system collapses. In the digital era, the history of failure is repeating where the system calls for prediction techniques to identify warning signals of financial distress much before it comes into existence.

3   Objectives of the Study

- To know the concepts of the soft computing
- To realize its rationale in financial world
- To review the applications of techniques of soft computing

4   The Concept: What is Soft Computing?

Soft computing is defined as the process of providing imprecise solutions by use of approximate calculations to complex computational problems. Soft computing is used to achieve traceable, robust and low-cost solution to imprecise, uncertain, partial truth and approximate models.

For example, financial turmoil (financial crisis) is a real-life problem with higher degrees of complexity. Traditional computing proved ineffective in solving complexity where the soft computing (approximate calculations) showed more effectiveness in handling complexity. Soft computing use approximate models—approximate reasoning and approximate modelling (Fig. 1 Hard Computing Vs Soft computing given by Dogan Ibrahim [3]).

The techniques of soft computing include (Fig. 2) suggested by Roberto Falcone, CarmineLima and Enzo Martinelli [4]:

4.1   Fuzzy Logic

Human knowledge by nature is imprecise. Zadeh introduced the fuzzy logic concept. The system comprises of three steps:

Step 1: deals with fuzzification interface which accepts crisp input and converts into fuzzy linguistic value.

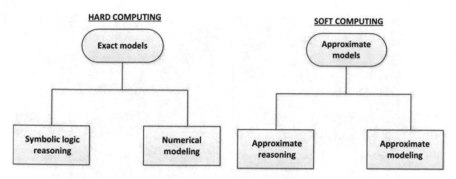

Fig. 1 Hard computing versus soft computing

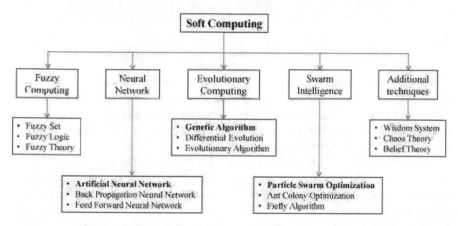

Fig. 2 Soft computing techniques

Step 2: deals with interface engine generates fuzzy output from fuzzy inputs based on fuzzy rule base ("IF–THEN").

Step 3: deals with defuzzification interface—gives crisp output actions nonlinear control systems can be easily designed using this technique (Fig. 3 shows the architecture of Fuzzy logic system suggested by Sayantini [5]).

Neural networks (NN) [6] are based on biological nervous system and brain. It is highly interdisciplinary field involving information processing system based on many other fields, such as computer science, neuroscience, psychology, philosophy, mathematics and linguistics. The specialized applications of NN include data recognition, image processing, image compression, pattern recognition, weather prediction, stock market prediction and security and loan applications.

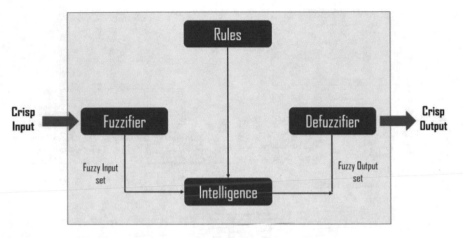

Fig. 3 Fuzzy logic system architecture

For highly nonlinear inputs and outputs, neural networks can be applied. It is applicable for solving problems which have no specific rule sets or algorithms. A network consists of interconnected large network of elements similar to neurons in humans. Each element in the network performs its little operation whose weighted sum gives the overall operation.

The process of the system is that the network is trained with input sets to produce desired outputs. Training is carried out by feeding the network with patterns based on defined learning rules. Generally, learning can supervised or unsupervised. Supervised learning: specific inputs with matching output patterns; unsupervised learning: based on input patterns—response of output comprised of three layers.

- Layer 1: Input layer
- Layer 2: Hidden layer
- Layer 3: Output layer.

It works on backpropagation algorithm; the mechanism involves training of the network for desired outputs. The output generated is compared with the desired output, if deviation is there then through backward propagation weights are calculated and modified to reduce the error values. This process of forward and backward is carried on till desired output is generated (Fig. 4 given by Diana Ramos [7]).

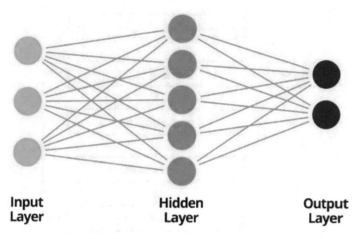

Input Layer   **Hidden Layer**   **Output Layer**

Fig. 4 Neural networks architecture

4.2 Evolutionary Computation (EC)—Genetic Algorithms

Genetic algorithm (GA)—a model developed based on evolutionary theory of the fittest. GA is a combination of artificial intelligence, machine learning and fuzzy computing. It is basically an algorithm used for real-world applications where optimization is required. Today, its applications are still at the stage of infancy. It can be applied to complex problems including climatology, code-taking, game theory, automated manufacturing and design, etc.

The process includes five steps:

Step	Title	Activity
1	Initialization	Creating initial population randomly
2	Evaluation	Assessing fitness of each member of the population with respect to desired requirements
3	Selection	Selecting parents who are fit with desired requirements
4	Crossover	Swapping tails of two randomly selected parents—new off-springs are produced
5	Mutation	Creating individuals satisfying desired requirements

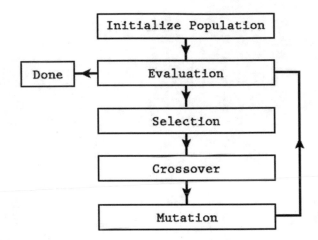

Fig. 5 Genetic algorithm flowchart

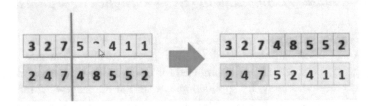

Fig. 6 Crossover in genetic algorithm

The process is iterated from step 2 till the desired condition is accomplished (Figs. 5 referred from GitHub [8] and Fig. 6 given by Faizan [9]).

4.3 Expert Systems

Expert system or knowledge-based system is a rule-based system for making intelligent decisions which works on the principle of artificial intelligence. In application, expert system is designed either to replace humans or to aid the humans. The followings are the main parts (Fig. 7 referred from ryan_blogwolf [10]):

Part	Name of the part	Activity
1	Knowledge base	Storing intelligence in the form of IF-THEN-ELSE statements
2	Interface engine	Making decisions based on conditions and requirements
3	User interface	Query/advice—user uses natural language to interact

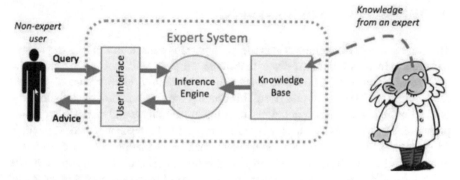

Fig. 7 Architecture of expert system

5 Rationale in Financial World

The life blood of the economic system is finance. The economic system circulates funds from surplus segment to deficit segment. The degree of complexity increases with the large amounts of risk associated with for smooth circulation. Risk is the result of uncertainty. Uncertainty is a probabilistic situation with 50–50 chances of success and failure. This situation depends on multiple factors—internal and external. Success position is desired but a failure position is not desired because it results in a contagion that is chain reaction from one failure to another failure, thus, leading to failure of the total system called as a crisis or financial turmoil of financial crisis. For example, suppose A, B, C are three banks now A promises B to pay Rs. 10 crores in expectation of cash inflows in next one month, but due to adverse situations like political turmoil, strict regulations, high volatile markets, etc., it may not receive the cash inflows in time, then A is unable to pay to B, where based on A's commitment B might commit itself to pay C. when A is unable to pay B, then B cannot pay C, thus, leading to chain of failures. Here, A is exposed to credit risk, liquidity risk, market risk, etc., which all lead to insolvency risk and this carried on to the next bank, individual or corporates or the Government, thus, leading to total failure called as financial crisis. For this reason, financial world becomes a complex system with features of uncertainty, imprecision, partial truth and approximate models. This type of failures, if handled effectively then gives a robust system. If the probable situation is predicted well in advance, then preventive measures can be initiated to avoid failures. For this the solution, it is soft computing which provides traceable, robust and low-cost solution to the real-world complex problems of financial world. The followings are the areas of financial system where soft computing can be applied:

(a) Credit scoring or financial distress
(b) Stock and currency market prediction
(c) Trading
(d) Portfolio management, etc.

6 Application of Soft Computing

The complexity of financial world is based on various parameters calling for various techniques to avoid failures. Each technique predicts with respect to particular parameter. Thus, there should be a system which is hybrid in nature to address the complexity. The techniques of soft computing are combination of statistical and intelligence techniques (computational techniques) to analyse and model the complex principle. The complexity calls for cognitive processes like trial and error reasoning, computations which are inexact and decision making in a subjective form similar to human mind where the traditional computing is failing. Exactness is the pre-requisite of traditional computing but exactness is hard to find in complexity. For example, market risk is a systematic risk which is dependent on external factors which is hard

Table 1 Soft computing techniques applied in banking

Technique	Application
Intelligence modelling techniques	
Early warning systems (EWS)	Monitoring the bank risk
Intelligence techniques	
Backpropagation neural networks (BPNN) model	Prediction and classification of bank problems
Trait recognition technique develops	Voting and classifying failures of banks into failed or non-failed
Support vector machine (SVM)	The margin is maximized between the two data sets
Decision Tree (DT) technique	Prediction problems
Rough Set technique	Modelling incomplete data based
Case-based reasoning (CBR)	Predicting failure of a bank based on past failures of other banks
K-nearest neighbour (K-NN)	Determining cut point and finding out the class—failed or non-failed banks
Soft computing technique	
Data envelopment analysis (DEA)	Measuring the organizational relative efficiencies in decision-making units
Multicriteria decision aid (MCDA)	Simultaneous analysis of preferred criteria
Multi-group hierarchical discrimination (M.H.DIS)	Determining alternatives based on the class of risk
Fuzzy cerebellar model articulation controller model (FCMAC)	Analysing financial distress patterns
Genetic algorithms	Optimizing decisions of lending by banks
Expert systems	In decision making with respect to financial loan/credit decisions

to predict with traditional computing. Table 1 shows the application techniques of soft computing in the area of banking.

The following is the review table of the application techniques of soft computing (Table 2).

Soft computing is at infancy stage in its application to financial distress but shows a hope of reliability to wave-off financial crisis.

Table 2 Application of soft computing techniques based on researcher

Researcher	Technique	Title
Altman [2]	Multilinear discriminant analysis (MDA)	Failure predictions of businesses
Korobow et al. [11]	Probabilistic approach	Failure predictions of businesses
Karels and Prakash [12]	Discriminant analysis (DA)	Conditions of normality in financial ratios
Pantalone and Platt [13]	Logistic regression	For bankruptcy prediction
Olmeda and Fernandez [14]	BPNN, multivariate adaptive splines (MARS), logistic regression, C4.5 and MDA	Predicted bankruptcy in Spanish banks
Alam et al. [2]	Self-organizing neural networks and fuzzy clustering	For bankruptcy prediction
McKee [15]	Rough set theory	For bankruptcy prediction
Ahn et al. [16]	BPNN-sample size reduction, selection and rough sets	For bankruptcy prediction
Swicegood and Clark [17]	Compared DA, BPNN and human judgement	Predicting bank failures
Shin and Lee [18]	Genetic algorithm(GA)—generated rules	For bankruptcy prediction
Park and Han [19]	K-NN applied with analytical hierarchy process (AHP)	Prediction of Korean firms bankruptcy
Cielen et al. [20]	Data envelopment analysis (DEA)—a combination of discriminant analysis and linear programming	Prediction of Belgian banks bankruptcy
Tung et al. [21]	New neuro-fuzzy system	Prediction of bankruptcy in banks
Andres et al. [22]	Additive fuzzy systems	Prediction Spanish commercial and industrial firms failure
Atiya [23]	Neural networks	Prediction of bankruptcy-credit risk
Shin et al. [24]	Support vector machines	In predicting corporate bankruptcy

(continued)

Table 2 (continued)

Researcher	Technique	Title
Pramodh and Ravi [25]	Auto associative neural network—modified and trained great deluge algorithm	For bankruptcy prediction
Ravi et al. [26]	Hybrid system-RBF, BPNN, probabilistic neural network (PNN), classification and regression techniques (CART), FRBC and PCA-based hybrid techniques	For bankruptcy prediction
Ravi and Pramodh [27]	Neural network of principal component (PCNN)	For bankruptcy prediction
Sun and Li	Weighted majority voting-based ensemble of classifiers serial combination of classifiers	For bankruptcy prediction
Li and Sun [28]	Ensemble of case-based reasoning classifiers	To bankruptcy prediction
Madireddi Vasu and Vadlamani Ravi [27]	PCA and a modified TAWNN (PCA-TAWNN)	For bankruptcy prediction in banks

7   Conclusion

Financial intermediation, the key operation of financial institutions is always associated with the exposure of different types of risks at different levels where the traditional computing is unable to offer a reliable model to handle the complex situation. Soft computing techniques offer a cognitive process of hybrid models which provide approximation models relevant to the complexity for predicting the financial stress conditions. Thus, it can be concluded that the new computing techniques provide approximation models of approximate situations leading to failure of total banking system in particular and total economic system in general.

References

1. https://www.britannica.com/list/5-of-the-worlds-most-devastating-financial-crises
2. E. Altman, Financial ratios, discriminant analysis, and the prediction of corporate bank-ruptcy. J. Finance **23**, 589–609 (1968)
3. D. Ibrahim, *An overview of Soft computing*, 12th International Conference on Application of Fuzzy Systems and Soft Computing, ICAFS 2016, pp. 29–30 August 2016, Vienna, Austria, in Procedia Computer Science, vol. 102, Science Direct (Elsevier, 2016), pp. 34–38
4. R. Falcone, C. Lima, E. Martinelli, Soft computing techniques in structural and earthquake engineering: a literature review. Eng. Struct. **207**, 110269 (15 March 2020). https://doi.org/10.1016/j.engstruct.2020.110269
5. Sayantini, What is fuzzy logic in AI and what are its applications? Published on Dec 10, 2019. https://www.edureka.co/blog/fuzzy-logic-ai/

6. P. Alam, D. Booth, K. Lee, T. Thordarson, The use of fuzzy clustering algorithm and self-organization neural networks for identifying potentially failing banks: an experimental study. Expert Syst. Appl. **18**, 185–199 (2000)
7. D. Ramos, Real-life and business applications of neural networks. https://www.smartsheet.com/neural-network-applications, Published on Oct 17, 2018
8. https://www.google.com/url?sa=i&url=%3A%2F%2Fgithub.com%2F2black0%2FGA-Python&psig=AOvVaw35Ky20QDKamJMMW3JLYKz2&ust=1619852234840000&source=images&cd=vfe&ved=0CA0QjhxqFwoTCJDYw4KypfACFQAAAAAdAAAAABAD
9. Faizan, Genetic algorithm | Artificial intelligence tutorial in hindi urdu | Genetic algorithm example. https://www.youtube.com/watch?v=frB2zIpOOBk. Uploaded on 26 April, 2018
10. Ryan_blogwolf, Neural networks and gradient descent. https://wp.wwu.edu/blogwolf/. Posted on January 29, 2017
11. L. Korobow, D. Stuhr, D. Martin, A probabilistic approach to early warning changes, in *Bank Financial Condition* (Federal Reserve Bank of New York, 1976), pp. 187–194 (Month-ly Review).
12. G.V. Karels, A.J. Prakash, Multivariate normality and forecasting of business bankruptcy. J. Business Finance Acc. **14**, 573–593 (1987)
13. C.C. Pantalone, M.B. Platt, Predicting bank failure since deregulation. N. Engl. Econ. Rev., Federal Reserve Bank of Boston, 37–47 (1987)
14. E. Olmeda, Fernandez, Hybrid classifiers for financial multicriteria decision making: the case of bankruptcy prediction. Comput. Econ. **10**, 317–335 (1997)
15. T.E. McKee, Developing a bankruptcy prediction model via rough set theory. Int. J. Intell. Syst. Acc. Finance, Manage. **9**, 159–173 (2000)
16. B.A. Ahn, S.S. Cho, C.Y. Kim, The integrated methodology of rough set theory and artificial neural network for business failure prediction. Expert Syst. Appl. **18**, 65–74 (2000)
17. P.G. Swicegood, Predicting poor bank profitability: a comparison of neural network, discriminant analysis and professional human judgement, Ph.D. Thesis, Department of Fi-nance, Florida State University, 1998.
18. K.-S. Shin, Y.-J. Lee, A genetic algorithm application in bankruptcy prediction modeling. Expert Syst. Appl. **23**(3), 321–328 (2002)
19. C.-S. Park, I. Han, A case-based reasoning with the feature weights derived byanalytic hierarchy process for bankruptcy prediction. Expert Syst. Appl. **23**(3), 255–264 (2002)
20. L. Cielen, K. Peeters, Vanhoof, Bankruptcy prediction using a data envelopment analysis. Eur. J. Oper. Res. **154**, 526–532 (2004)
21. W.L. Tung, C. Quek, P. Cheng, GenSo-EWS: a novel neural-fuzzy based early warning system for predicting bank failures. Neural Networks **17**, 567–587 (2004)
22. J. Andres, M. Landajo, P. Lorca, Forecasting business profitability by using classification techniques: a comparative analysis based on a Spanish case. Eur. J. Oper. Res. **167**, 518–542 (2005)
23. A.F. Atiya, Bankruptcy prediction for credit risk using neural networks: a survey and new results. IEEE Trans. Neural Networks **12**, 929–935 (2001)
24. K.-S. Shin, T.S. Lee, H.-J. Kim, An application of support vector machines in bankruptcy prediction model. Expert Syst. Appl. **28**, 127–135 (2005)
25. V. Ravi, P.J. Reddy, H.-J. Zimmermann, Fuzzy rule base generation for classification and its optimization via modified threshold accepting. Fuzzy Sets Syst. **120**(2), 271–279 (2001)
26. V. Ravi et al., Soft computing system for bank performance prediction. Appl. Soft Comput J. (2007). https://doi.org/10.1016/j.asoc.2007.02.001
27. V. Ravi, H.-J. Zimmermann, A neural network and fuzzy rule base hybrid for pattern classification. Soft Comput. **5**(2), 152–159 (2001)
28. L. Yu, S.Y. Wang, K.K. Lai, A novel non-linear ensemble forecasting model incorporating GLAR and ANN for foreign exchange rates. Comput. Oper. Res. **32**(10), 2523–2541 (2005)

Lesion Detection and Classification Using Sematic Deep Segmentation Network

Anil B. Gavade, Rajendra B. Nerli, and Shridhar Ghagane

Abstract Innovation of modern algorithms directed to development of effective segmentation and classification for liver segments. The lesion segmentation is essential in pre-operative liver surgical planning. In addition, the segmentation of liver lesions is considered as a crucial step for deriving qualitative biomarkers for precise clinical diagnosis. The aim of the research is to model an effective method that comprises of the extraction of the liver vessels and classification of liver segments for which Computed Tomography (CT) images is utilized. In classification module, the lesion detection and classification are performed using newly designed optimization algorithm named Adaptive Adam optimization algorithm along with DeepSegNet model. The proposed Adaptive Adam is designed by integrating adaptive concepts in the Adam algorithm. The liver image undergoes segmentation, wherein the tumors present in the liver are extracted based on an optimization strategy. Moreover, the accuracy is attained based on the training of the DeepSegNet using proposed Adaptive Adam algorithm. In this strategy, the probability for lesion or non-lesion is identified effectively. The performance of proposed Adaptive Adam algorithm found superior in terms maximal accuracy 0.846, sensitivity value of 0.881, and specificity value of 0.828, respectively.

Keywords Lesion detection · DeepSegNet · Liver vessels · Segmentation · Computed tomography (CT)

A. B. Gavade (✉)
The Department of Electronics and Communication Engineering, KLS Gogte Institute of Technology, Belagavi, Karnataka, India

R. B. Nerli · S. Ghagane
Department of Urology and Radiology, JN Medical College, KLE Academy of Higher Education and Research (Deemed-To-Be-University), Belagavi, Karnataka, India

1 Introduction

CT is the well-known modality utilized for detecting the lesions of liver, diagnosing liver lesions, and staging of lesions. Focal liver lesions are categorized into different sets like benign lesions and malignant lesions, with substantial deviations in shape, size, location, and contrast. Precise bias between the different lesions is of great significance. Manually segmenting and classifying the liver lesions using the CT images is time consuming and leads to perplexity, which makes it a multifaceted task. Thus, the usage of automated tools assists radiologists to diagnose the liver lesions using CT images [1]. The CT images offer high spatial and temporal resolution by analyzing the pulmonary structures using three-dimensional (3-D) human thorax. Numerous techniques are devised for identifying the precise lung boundaries [2]. Computer-aided diagnostic methods are utilized for detecting the lung cancers using CT images. In CT images, it is acceptable to prohibit the pre-processing step, but the application of enhancement filter is beneficial for determining nodules in the CT images as the thresholding of the CT images may tackle the complexities [3].

Separating blood vessels from CT images and separation of liver into number of segments is essential to plan the liver transplantation [4]. Different liver segmentation algorithms are categorized into semi-automatic, automatic, and manual methods. Sketching the outline of liver manually using CT images offers more precise segmentation, but it is a burdensome and lengthy with respect to the users. The automated techniques are used for segmenting the liver without user interaction but failed to uphold accuracy [4]. In [5], an automatic segmentation method is devised based on supervoxel-based graph cuts. Then, a semi-automatic technique is applied for extracting the liver from the regions. In [6], multiregional-appearance-based approach is devised for segmenting the liver using manual initialization considering various CT slices. The vessel segmentation algorithms are mainly divided as semi-automatic and fully-automatic. In [7], region growing method was devised to segment the liver vessels by altering the thresholds. However, the method was inapplicable with smaller vessels [8]. In [9], locally adaptive region growing method was devised for segmenting the vessel trees wherein the locally adaptive analysis was performed for identifying small vessels. A level-set method was devised in [10] for segmenting the vessels using statistical histogram analysis. In [11], the thresholds are evaluated for extracting vessels in order to refine the counterfeit branches by examining the skeletonized vessel structures. In [12], a matched filter was utilized for enhancing the vessel structures and finally a generic algorithm are applied for liver lesion detection.

In this paper, we implemented a technique, called Adaptive Adam algorithm-based DeepSegNet, for detecting liver lesions and classification using CT images. Implementation is divided into three steps, which comprises thresholding, segmentation, and classification for operative liver lesion detection. Initially, the input CT images are progressed to the thresholding process, wherein the thresholding is carried out for separating pixels. After segmentation, the classification is done by the

DeepSegNet model, which is trained using the proposed Adaptive Adam optimization algorithm. Here, the proposed Adaptive Adam is developed by incorporating the self-adaptive concept into Adam algorithm. In this strategy, the probability for lesion and non-lesion is identified effectively.

This paper contributes toward the optimization algorithm, viz. Adaptive Adam, by altering the Adam algorithm and depends on self-adaptive concept. The major role of the proposed Adaptive Adam is to train the DeepSegNet for lesion detection and classification. The paper is divided into five sections, as introductory part of the liver lesion detection and classification, literature review of liver lesion classification, implemented Adaptive Adam algorithm-based DeepSegNet model, experimental results, and finally conclusion of the implementation.

2 Motivation

In this section, we provide overview of literature review on lesion detection, with limitation of available methods, about eight different techniques are considered and final implementation carried out.

2.1 Literature Review

The four existing literary techniques based on lesion detection techniques are discussed below. Yang et al. [4] developed a method named automatic seed point identification method using automated thresholding method. The method was utilized for extracting liver, hepatic vein (HV), and portal vein (PV), using the CT images. Further, a semi-automatic method was designed for separating HV and PV. Then local searching method was used to categorize the liver segments. The method was effective in segmenting the livers but failed to include other methods like geometrical analysis, graph cuts and histogram analysis. Yan et al. [13] designed a paradigm named DeepLesion for collecting the annotations of liver and then construct a huge scale lesion datasets with less manual effort. A Picture Archiving and Communication System (PACS) are radiological imagery platform they deliver cost effective storage, retrieval, managing, distribution and presentation of medical images. Hospital PACS have collected large amount of clinical annotation, meanwhile extracting, harvesting building large annotated radiological imagery datasets is a challenging process. The method poses many advantages and was considered as a widespread detector of lesions, which could determine all types of lesion in a single framework. However, this model is inapplicable with 3-D images. Diamant et al. [14] developed a technique for automating the diagnosis of liver lesions based on the CT images that enhances the image patch representation of boundary lesions. This method captured the features of the lesion periphery using the lesion interior. The method improved the features based on visual word histograms and ROI. However, the method was unable to analyze the 3-D volumetric data. Frid-Adar et al. [15] developed a Generative Adversarial Networks (GANs) using synthetic medical images for determining the lesions by analyzing data and to improve the performance of CNN. The method synthesized improved quality liver lesion using ROI. The method was not applicable with other medical domains for enhancing the synthesis of liver lesions.

2.2 Challenges

The challenges of the research are deliberated below:

- In [4], a semi-automatic method is designed for distinguishing hepatic vein (HV) and portal vein (PV) using CT images. However, the recognition of PV branches using eight different liver segments is a major obstacle because of complexity in choosing a basis point for recognition of PV branches from 3-D images. In addition, this method faces complexities while dealing with local disturbances like connected structures or low contrast with intensity values.
- The major issue in the context of radiological imaging is to manage with diminutive datasets and the fewer amounts of processed samples particularly, while adapting supervised machine learning techniques that necessitate labeled data and outsized training. However, the annotations of CT images are a lengthy process and require more time for processing [15].
- Analysis of liver lesions is major challenge due to inconsistencies occurred with respect to the appearances like differences in shape of lesions, margin sharpness, size, and interior textures. In addition, the manifestation of many lesions overlaps and thereby leads to lots of variation [14].

3 Proposed DeepSegNet Model for Liver Lesion Detection

Figure 1 portrays the architecture of DeepSegNet model for detecting and classifying the lesion present in the CT image. The aim of implementation is to model an effective planning method that comprises of the extraction stage of the liver, extraction of vessels, and classification of the liver segments for which the CT images is employed. At first, the CT image is taken as an input, which is then fed to the liver extraction module for extracting the interesting regions from the image. The ROI from the CT image is extracted using the thresholding process. Here, the region covering the liver area is subsequently extracted from the input CT image. Once the ROI is extracted then, the extracted regions are fed to the proposed DeepSegNet model for performing deep segmentation and classification using the Adam optimization [16] algorithm.

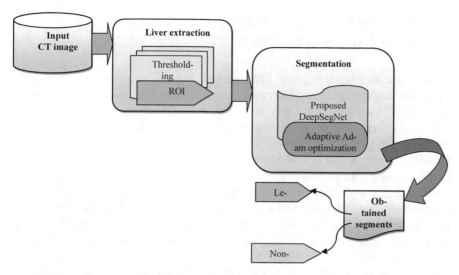

Fig. 1 Framework of proposed adaptive Adam-based DeepSegNet model for liver lesion detection

Based on the detected regions, the classification is done to determine whether the detected region is lesion or non-lesion.

Here, D is database with 'f' number of CT images and is given as,

$$D = \{I_1, I_2, \ldots, I_g, \ldots, I_f\} \tag{1}$$

where I_g represents the gth input image.

3.1 Thresholding to Extract the Interesting Regions

For qualitative valuation of CT image, the ROI is evaluated for estimating the noise variance. Generally, the CT images are assumed to hold additive noise along with the image. Thus, the thresholding is employed for performing ROI extraction in the CT images, wherein the pixels are normalized and set to one or zero. Thus, the thresholding result is a binary image, which can offer a mask for extracting uniform ROI. The thresholding selects the ROI on the basis of pixel intensity. The goal of image thresholding is to provide segments of the image based on typical features of the image and characteristics are devised on the basis of pixels intensity level. The thresholding process represents the image in the digital format by categorizing the image as light or dark. In order to overwhelm the changes for brightness, the thresholding is utilized that utilizes different values of thresholds fir each pixel in the image. The thresholding of image offers robustness with respect to the changes in the illumination. The merits of the thresholding are that the method is simple, upfront, and result is assumed to be same, if the steps taken for the comparison are different.

3.2 Proposed DeepSegNet Model for Extracting Lesions of the Liver from CT Image

Liver lesion detection is employed for segmenting the liver to detect the unhealthy tissues for diagnosing hepatic diseases and access the medical practitioner for earlier treatment. The segmentation of liver and lesion using CT images helps the oncologists to precisely diagnose liver cancer to spot the presence of lesion. The DeepSegNet [17] model is employed for segmenting the liver lesions to detect the existence of lesion from the CT image. The DeepSegNet is an encoder-decoder network that contains a hierarchy of decoder each linking to a specific encoder. The DeepSegNet is an architecture used for performing pixel-wise semantic segmentation. The DeepSegNet is efficient with respect to computation time and memory and the DeepSegNet is flexible to implement. This section shows the DeepSegNet employed for the classification to detect the liver lesion from the CT images. The training of the DeepSegNet model is performed with proposed Adaptive Adam, wherein Adam algorithm [16] is modified using the self-adaptive concept to enhance the convergence and optimization behavior. The architecture of DeepSegNet model and the proposed Adaptive Adam algorithm is illustrated in the subsections.

3.2.1 Architecture of DeepSegNet Model

DeepSegNet is new segmentation technique, this section explains, segmenting objects by means of DeepSegNet. The CT images are subjected to the DeepSegNet [17] for the liver lesion detection. In DeepSegNet, the decision is taken for each pixel contained in the image. Semantic segmentation is defined as a method, which can understand image at its pixel level. Moreover, the model classifies each pixel according to the pre-determined class. DeepSegNet consists of decoder and encoder network and classification layer. The structural design of DeepSegNet model is demonstrated in Fig. 2.

Here, the encoder network contains convolutional layers for performing the liver lesion classification. Thus, the training process is initiated from the weights trained for classification using CT images. Each layer of encoder consists of a equivalent decoder layer for the reconstruction process. The output generated by the decoder is given to the multi-class soft-max classifier for generating the probabilities of class considering each pixel. Encoder performs convolution by means of a filter bank and generates feature maps. Then, an element-wise Rectified Linear Non-Linearity (ReLU) is applied and the resulting output is sub-sampled. ReLU is used for ensuring effectiveness, simplicity, and ReLU layers work faster while dealing with large networks. Moreover, POOL layers reduce the complication and there is no weight/bias for training rather it process the CT input images in combining the local regions in the filter. Then, the suitable decoder is employed for upsampling feature maps using the learned max-pooling indices from the corresponding encoder feature maps. Then, the feature maps are convoluted using a filter bank for generating dense feature maps.

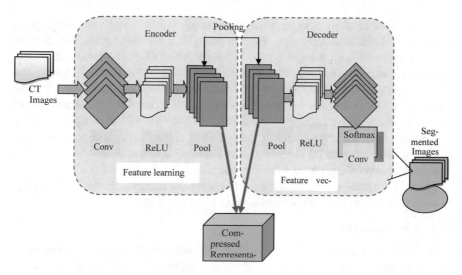

Fig. 2 Architecture of DeepSegNet model

The output produced by the soft-max classifier is K, where K indicates the count of classes. Thus, the DeepSegNet is responsible for generating the segments for liver lesion classification. Thus, the liver lesions are determined using the DeepSegNet module.

The segments are generated using the proposed DeepSegNet model, which is given as,

$$S^g = \left\{ S_1^g, S_2^g, \ldots, S_k^g, \ldots S_p^g \right\} \tag{2}$$

where S_k^g specifies the kth segment of the gth image, and 'p' refers the total number of the segments obtained from the gth image.

3.2.2 Proposed Adaptive Adam Algorithm for Training DeepSegNet

The proposed Adaptive Adam is utilized for training the DeepSegNet model to extract the lesion regions. In addition, the proposed Adaptive Adam is designed by integrating the merits of Adam optimization algorithm [16] along with the self-adaptive concepts. Adam [16] is the first-order stochastic gradient-based optimization that well-suits for the objective function that varies with respect to the parameters. The hitches associated with the non-stationary objectives and issues in the presence of the noisy gradients are handled effectively. This algorithm utilizes minimum memory by improving computational efficiency, the hitches associated non-stationary objectives and issues in the presence of noisy gradients are handled efficiently. Additionally, Adam witnesses the following advantages: the magnitudes of the updated parameters are invariant with respect to the rescaling of the gradient, and the step-sizes are approximated using a hyper-parameter, which works effectively using sparse gradients and non-stationary objectives. In addition, Adam is effective for step size annealing. To obtain global optimal solutions, the Adam algorithm incorporates self-adaptive concept. The self-adaptive parameter minimizes the impacts of conventional parameters by balances the efficiency and accuracy for segmenting the liver lesions. Thus, the accuracy and the average time of the method becomes faster and higher. Generally, the adaptive parameter is utilized for making the method highly efficient and precise for solving optimization issues. Thus, the Adam optimization incorporates adaptive parameter for obtaining global optimal solution while classifying liver lesions. Moreover, the propose Adaptive Adam assists to jump out from local optimal and provides a trade-off between local and global optimization strategies. The steps of proposed Adaptive Adam are illustrated in this subsection.

Step 1: Initialization

Initialization of bias corrections is the first step, wherein $\hat{q}_l$ is the bias corrected of first moment estimate and $\hat{m}_l$ indicates the corrected bias of second moment estimate.

Step 2: Evaluation of the fitness function

Fitness of bias is calculated to choose optimal weights for training DeepSegNet model, fitness function is treated as error function which results global optimum solution. The function is a minimization function and it is given by,

$$J = \frac{1}{r} \sum_{k=1}^{r} \left(H_k - H_k^* \right)^2 \tag{3}$$

where r represents the total image samples, H_k referees output produced by DeepSegNet classifier, H_k^* indicates ground truth value.

Step 3: Determination of updated bias

Adam algorithm is employed for improving the optimization and convergence behavior, which produces smoother distinction with better computational efficiency at minimal memory utilization. According to the Adam [16], the bias is given as,

$$\theta_l = \theta_{l-1} - \frac{\alpha \hat{q}_l}{\sqrt{\hat{m}_l} + \varepsilon} \tag{4}$$

where α is the step size, $\hat{q}_l$ is the bias corrected, $\hat{m}_l$ indicates the bias corrected second moment estimate, ε specifies the constant, θ_{l-1} indicates the parameter at previous time instant $(l-1)$. The corrected bias of first-order moment is formulated as,

$$\hat{q}_l = \frac{q_l}{(1 - \eta_1^l)} \tag{5}$$

$$\hat{q}_l = \eta_1 q_{l-1} + (1 - \eta_1) G_l^1 \tag{6}$$

The corrected bias of second-order moment is expressed as,

$$\hat{m}_l = \frac{m_l}{(1 - \eta_2^l)} \tag{7}$$

$$\hat{m}_l = \eta_2 m_{l-1} + (1 - \eta_2) G_l^2 \tag{8}$$

where,

$$G_l = \nabla_\theta \text{loss}(\theta_{l-1}) \tag{9}$$

While update, algorithm is made self-adaptive, by adjusting the step size without any user intervention. The standard parameter update equation of the Adam optimization is modified using the adaptive parameter for which α is considered. Here, the α of the Adam optimization algorithm is made self-adaptive and is formulated as,

$$\alpha = \left(1 + \frac{(1 - n)}{1 + n_{\max}}\right) \tag{10}$$

where n is current iteration, and $n_{\max}$ indicates the maximum iteration.

Step 4: Stopping criterion

Optimal weights are developed in and reiterative way until final iterations is attained.

4 Results and Discussion

The results of implemented Adaptive Adam algorithm devised for the lesion detection is illustrated. The effectiveness of the Adaptive Adam algorithm is estimated by varying the training data percentages with respect to accuracy, sensitivity, and specificity parameters.

4.1 Experimental Setup

The implementation is done using MathWorks MATLAB tool on Windows platform, with Titan-Nvidia GPU hardware.

4.2 Dataset Description

The DeepLesion dataset [18] is created in November 27, 2020, by National Institutes of Health—Clinical Center and donated by Ke Yan. The dataset comprised of certain attributes like patient_index, study_index, series_index, and slice_index. The DeepLesion dataset was collected on the basis of radiologists' annotations named as 'bookmarks.' The dataset accumulated the key slices, which comprise of the bookmarks and 60 mm contexts with extra slices above and below the key slice for enabling usage of the 3-D information.

4.3 Evaluation Metrics

The performance of the implementation is computed by means of three parameters accuracy, sensitivity, and measures.

(a) *Accuracy*: Accuracy is one of the most significance parameter in classification of objects, this is the degree of familiarity of assessed value with respect to it unique value and accuracy is represented as,

$$A = \frac{(TP + TN)}{(TP + FP + FN + TN)} \tag{11}$$

where TP is true positive, TN is true negative, FP and FN denotes false positive and indicates false negative.

(b) *Sensitivity*: Represents as ratio of the positives that are recognized appropriately.

$$TPR = \frac{TP}{(TP + FN)} \tag{12}$$

(c) *Specificity*: The specificity is utilized to produce a result which indicates that a given condition is present when it is not.

$$FPR = \frac{FP}{(FP + TN)} \tag{13}$$

4.4 Experimental Results

The experimental results of liver lesion segmentation using proposed Adaptive Adam-based DeepSegNet model are illustrated in Fig. 3. Figure 3a portrays the input CT images of liver disease patient, and Fig. 3b portrays the segmented CT

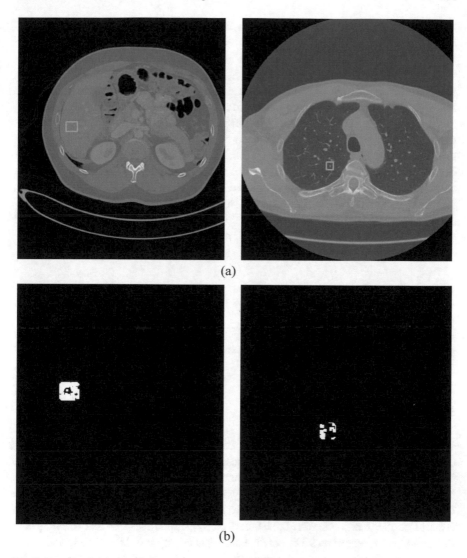

Fig. 3 Results of Adaptive Adam algorithm. **a** Input image, **b** segmented image

image for detecting the presence of liver lesions. The white indicates the detected lesions whereas the black depicts the non-lesion regions.

4.5 Performance Analysis

Implementation analysis of proposed algorithm is carried by changing the convolutional layers ranging from 2 to 10, showing in Fig. 3. Figure 3a indicates the analysis based on accuracy, with 50% training data, the calculated vales for different convolutional layer 2, 4, 6, 8, and 10 are 0.603, 0.688, 0.726, 0.747, and 0.771, respectively. When the training data increased to 90% corresponding accuracies increased drastically, with convolutional layer 2, 4, 6, 8, and 10 are 0.651, 0.733, 0.774, 0.796, and 0.822, respectively. Figure 3b indicates the analysis based on sensitivity, with 50% training data, the calculated vales for different convolutional layer 2, 4, 6, 8, and 10 are 0.586, 0.668, 0.705, 0.726, and 0.749, respectively. When the training data increased to 90% corresponding accuracies increased drastically, with convolutional layer 2, 4, 6, 8, and 10 are 0.648, 0.730, 0.770, 0.793, and 0.818, respectively. Figure 3c indicates the analysis based on specificity, with 50% training data, the calculated vales for different convolutional layer 2, 4, 6, 8, and 10 are 0.817, 0.714, 0.686, 0.589, and 0.573, respectively. When the training data increased to 90% corresponding accuracies increased drastically, with convolutional layer 2, 4, 6, 8, and 10 are 0.891, 0.735, 0.705, 0.598, and 0.584, respectively.

4.6 Comparative Methods

Implemented Adaptive Adam DeepSegNet is compared with Fuzzy-c-means (FCM) [19], Adaptive threshold [20] with parameters accuracy, specificity, and sensitivity is elaborated further.

4.6.1 Comparative Analysis

Implementation evaluation of proposed Adaptive Adam DeepSegNet algorithm is carried using accuracy, sensitivity, and specificity metrics, represented in Fig. 4. We have observed when the training data dimension is large; we get better accuracy for classification. Figure 5a indicates the analysis based on accuracy parameter with 50% training data, obtained results of FCM, Adaptive threshold, and proposed Adaptive Adam DeepSegNet are 0.671, 0.819, and 0.797, respectively. When training data dimension is increased to 90%, corresponding accuracy are 0.81, 0.831, and 0.846 respectively. Figure 5b represents analysis based on sensitivity parameter with 50% training data; obtained results of FCM, Adaptive threshold, and proposed Adaptive Adam DeepSegNet are 0.672, 0.831, and 0.850 respectively. When training data dimension is increased to 90%, corresponding accuracy are 0.839, 0.842 and 0.881 respectively. Figure 5c represents analysis based on specificity parameter with 50% training data, obtained results of FCM, Adaptive threshold, and proposed Adaptive Adam DeepSegNet are 0.654, 0.600, and 0.676 respectively. When training data dimension is increased to 90%, corresponding accuracy are 0.800, 0.726, and 0.828 respectively.

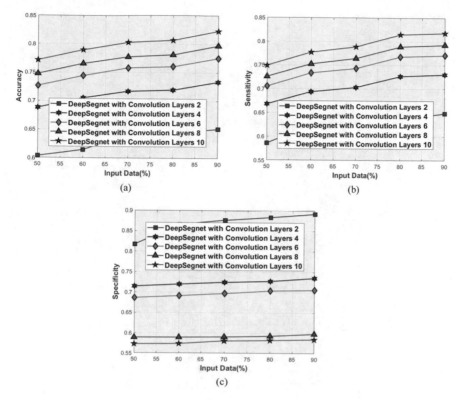

Fig. 4 Performance analysis of proposed DeepSegNet based on convolution layers with **a** accuracy, **b** sensitivity, **c** specificity

4.7 Comparative Discussion

Table 1 describes the comparative analysis of proposed Adaptive Adam DeepSegNet model with the available methods like FCM, Adaptive threshold. The performance analysis is carried out by the training data dimensions of 90%, with classification assessing parameters such as accuracy, sensitivity, and specificity measures matrices. The maximal performance is attained with accuracy value of proposed Adaptive Adam DeepSegNet as 0.846, whereas the accuracy values of available FCM, Adaptive threshold as 0.81, and 0.831, respectively. Similarly, for 90% training data, the sensitivity values calculated by proposed DeepSegNet as 0.881, whereas the sensitivity values of existing FCM, Adaptive threshold as 0.839, and 0.842, respectively. Similarly, for 90% training data, the specificity values computed by proposed DeepSegNet as 0.828, whereas the specificity values of existing FCM, Adaptive threshold as 0.800, and 0.726, respectively.

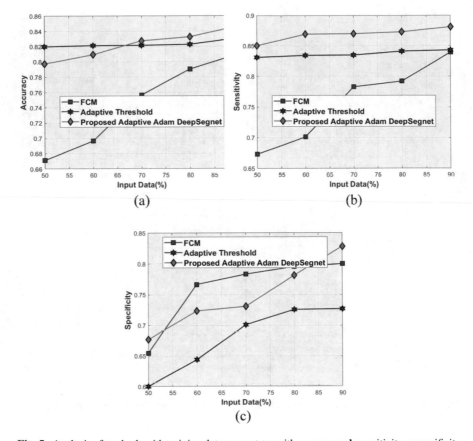

Fig. 5 Analysis of methods with training data percentage with **a** accuracy, **b** sensitivity, **c** specificity

Table 1 Comparative discussion

Metrics	FCM	Adaptive threshold	Proposed Adaptive Adam DeepSegNet
Accuracy	0.81	0.831	0.846
Sensitivity	0.839	0.842	0.881
Specificity	0.800	0.726	0.828

5 Conclusion

This paper proposes an effective optimization strategy for extracting the stage of the liver, vessel extraction, and classification of the liver segments for which the CT images is utilized. In the classification module, the lesion detection and classification is done, which is the goal of this paper. The liver with tumors is segmented through including the tumors to the extracted liver based on a strategy that work on the deviation of the intensity variations of the image with respect to the tumor and tumor regions. Once the liver and vessel extraction are done, lesion detection and classification are progressed for which the DeepSegNet model is utilized. Here, the training of the DeepSegNet classifier is done by proposed Adaptive Adam algorithm. The proposed Adaptive Adam is designed by making the Adam algorithm self-adaptive. In this strategy, the probability for lesion or non-lesion is identified effectively. The performance of the proposed Adaptive Adam is superior to the existing methods with maximal accuracy value of 0.846, sensitivity value of 0.881, and specificity value of 0.828, respectively.

References

1. M. Heker, A. Ben-Cohen, H. Greenspan, Hierarchical fine-tuning for joint liver lesion segmentation and lesion classification in CT (2019). arXiv preprint arXiv:1907.13409
2. S. Hu, E.A. Hoffman, J.M. Reinhardt, Automatic lung segmentation for accurate quantitation of volumetric X-ray CT images. IEEE Trans. Med. Imaging **20**(6), 490–498 (2001)
3. Q. Li, S. Sone, K. Doi, Selective enhancement filters for nodules, vessels, and airway walls in two-and three-dimensional CT scans. Med. Phys. **30**(8), 2040–2051 (2003)
4. X. Yang, J. Do Yang, H.P. Hwang, H.C. Yu, S. Ahn, B.W. Kim, H. You, Segmentation of liver and vessels from CT images and classification of liver segments for preoperative liver surgical planning in living donor liver transplantation. Comput. Methods Programs Biomed. **158**, 41–52 (2018)
5. W. Wu, Z. Zhou, S. Wu, Y. Zhang, Automatic liver segmentation on volumetric CT images using supervoxel-based graph cuts. Comput. Math. Methods Med. 14 (2016)
6. J. Peng, P. Hu, F. Lu, Z. Peng, D. Kong, H. Zhang, 3D liver segmentation using multiple region appearances and graph cuts. Med. Phys. **42**, 6840–6852 (2015)
7. D. Selle, B. Preim, A. Schenk, H.O. Peitgen, Analysis of vasculature for liver surgical planning. IEEE Trans. Med. Imag. **21**, 1344–1357 (2002)
8. S. Esneault, C. Lafon, J.-L. Dillenseger, Liver vessels segmentation using a hybrid geometrical moments/graph cuts method. IEEE Trans. Biomed. Eng. **57**, 276–283 (2010)
9. J. Yi, J.B. Ra, A locally adaptive region growing algorithm for vascular segmentation. Int. J. Imaging Syst. Technol. **13**, 208–214 (2003)
10. L.M. Lorigo, O.D. Faugeras, W.E.L. Grimson, R. Keriven, R. Kikinis, A. Nabavi, C.-F.Westin, CURVES: curve evolution for vessel segmentation. Med. Image Anal. **5** (2001)
11. L. Soler, H. Delingette, G. Malandain, J. Montagnat, N. Ayache, C. Koehl, O. Dourthe, B. Malassagne, M. Smith, D. Mutter, J. Marescaux, Fully automatic anatomical, pathological, and functional segmentation from CT scans for hepatic surgery. Comput. Aid. Surg **6**, 131–142 (2001)
12. O.C. Eidheim, L. Aurdal, T. Omholt-Jensen, T. Mala, B. Edwin, Segmentation of liver vessels as seen in MR and CT images. Int. J. Comput. Assisted Radiol. Surg. **1268**, 201–206 (2004)
13. K. Yan, X. Wang, L. Lu, R.M. Summers, DeepLesion: automated mining of large-scale lesion annotations and universal lesion detection with deep learning. J. Med. Imaging **5**(3), 036501 (2018)
14. I. Diamant, A. Hoogi, C.F. Beaulieu, M. Safdari, E. Klang, M. Amitai, H. Greenspan, D.L. Rubin, Improved patch-based automated liver lesion classification by separate analysis of the interior and boundary regions. IEEE J. Biomed. Health Inf. **20**(6), 1585–1594 (2015)
15. M. Frid-Adar, I. Diamant, E. Klang, M. Amitai, J. Goldberger, H. Greenspan, GAN-based synthetic medical image augmentation for increased CNN performance in liver lesion classification. Neurocomputing **321**, 321–331 (2018)
16. D.P. Kingma, J. Ba, Adam: a method for stochastic optimization (2014). arXiv preprint arXiv: 1412.6980

17. S. Kwak, S. Hong, B. Han, Weakly supervised semantic segmentation using superpixel pooling network, in *AAAI*, Feb 2017, pp. 4111–4117
18. DeepLesion Dataset. https://nihcc.app.box.com/v/DeepLesion/folder/50715173939. Accessed on Nov 2020
19. A.R. Ali, , M. Couceiro, A.E. Hassanien, M.F. Tolba, V. Snášel, Fuzzy c-means based liver CT image segmentation with optimum number of clusters, in *Proceedings of the Fifth International Conference on Innovations in Bio-Inspired Computing and Applications* (Springer, Berlin, 2014), pp. 131–139
20. M. Anju Krishna, D. Edwin, Liver tumor segmentation using adaptive thresholding. Int. J. Innov. Res. Electr. Electron. Instrum. Control En. **5**(1) (2017)

Automatic Gland Segmentation for Detection of CRC Using Enhanced SegNet Neural Network

Mohan Mahanty⊙, Debnath Bhattacharyya⊙,
and Divya Midhunchakkaravarthy⊙

Abstract Colorectal cancer (CRC) is a sort of cancer that begins in the rectum or large intestine, generally showing more impact on adults. It generally starts as tiny, non-cancerous (benign) lumps of cells called polyps that form on the colon's inward lining. Detection of colorectal cancers in early stages can reduce the risk of mortality rate. Detection of polyps through screening tests may fail because of the length of the rectum and the tiny size of the polyps. Computer-based biomedical imagining might help the radiologists sift through the information and analyse clinical pathology images better. The majority of research focuses on gland segmentation using CNN architectures, but accuracy is still challenging. In this work, encoder-decoder-based deep learning model was proposed for morphological colon gland segmentation. This research of enhanced SegNet architecture improved the system's overall performance and, when tested on the standard benchmark Warwick-QU dataset for colonoscopy images. We trained the system by the training dataset consisting of 7225 images, and evaluation of the model is done using a test dataset divided into two parts: A (60 images), B (20 images). The results show that the proposed model is comparable to all the existing models regarding the same dataset and attain segmentation accuracy of 0.924 on Part A, 0.861 on Part B. The shape similarity of 83.145 on Part A, 88.642 on Part B, is considered adequate compared with the existing methods.

Keywords CRC · Colonoscopy images · Deep learning · Image segmentation · CNN · SegNet

M. Mahanty (✉) · D. Midhunchakkaravarthy
Department of Computer Science and Multimedia, Lincoln University College, Kuala Lumpur, Malaysia
e-mail: mohan.mahanty@lincoln.edu.my

D. Midhunchakkaravarthy
e-mail: divya@lincoln.edu.my

D. Bhattacharyya
Department of Computer Science and Engineering, K L Deemed To Be University, KLEF, Guntur 522502, India
e-mail: debnathb@kluniversity.in

337

1   Introduction

The bowel is the lower portion of the body's gastrointestinal structure, a complicated part of the digestive system, which carries the ingested nutrients. Many millions of organs in the large intestine and rectum take care of the absorbing of water and minerals and secreting mucous for the regrowth of epithelial cells. Still, precise reasons for colorectal cancer are unknown. The abnormal growth of tissues in the colon, named polyps, can change into cancer over time, typically three to seven years. Most of the polyps are hyperplasic, which are non-cancerous. Suspicious colorectal polyps may invade the surrounding tissues and diffusion to various other body organs and harm them. Still, precise reasons for colorectal cancer are unknown. If the polyp's size is more significant than 1 cm or two, polyps found in the interlining of the colon cells appear adenomatous. In sporadic cases of exposure to carcinogen agents in the environment, specific genetic reasons may be the reasons. The stage of disease describes just how much it has spread, and the grading stage of cancer helps to choose the best treatment.

According to the American Cancer Society, Surveillance Research, 2020 report [1], colon cancer is the succeeding prominent reason for mortality in males and females. The statistical analysis estimated that the cases are also increasing rapidly, represented in Table 1. As per the survey, this is the second leading cancer in men and even third in women. Gastroenterologist often fails to explain why one person develops this disease, and another does not. Nevertheless, an age far more than 50, a family history of colorectal cancers, personal history of uterine, breast cancer, or maybe ovarian cancer can improve the possibility to affect by colorectal cancer.

The recent advances in deep learning frameworks have allowed faster and more accurate detection. After exploring GPUs and modern computer architectures, researchers effectively applied the deep learning models in biomedical imagining. Computer-based medical imaging diagnosis mainly focuses on gland segmentation, which plays an essential role in diagnosing malignant tissues. Deep neural networks are traditional neural networks that can effectively segment the objects in the images. The introduction of CNN changed image data processing in deep learning models as the increased CPU and GPU processing capacity available allows radiologists to scale their diagnostic efforts.

Table 1 Estimated cases of colon cancer in the USA

Estimated cases of colon caner	All ages	≤ 45	≥ 45	≤ 65	≥ 65
Male	78,300	4830	73,470	37,990	40,310
Female	69,650	4730	64,920	29,950	39,700

2 Related Work

Jing Tang et al. proposed SegNet [2] for gland segmentation. They evaluated their model over the Warwick-QU dataset, and training dataset consists of 85 images and the testing performed on two sets of test dataset. By consider the object-level Dice distance ($Dice_{obj}$) and object-level Hausdorff distance (H_{obj}) as metrics, Part A has achieved 0.8820. Moreover, the shape similarity on section A is 106.6471, and for Part B, the model has achieved $Dice_{obj}$ and H_{obj} of 0.8636 and 102.5729.

Xin Yan et al. proposed an image detection method for cell pathology [3]. After pre-processing, the pathology images, processed on the proposed system, can accurately extract their hidden features. The model was processed on the Warwick-QU dataset. The researchers claimed that their model yields classification accuracy of 73.94%, 80.00%, and 71.52% in the healthy, adenomatous, moderately differentiated, categories, respectively. And it attains an accuracy of 87.88% for moderately to poorly differentiated, and 84.24% in the poorly differentiated categories, respectively.

Khvostikov et al. proposed a CNN architecture inspired from U-Net for "automatic mucous glands segmentation in histological images" [4]. Their proposed CNN to get segmentation is patch-oriented. When an image is fed into the model, it is split up into patches, and then each patch processes through the model, and finally merged to the final result. They evaluated their model on the standard benchmark Warwick-QU and obtained the Dice score of 0.92 and D_{obj} of 0.88.

Rezaei et al. suggested a modified version of LinkNet [5] for gland segmentations. They considered the handcrafted features and deep networks are the key points in the segmentation process. They claimed their proposed architecture of link net consists of the parameters that are 1/3 of the parameters of the U-Net model; hence, the training and testing can be performed effectively compared to U-Net. After pre-processing, the training dataset composed of 85 colour images is fed into the LinkNet model. They evaluated their system on two separate parts (60, 20 images) of the Warwick-QU test set. Part A has achieved segmentation accuracy of 0.897, $F1$ Score of 0.912, and shape similarity of 45.418, and for Part B, the segmentation accuracy, $F1$ Score and shape similarity of 0.822 and 0.750 and 108.208.

Ding et al. suggested a multi-scale FCN for gland segmentation [6]. Their proposed MSFCN model additionally follows the complete fully 210 convolutional frameworks, and when it is evaluated over the Warwick-QU dataset, attain $F1$ Score, D_{obj}, H_{obj} as 0.914, 0.913, 39.848 on Part A and on Part B as 0.850, 0.858, 93.244. Safiyeh Rezaei et al. proposed their work based upon LinkNet [7] and based upon LinkNet-based network. The authors claimed that their proposed method consists of three components, the first unit, pre-processing unit, the second unit is the proposed LinkNet used for segmentation of glands and the post-processing module. They evaluated their model on the benchmark Warwick-QU dataset (training dataset 85 images and testing two sections 60 images and 20 images) and attain D_{obj}, $F1$ Score, H_{obj} distances as 0.872, 0.832, 55.51 on Part A and Part B 0.82, 0.73, 109.72.

Tianyu Shi et al. proposed SG-U-Net architecture [8], which uses hybrid features and the multiple sub-modules, they can segment ROI and effectively detect the diseases. The features extracted at lower-level and higher-level are directly represented from one convolution block through dense connections. They evaluated their proposed algorithm on the Warwick-QU dataset and achieved 0.911 object Dice, 42.192 object Hausdorff distances, and 0.843, 89.677 object Dice and object Hausdorff distances, respectively.

However, gland segmentation plays a crucial role in diagnosing CRC detection. But accurate detection of the polyps through image segmentation is still a challenge in medical diagnosis. A very deep neural network increases the computational time and model execution complexity. To overcome this, we propose an enhanced SegNet-based encoder-decoder model for accurate segmentation of the polyps.

3 Proposed Architecture for Gland Segmentation

Analysis of pathological images plays an important role in the recognition of cancer. With the extraordinary improvement in computer-aided medical diagnosis in recent years, full-slice scanning equipment technology can quickly digitize the pathological images. The computer-based diagnosis (CAD) of digitized medical pathological images became one of the hotspots in medical image diagnosis. Polyp segmentation from the colonoscopy images is a bottleneck for any system and is remained as a difficult problem. In this present work, we proposed a novel model for accurate polyp detection from the colonoscopy images. Figure 1 represents the architecture of the enhanced SegNet architecture.

Our proposed model of SegNet is an enhanced variation of the primary SegNet architecture. Specifically, we recommended to scale down the down-sampling and up-sampling layers to make the original SegNet architecture computationally inexpensive. And the reduced convolutional layers affect the total learnable parameters of the primary SegNet model, whereas in the original model encoder network had 14.7 million learnable parameters, and is reduced in the proposed model. The proposed model reduced the complexity of computation and accomplished

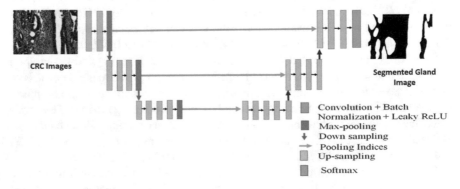

Fig. 1 Proposed modified SegNet architecture for gland segmentation

outstanding outcomes on the Warwick-QU dataset. Generally, CNN models are finally attached to fully connected layers, but here we discarded the fully connected (FCN) layers to minimize the proposed model's evaluation complexity. The pooling indices of the higher resolution feature maps (channels) are saved before sub-sampling. This additionally decreases the total number of parameters (learnable) of the proposed SegNet model.

The model is classified into two parts, as encoder and decoder networks. The encoder network (left side) of the model comprises nine convolutional (Conv) layers, as well as the decoder network (right side) has nine corresponding convolutional layers. For every convolutional layer of the encoder, convolutions are applied using kernels of size 3×3, one-pixel stride, to generate the featured maps. For the generated feature channels, batch normalization (BN) is applied, and then leaky ReLU (LReLU) [9] is applied.

In the existed deep learning models, ReLU is applied; however, it is not continuously differentiable. At some $x = 0$, the breakeven point between x and 0, the activity cannot be computed. This is not as well bothersome, but in real-time, it can slightly influence training performance. And ReLU sets the values to zero when a neuron obtained the negative values. When the neurons arrive at large negative values, the neuron effectively dies and cannot recover from being stuck at 0, known as the dying ReLU problem. If the neurons are not initialized appropriately or when data passed to them is not normalized very well, it can create considerable weight swings during the first phases of optimizing the model. Because of this problem, the network model may essentially stop learning and underperforms.

Leaky ReLU (LReLU) can fix the "dying ReLU" problem and additionally attempts to minimize one's level of sensitivity to the dying ReLU problem. Mathematically, Leaky ReLU can be represented as

$$f(x) = \begin{cases} 0.01x & \text{if } x < 0 \\ x & \text{otherwise} \end{cases} \tag{1}$$

The working model of ReLU and Leaky ReLU is graphically shown in Fig. 2. As shown in Eq. (1) for all the inputs less than zero, ReLU sets the values to zero, and in Leaky ReLU, the outputs are slightly descending, which indicates that these small numbers reduce the death of ReLU activated neurons.

After a series of convolutional layers, max-pooling is applied to subsample the feature maps. A non-overlapping 2×2 sized window is used for max-pooling, with stride 2. Totally three times, sub-sampling is performed at the encoder network (shrinking path), making the spatial dimensions of features divided into two. Nonetheless, by sub-sampling (using max-pooling), the significant loss of the feature channels' spatial resolution causes an adverse outcome for segmentation when contour information is vanishing. So, there is a necessity to capture and save the encoder feature map's contour information before sub-sampling.

A variant of the model, where instead of pooling indices if the feature maps are transferred to corresponding decoder network definitely consumes even more

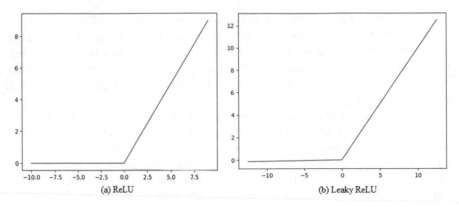

Fig. 2 Working of **a** ReLU and **b** leaky ReLU

memory, and it also takes a lot of computational time. So, we can apply a straight-forward method as well as save the max-pooling indices and transferred those to the corresponding decoder network. Therefore, the pooling indices for each feature map are memorized using a minimal amount of memory. The corresponding decoder placed in the (right side) decoder network up-samples the received input feature channels by utilizing the memorized pooling indices obtained from the corresponding encoder network. Each up-sampling layer's outcome is incorporated with the matching feature map (received from the encoder network) to generate a sparse feature map(s).

For each up-sampling, the number of kernels used for convolutional operations is doubled, used for deconvolutional operations, to generate the non-lossy boundary information of the objects. And the pooling indices obtained from the corresponding contracting path (encoder network) are used to generate the sparse feature maps. After the decoder network convolutions, the sparse feature maps are batch normalized like the encoder network. In up-sampling, fixed bicubic interpolation weights are used because no learning is there in the up-sampling process. Previously, bilinear interpolation is used in the up-sampling process, considers the 2×2 (Four pixels), and uses the weighted average of those block of four pixels for each missed pixel value, yields to blur edges, whereas bicubic interpolation considers 4×4 (16 pixels) for each missed pixel value. Images resembled with bicubic interpolation are smoother and have less interpolation distortion. The final decoder network result is a dense feature map (channel), which has a similar resolution and the same number of channels as the input image. Finally, these feature map(s) are given as input to softmax (multi-class) classifier to separately classify each pixel.

 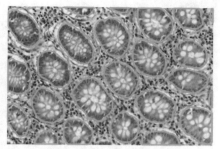

(a) Typical histopathology image   (b) Segmented gland by a pathologist

Fig. 3 Examples of typical histopathology and segmented glands

Table 2 Training and testing samples of Warwick-QU dataset

Dataset (total images 165)	Warwick-QU	
Training (85)	Benign (37)	Malignant (48)
Testing (80)	Benign (37)	Malignant (43)

4 Experimental Results and Analysis

4.1 Dataset

We consider the standard benchmark Warwick-QU dataset [10, 11] to evaluate the proposed model. The slides have been generated using a Zeiss MIRAXMIDI Slide Scanner to get a 20× magnification of digital medical image scanning. The dataset consists of 165 colon cancer cell tissues images extracted from 52 visual fields and hand-marked ground truth annotations. As shown in Fig. 3, ground truth annotations of every image in training and testing datasets are provided by an expert pathologist.

As shown in Table 2, the dataset is then categorized into training and testing. The primary training dataset comprises 85 images, of which 37 are benign, and 48 are malignant. And the testing dataset is categorized into two parts, as Part A, Part B consists of 60, 20 images, respectively.

4.2 Data Augmentation

In the real-world circumstance of medical imagining, we may have a limited set of images. To deal with various circumstances like a different orientation, location, scale, brightness, etc. To overcome this, synthetically modified data is generated from the original dataset by using data augmentation techniques. It can help raise the amount of relevant data according to the dataset and lower the over-fitting. The primary dataset consists of 85 images that have various sizes {(581 × 442), (589 ×

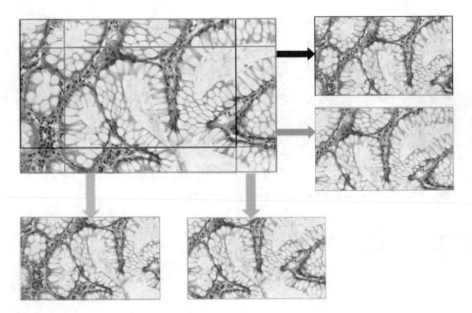

Fig. 4 Augmented images from the original image

453), (775 × 522)}, we initially resize all of them to 256 × 256 to be appropriate for successive down-sampling layers of proposed enhanced SegNet. Data augmentation is performed on the training data, where each image in the dataset is rotated by 90°, 180°, and then flipped in vertical and horizontal directions. Then, 4 overlapping massive patches (¾ of original size) are cropped and again resized to 256 × 256, as shown in Fig. 4.

Finally, again all the images are rotated by 90°, 180°, and then both vertical and horizontal flip is applied on the total images. By this, a total of 7225 images are originated from the primary dataset (training) images.

4.3   Model Evaluation

Before initiating the enhanced SegNet model's training procedure, all the kernels used to perform convolutions at the encoder network are fixed to 3 × 3, and max-pooling is performed with a window size of 2 × 2. We trained the proposed enhanced SegNet model with training dataset and GTs and then tested with the two sets of test dataset images. The proposed enhanced SegNet model is evaluated using shape similarity and segmentation accuracy. The shape similarity can be measured by the Dice index [10], between ground truth (G) objects and the segmented (S) objects are shown in Eq. (2).

$$\text{Dice}(G, S) = \frac{2 * |G \cap S|}{|G| + |S|} \tag{2}$$

If the Dice index near to 1 indicates perfect segmentation, Dice (G, S) finds the accuracy at pixel level only and cannot calculate the segmentation accuracy at gland level. So the object-level Dice index is defined as in Eq. (3), as

$$\text{Dice}_{\text{obj}}(G, S) = \frac{1}{2} * \left[\sum_{i=1}^{n_G} \gamma_i * \text{Dice}(G_i, S_*(G_i)) + \sum_{j=1}^{n_S} \sigma_j * \text{Dice}(G_*(S_j), S_j) \right] \tag{3}$$

$$\gamma_i = |G_i| \bigg/ \sum_{p=1}^{n_G} |G_p|, \quad \sigma_j = |S_j| \bigg/ \sum_{q=1}^{n_S} |S_q| \tag{4}$$

From Eq. (4), G_i describes the ith ground truth object,
G_j indicates the jth SO.
$S * (G_i)$ denotes a GTO(G) that maximally overlaps G_i, and
$G * (S_j)$ describes an object that overlaps S_j at maximum.
Hausdorff distance [10] measures the shape similarity between the shape of S and G, given by Eq. (5) as

$$H(G, S) = \max \left\{ \sup_{x \in G} \inf_{y \in S} d(x, y), \sup_{y \in S} \inf_{x \in G} d(x, y) \right\} \tag{5}$$

In Eq. (5), Euclidean distance denoted by $d(x, y)$ between the pixels x of G and y of S. If the obtained Hausdorff distance is small, it indicates the maximum similarity between the contours of S and G. if S equals to G, which indicates the $H(G, S)$ is zero.

$$H_{\text{obj}}(G, S) = \frac{1}{2} * \left[\sum_{i=1}^{n_G} \gamma_i * H(G_i, S_*(G_i)) + \sum_{j=1}^{n_S} \sigma_j * H(G_*(S_j), S_j) \right] \tag{6}$$

Object wise Hausdorff distance is implemented as Eq. (5) to find the distance between object wise boundary-based segmentation accuracy. Object-level Dice and Hausdorff distance are used to evaluate the efficiency of the proposed model. Where D_{obj} is used to measure the segmentation accuracy, and H_{obj} is used to measure the shape similarity.

4.4 Experimental Setup

All the experiments were carried out on a PC with a 2.4 GHz Intel(R) I7-7th gen processor, CPU (32 GB of RAM), and NVIDIA GeForce Titan X GPU. The model execution has been conducted with Python 2.7 on TensorFlow [12]. Since we do not use a pre-trained model, we augmented the training data. The proposed architecture is trained by using the 7225 augmented images is conducted over 100 epochs.

4.5 Results and Analysis

After designing the network model, training was initiated using the Warwick-QU dataset. The images of the training data and images generated from data augmentation are fed to the training algorithm. However, CNN training consumes some time, but the features can be extracted accurately, compared to conventional textural methods. After completion of training, testing is performed on both parts of the test dataset to evaluate the proposed model's performance. After segmented the glands from the test images, we can differentiate the glands in benign and malignant images. The model is evaluated and compared with other models using the segmentation accuracy and shape similarity results. Table 3 describes various deep learning models' performance on the benchmark Warwick-QU dataset (Part A).

The quantitative results from Table 3 demonstrate that enhanced SegNet model outperforms all the other deep learning models. The segmentation accuracy and shape similarity on Part A have achieved 0.924 and 43.145, respectively. With regard to the object-level Dice score, our model achieves a better result.

Table 4 demonstrates various deep learning models' performance on the benchmark Warwick-QU dataset (Part B). Our proposed model achieved the segmentation

Table 3 Segmentation results on Part A

Method	$Dice_{obj}$	H_{obj}
Tang et al. [2]	0.882	106.647
Safiyeh Rezaei et al. [5]	0.897	45.418
Huijun Ding et al. [6]	0.913	39.848
Safiyeh Rezaei et al. [8]	0.872	55.51
Tianyu Shi et al. [9]	0.911	42.192
Our proposed model	0.924	83.145

Table 4 Segmentation results on Part B

Method	$Dice_{obj}$	H_{obj}
Jing Tang et al. [2]	0.863	102.572
Rezaei et al. [5]	0.822	108.208
Ding et al. [6]	0.858	93.244
Rezaei et al. [8]	0.82	109.72
Shi et al. [9]	0.843	89.677
Our proposed model	0.861	88.642

accuracy. Similarity on Part *B* has achieved 0.861 and 88.642, respectively, indicating that the proposed model outshined all the other advanced methods with respect to object-level Hausdorff distance. In summary, the proposed enhanced model has achieved a much better D_{obj} and H_{obj} on the whole.

5 Conclusion and Future Work

This paper has presented an enhanced SegNet-based pixel-wise semantic segmentation method for effective segmentation of the polyps (gland structures) in CRC pathology images. Initially, the training dataset for the enhanced SegNet model is acquired by data augmentation. Then, the parameters of the model were adjusted for optimized training results. Testing is performed on the trained model by using the two parts of the test dataset. And the model is evaluated by using the two metrics, object-level Dice index (D_{obj}) and object-level Hausdorff distance (H_{obj}). Our results proved that the proposed model could accomplish the highest shape similarity and segmentation accuracy on the whole. Our work additionally suggests that the advanced deep learning models can effectively segment the histology images' glands or polyps and help the pathologists diagnose the diseases.

References

1. https://www.cancer.org/content/dam/cancer-org/research/cancer-facts-and-statistics/annual-cancer-facts-and-figures/2020/estimated-number-new-cases-by-sex-and-age-group-2020.pdf. Accessed 06 Nov 2020
2. J. Tang, J. Li, X. Xu, Segnet-based gland segmentation from colon cancer histology images, in *Proceedings - 2018 33rd Youth Academic Annual Conference of Chinese Association of Automation, YAC 2018*, July 2018, pp. 1078–1082. https://doi.org/10.1109/YAC.2018.8406531
3. X. Yan, L. Wang, An effective cell pathology image detection method based on deep stacked auto-encoder combined with Random Forest. J. Phys. Conf. Ser. **1288**(1) (2019).https://doi.org/10.1088/1742-6596/1288/1/012004
4. A. Khvostikov, A. Krylov, I. Mikhailov, O. Kharlova, N. Oleynikova, P. Malkov, Automatic mucous glands segmentation in histological images, ISPRS Int. Arch. Photogramm. Remote Sens. Spat. Inf. Sci. **XLII-2/W12**(2/W12), 103–109, May 2019. https://doi.org/10.5194/isprs-archives-XLII-2-W12-103-2019
5. S. Rezaei et al., Gland segmentation in histopathology images using deep networks and hand-crafted features, in *Proceedings of the Annual International Conference of the IEEE Engineering in Medicine and Biology Society, EMBS*, July 2019, pp. 1031–1034. https://doi.org/10.1109/EMBC.2019.8856776
6. H. Ding, Z. Pan, Q. Cen, Y. Li, S. Chen, Multi-scale fully convolutional network for gland segmentation using three-class classification. Neurocomputing **380**, 150–161 (2020). https://doi.org/10.1016/j.neucom.2019.10.097
7. S. Rezaei, A. Emami, N. Karimi, S. Samavi, Gland segmentation in histopathological images by deep neural network, in *2020 25th International Computer Conference, Computer Society of Iran, CSICC 2020*, Jan 2020, pp. 1–5. https://doi.org/10.1109/CSICC49403.2020.9050084

8. T. Shi, H. Jiang, B. Zheng, A stacked generalization U-shape network based on zoom strategy and its application in biomedical image segmentation. Comput. Methods Programs Biomed. **197**, 105678 (2020). https://doi.org/10.1016/j.cmpb.2020.105678
9. B. Xu, N. Wang, H. Kong, T. Chen, M. Li, Empirical evaluation of rectified activations in convolution network. Accessed 16 Nov 2020. Available https://github.com/
10. K. Sirinukunwattana et al., Gland segmentation in colon histology images: the GlaS challenge contest (2016). Available https://www.warwick.ac.uk/bialab/GlaScontest
11. K. Sirinukunwattana, D.R.J. Snead, N.M. Rajpoot, A Stochastic Polygons model for glandular structures in colon histology images. IEEE Trans. Med. Imaging **34**(11), 2366–2378 (2015). https://doi.org/10.1109/TMI.2015.2433900
12. M. Abadi et al., TensorFlow: large-scale machine learning on heterogeneous distributed systems. Accessed 16 Nov 2020. Available www.tensorflow.org

Smart Cyclones: Creating Artificial Cyclones with Specific Intensity in the Dearth Situations Using IoT

G. Subbarao, S. Hrushikesava Raju, Lakshmi Ramani Burra, Venkata Naresh Mandhala, and P. Seetha Rama Krishna

Abstract Nowadays, there are certain occasions in the world to face scarcity of water due to lack of groundwater levels. During these modern days, everything is possible with smart and hard ideas. The proposed system is developing a simulated device with similar intensity and features of a cyclone. Incorporate the customized factors such as the direction of cyclone movement, and the amount of rainfall from beginning till weakening of the cyclone intensity. This model analyzes the various cyclone intensity and impact. The analysis helps to do a simulated device and that may be injected into the sea. The operating of it from a remote place helps for the creation of the artificial cyclone. The customization factors are provided at the time of generation of the cyclone. The artificial cyclone may be a little disaster to the society of a particular region but the scarcity of water will be removed. The result will be making groundwater levels improved and society to be benefited. The development of the simulated device will be done using suitable sensors embedded in it. The integration of predicted factors and their simulation over a device would help to create a virtual cyclone during water scarcity times namely dearth's time.

Keywords Pressure · Temperature · Sensors · Networked · IoT and dearth's

The original version of this chapter was revised: Author provided belated corrections have been incorporated. The correction to this chapter is available at https://doi.org/10.1007/978-981-16-1773-7_33

G. Subbarao (✉) · S. Hrushikesava Raju · V. N. Mandhala · P. Seetha Rama Krishna
Department of Computer Science and Engineering, Koneru Lakshmaiah Education Foundation, Vaddeswaram, Guntur, India
e-mail: hkesavaraju@kluniversity.in

L. R. Burra
Department of Computer Science and Engineering, PVP Siddhartha Institute of Technology, Kanuru, Vijayawada, India

S. K. Saha et al. (eds.), *Smart Technologies in Data Science and Communication*, Lecture Notes in Networks and Systems 210, https://doi.org/10.1007/978-981-16-1773-7_28

349

1 Introduction

The regions name the cyclones where hurricanes and typhoons in the northern hemisphere and cyclones in the southern hemisphere. The cyclones show their impact based on distance, size, and wind speed. There are many situations demanding the brought up of proposed system. As technology is upgrading, the scenarios are becoming simpler but it is cost oriented. For example, a baby also to be developed from the sperm through artificial approach but not achieved through the natural approach. There is also another kind to make rains to fall though artificial forming of clouds and getting merging. Similarly, anything is possible and nothing is impossible to get when there will be strong will to achieve it. Another example, communication becomes simple through the internet and through computers. Likewise, artificial cyclones are to be generated through latest technology called Internet of Things. For generating a cyclone, high temperature is found on the surface of the sea and should move to low-pressure regions. The information for generating the cyclones through technology available is provided and much features to be provided in necessary components. Integrate such components through Internet of Things. Here, the understanding of how a cyclone to be formed gives a backbone of the preparing the proposed system. At near the equator below the 5 degrees Celsius, cyclone will not be formed at any cost. When warm air is found on the surface of the sea, that warm air will move up makes surface of the sea is having the low-pressure air. This activity requires warm air to be brought at that surface of the water, and this warm air moving up. This process of moving the warm air up and bringing the warm air from surroundings cause low pressure on the surface of the sea. This creates the eye on the surface of the water and risen warm air makes clouds. This process creates carioles' impact toward right in the northern hemisphere and left in the southern hemisphere. This creates a wind spiral at the low-pressure area which creates a eye from which cyclone impact actually begins. The kinds of cyclones to be formed according certain factors are listed in Table 1. It lists more chaos cyclones to normal cyclones.

The important aspect is to create a warm air above 26.5 °C on the surface of the sea, which may cause cyclone. That warm air when moving up results sea surface is having low pressure. The cool air when moving downwards will become clouds. The surrounding areas of the sea surface extract the warm air and this air goes up results low pressure on the sea surface. This process creates wind spiral with specific features. Based on its factors, the impact may happen. Forming the cyclones is sometimes required in the future where dearth of water areas will be identified. Hence, this study became significant to avoid water level scarcity and improve greenery on the earth which avoids global warming also.

Table 1 Kinds of cyclones

Cyclone to be named	Wind speed (mph)	Damage	Surge surface (feet)
Category5 (Catastrophic)	>155	Too much	19+
Category4 (Extreme)	131–155	Less compared to category5	13–18
Category3 (Extensive)	111–130	Less compared to category4	9–12
Category2 (Moderate)	96–110	Less compared to category3	6–8
Category1 (Normal)	74–95	Less compared to category2	4–5

2 Literature Review

In this, the various cyclones are described and post-survey results are mentioned for preparedness of any upcoming cyclone. But the proposed system aims for generating own intensity cyclone. The detained steps and architecture are to be demonstrated in the proposed methodology. The studies related to cyclones are listed in the references and are described in here through their noted articles. According to source [1] mentioned by un-habitat, Myanmar, the manual on especially cyclones is developed and is released where the information from fundamental such as cyclone definition, how a cyclone to be created naturally, its impact on the regions, etc. are predicted. The concepts related to cyclones are described in this article but it will not tell how to create user oriented kind of cyclone. With respect to sources [2–4], the description is specified in terms of formation of tropical cyclones, its impact in terms of loss of human lives as well loss of property. This work acts as the supporting guide but not mentioned how to create customized cyclones during water dearth times. In the source [5], the description is about how the cyclone to be formed and categories of cyclones based on specific factor such as wind and its range of damage over the society. In the resource specified in [6], the information is provided in terms of cyclone formation, types of cyclones, and past history of cyclones that are occurred. In the source mentioned in [7], the description is based on cyclones and their impact since 21 years. That analysis is done based on remote sensing and spatial guided theory. This study helps to minimize the cyclone disaster management in the future. It is done on two major continents such as Asia and America. Using Analytical Hierarchy Process, the cyclone's risk management is done through spatial multi-criteria factors as well as based on mitigation capacity. In the resource information demonstrated in [8], the various kinds of cyclone structures such as axi-symmetric and asymmetric kinds, spiral rain-bands, concentric eye-wall cycle, annular hurricane structure, inner-core size of the tropical cyclones. This also summarizes the factors that limit the maximum intensity of tropical cyclones. In the resource mentioned in [9], the description is covering vortex boundary changes such as unsteady cat's eye flow, more complex interaction of multiple waves, vortices, and shearing of background leads to a method called topological rearrangement. Here, there is an agreement exists between evolution of circulation and time integral of time tendency of circulation. This leads to same shape by material curves but lobes are transported across the vortex boundary. The transport of the boundary is measured by advective and non-advective fluxes across the time. The description given in [10] would demonstrate on comparison between ANN technique and ERA interim ECMWF that proves former method is good based on KI and DCI parameters. This analysis is done over Anakapalli, Vishakhapatnam district in the period between 2001 and 2010 in order to know thunderstorm details during the pre-monsoon's season. In the data given in [11], the numerical methods were applied on the data obtained from radiances of ATOVS and ATMS as well as with and without assimilation of scattero-meter winds. Here, the methods used for predicting the cyclone so that disaster to be known in advanced based on the intensity of the cyclone formed. It proved that certain method is best prediction with least errors for cyclones less than 100 km and other methods are on the good track for cyclones but with errors of 150–200 km intensity. In the information provided in [12], the demonstration is on various attenuation models that would predict the rainfall and its impact in the two cities Hyderabad and Visakhapatnam over a KU band scale from 10.99 to 14.2 GHz. Here, comparison is done over the different models in the prediction of rainfall and its impact. In [13] mentioned, the factors to be predicted such as vertical rain rate, liquid water content, path integrated attenuation, as well as fall velocity is up to 4.4 km in the atmosphere with respective rain intensity. In the source defined w.r.to [14], the parameters of the strati-form and convective rain fall are predicted over a Ka band scale with respect to location 16.24 latitude, and 80.45 longitude. This study is taken for four years where year is considered as a component. Hence, the rainfall measure and the cyclone predictions are demonstrated in the existing studies and also they guided the fundamentals of the cyclones. These studies are useful for analysis and aware of various details about the cyclone formation, cyclone movement, the impact of the cyclone based on the type of it, and other details. This might help to take measures towards the loss of lives as well as property. All these studies show losses and damages but the inherent factor considered is water is surplus and levels of water in the ground to be increased. Hence, when normal rail-fall is not happening for 1–2 years, the people and all living animals, as well as plants, must have to face big challenges. At that time, artificial cyclones play a vital role although they cause loss and damage. The smart cyclones are nothing but creating cyclones artificially according to reinforcement knowledge using the Internet of Things.

3 Implementation

The proposed methodology is demonstrated though a defined architecture where modules are available and are described with respect to flowcharts and pseudocodes. In the proposed approach, the sensors with wide range capability are designed for making surface of the sea's temperature above more than 26.5 °C and attracting the high-pressure wind towards lower pressure area. This will cause the cyclone to be formed.

The following is the architecture of the smart cyclones where modules are specified in rectangular boxes and their activities in ovals, arrows are used for connecting activities and modules (Fig. 1).

In this, two modules used are such as wide range temperature sensor and versatile pressure sensor. In this, temperature sensor is first module search for few locations based on past locations from which old cyclones were started. These locations are base for temperature sensor in terms of longitude and latitude. Based on the past cyclone impact and the factors, new cyclone to be generated by triggering the temperature sensor over the possible location of the sea. At that location, temperature sensor is useful for not only reading the atmospheric temperature but also for making the specified area more hot which is above 26.5 °C. These activities are mentioned in Table 2.

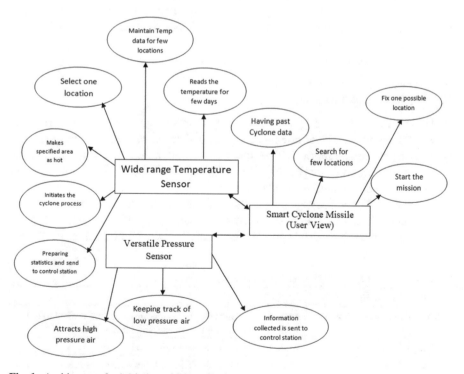

Fig. 1 Architecture for initiation of Smart Cyclone

Table 2 Activities of module named wide range temperature sensor

Name of the activity	Purpose
Reading temperature	Its main objective is to read atmospheric temperature for certain days, identify the change in temperature is possible from read data and return yes if case of change in temperature
Maintain temperature data for certain locations	User limits number of specific locations. Based on the given locations, temperature is recorded daily and generates a report for a week
Select one location	User selects the location based on reports of the locations
Activate heat temperature sensor	Once the location is fixed, activate hot sensor in the temperature module which increases the temperature more than 26.5 °C
Prepare statistics and send info to control center	Statistics of the location about the volume of temperature that turned to hot climate on the sea surface

Similarly, the second module such as versatile pressure sensor when finds the low-pressure air over the sea surface would record the statistics happening on the surface of the sea. If the surrounding air is found having the high pressure, would be naturally attracted and the designated pressure sensor also helps to attract the winds of HP to the spotted LP location. The activities used for second module are mentioned in Table 3.

Hence, the modules identified in this proposed approach are:

A. Wide range temperature sensor
B. Versatile pressure sensor

These two modules long with necessary components and resources such as network oriented and power oriented are provided in the customized missile prototype. Like preparing a missile to the space or making a nuclear missile for testing and for achieving some target, this kind of missile with low cost will be prepared and sent to selected longitude and latitude.

The activities of wide range temperature sensor are shown in Table 2.

The activities of versatile pressure sensor are shown in Table 3.

The flow chart of proposed approach is defined as shown in Fig. 2.

Table 3 Activities of module named versatile pressure sensor

Name of the activity	Purpose
Air from high-pressure area	Automatically air from high-pressure areas are moved to low pressure area. If not, this sensor will attract such air
Tacking of low pressure area	Volume of low pressure air is tracked
Sending of statistics report	The details such as air from both pressure areas are tracked in a report and communicated to the control center

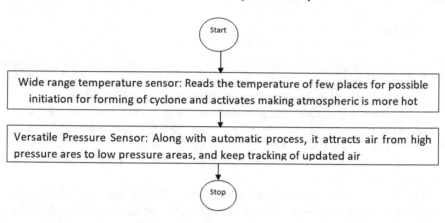

Fig. 2 Flow chart of proposed approach

The pseudocode of proposed approach is defined as follows for initiating the cyclones:

Pseudo_Procedure(location,   cut_off_temp,   past_history_cyclone_data, towards_direction):

Step1: Locate longitude and latitude for the missile to travel over the sea based on past cyclone locations.

Step2: Call Wide_range_temperature_sensor(cut_off_temp, locations_suitable[]):

2.1 Missile would travel to the specified location over the sea surface

2.2 Search for possible locations that support change in temperature based on certain days (By default one week)

2.3 Location of maximum varying temperature during that observed week is selected by the user or automatically by the sensor

2.4 if atmospheric_temp<cut_off_temp:

Activate normal temperature sensor over a specific range

else:

Track the volume of the warm air temperature and submit the report

Step3: Call Verstile_Pressure_sensor(towards_direction):):

3.1 Fixes the direction to which the cyclone to be sent using GPS

3.2 Activates attracting the air of high pressure area

3.3 Tracking of low pressure air and sending the report

4 Results

The customized featured cyclone is influenced by certain factors. Also, few cyclones with specific intensity are to be listed along with their impacts. The following is the list of significant cyclones occurred over Indian states in 2020 (Table 4).

There were many other cyclones such as Nivar, Burevi, BOB3, ARB3, etc. occurred but their intensity is very less compared to category 3 and 5. The cyclones happened in 2019 are Fani, Hikka, and vayu are significantly considered.

The cyclone performance is purely depends on the range of pressure created during the cyclone formation.

The following is the graph of the cyclones that affect the landfall based on wind peak over landfall and pressure range over the surface of the sea (Fig. 3; Table 5).

Table 4 Significant cyclones in the 2020

Cyclone name	States over cyclone passed	Wind speed (peak)	Pressure range	Type of cyclone	Remarks
Super cyclonic storm: Amphan	Kerala, Srilanka	240 kmph (150 mph)	920 hPa (27.46 in Hg)	Catastrophic damage	Occurred in may, 2020
Very severe cyclonic storm Gati	Ras Binnah, Somalia	140 km/h (85 mph)	978 hPa (mbar)	Extensive damage	November, 2020

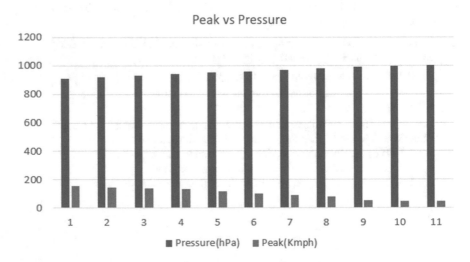

Fig. 3 Performance of cyclones based on pressure and peak factors

Table 5 Data sets for pressure and peak factors

Pressure (hpa)	910	920	930	940	950	960	970	980	990	995	1000
Peak (kmph)	155	145	140	135	120	100	90	80	55	48	45

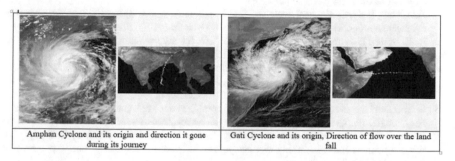

Amphan Cyclone and its origin and direction it gone during its journey	Gati Cyclone and its origin, Direction of flow over the land fall

Fig. 4 Significant cyclones over Indian states in 2020

The results vary based on pressure created on the surface of the sea and spiral wind strength and eye strength formed for that cyclone. The following are few snapshots of few significant cyclones occurred (Fig. 4).

5 Conclusion

There are times in the history, may rise dearth's for water. Such situations, the importance of artificial cyclones plays a vital role to avoid water scarcity and overcome from damage of many significant animals' existence over the earth. Hence, a user-friendly missile that consists of supporting sensors that may create a situation over the sea in order to generate a cyclone toward a specific direction. The sensors used here are advanced temperature sensor and pressure sensors with past cyclone histories are injected as a base. Once selected the possible suitable location, read temperature is made more hot above 26.5°, which causes bringing high-pressure area air to low-pressure area would cause wind spirals. The repeated process leads for the cyclone to be formed in certain duration and the resources in the hand crafted missile are networked. This will send track the necessary bases and communicates the statistics of the cyclone. Here, for activating few sensors and communicating the tracked information to the control center, Internet of Things is used and that plays a key role in this work.

References

1. Un-Habitat, Myanmar, Manual on Cyclone, Causes, Effects, and Preparedness, Mar 2015, https://unhabitat.org.mm/wp-content/uploads/2015/03/MANUAL-ON-CYCLONE-Causes-Effects-Preparedness_English.pdf
2. A.H. Fink, P. Speth, Tropical cyclones. Sci. Nat. **85**(10), 482–493, Jan 1998. https://www.res earchgate.net/publication/226073560_Tropical_Cyclones
3. R. K. Smith, Lectures on tropical cyclones (2006). https://www.meteo.physik.uni-muenchen.de/~roger/Lectures/Tropical_Cyclones/060510_tropical_cyclones.pdf
4. William M. Gray, Chapter 10, Tropical cyclones, The COMET® Program, Introduction to Tropical Meteorology, March, 2009, Version 1.3, pp. 1–248. https://www.meteo.physik.uni-muenchen.de/~roger/Mtheory/Ch10_Tropical_Cyclones.pdf
5. Storms and Cyclones, Cyclone Formation, Natural Disaster Management. April, 2021, https://sites.google.com/site/disasterportal/stroms_cyclones/cyclone formation
6. S. Anwar, Cyclone-formation: types and cyclone prone areas in India, May 2020. https://www.jagranjosh.com/general-knowledge/cyclone-formation-types-and-cyclone-prone-area-in-india-1556882714-1
7. M. Al-Amin Hoque, S. Phinn, C. Roelfsem, A systematic review of tropical cyclone disaster management research using remote sensing and spatial analysis, Sept 2017. https://doi.org/10.1016/j.ocecoaman.2017.07.001
8. Y. Wang, Recent research progress on tropical cyclone structure and intensity, May 2012. https://doi.org/10.6057/2012TCRR02.05
9. B. Rutherford, T.J. Dunkerton, Finite-time circulation changes from topological rearrangement of distinguished curves and non-advective fluxes. Trop. Cyclone Res. Rev. **9**(1) (2020). https://doi.org/10.1016/j.tcrr.2019.05.001
10. N. Umakanth, G.C. Satyanarayana, B. Simon, M.C. Rao, M.T. Kumar, N.R. Babu, Analysis of various thermodynamic instability parameters and their association with the rainfall during thunderstorm events over Anakapalle (Visakhapatnam district), India, Sept 2020. https://doi.org/10.1007/s11600-020-00478-1
11. V.B. Dodla, D. Srinivas, H.P. Dasari, C.S. Gubbal, Prediction of tropical cyclone over North Indian ocean using WRF model: sensitivity to scatterometer winds. ATOVS ATMS Rad. (2016). https://doi.org/10.1117/12.2223615
12. B.J. Philip, K.S. Kumar, K.C. Sri Kavya, C. Susmitha, C.N. Rao, S. Vasu, A. Madhulika, Comparison of different rain attenuation prediction models at Visakhapatnam and Hyderabad regions. J. Adv. Res. Dyn. Control Syst. 9(4), 220–229 (2019)
13. A. Kilaru, S.K. Kotamraju, N. Avlonitis, K. C. Sri Kavya, Vertical structure observations of precipitation using micro rain radar over indian region. Int. J. Innov. Technol. Explor. Eng. **8**(5), 1003–1007 (2019)
14. A. Kilaru, S. Kumarkotamraju, K.C. Sri Kavya, Stratiform and convective rain intensity effects on ka band links. Int. J. Innov. Technol. Explor. Eng. **8**(5), 146–149 (2019)

Machine Learning-Based Application to Detect Pepper Leaf Diseases Using HistGradientBoosting Classifier with Fused HOG and LBP Features

Matta Bharathi Devi and K. Amarendra

Abstract Pepper leaf disease detection is one of the interesting challenges in the field of machine learning. In this chapter, we propose a machine learning-based approach to extract texture features and use dimensionality reduction technique called Principal Component Analysis (PCA) and create composite feature descriptor. We use two different texture-based feature representations extracted by using HOG and LBP feature engineering techniques, from pepper leaf images and apply PCA to get reduced representations. These representations are fused and passed to machine learning models like Logistic Regression, Naïve Bayes, Decision Tree, Support Vector Machine, and HistGradientBoosting Classifier for classification. HistGradientBoosting Classifier achieved highest accuracy of 89.11% and outperformed other models.

Keywords Histogram of oriented gradients (HOG) · Local binary pattern (LBP) · Principal component analysis (PCA) · HistGradientBoosting Classifier (HGB) · Machine learning

1 Introduction

Detecting plant leaf diseases is one of the major challenges faced by farmers in agriculture. It is very important to identify the type of leaf diseases accurately for appropriate use of pesticides. Any mistakes in identifying diseases of plants leads to reduced yield. Plant diseases can be either biotic [1, 2] or abiotic. Primary cause behind the biotic diseases are various living organisms like bacteria, virus, and fungi. Biotic diseases are affected by viruses unlike abiotic diseases which are affected by inorganic conditions like weather changes, chemicals, etc. Identifying leaf diseases

M. B. Devi · K. Amarendra (✉)
Department of Computer Science and Engineering, Koneru Lakshmaiah Education Foundation,
Guntur, Andhra Pradesh, India
e-mail: amarendra@kluniversity.in

© The Author(s), under exclusive license to Springer Nature Singapore Pte Ltd. 2021
S. K. Saha et al. (eds.), *Smart Technologies in Data Science and Communication*,
Lecture Notes in Networks and Systems 210,
https://doi.org/10.1007/978-981-16-1773-7_29

accurately by observing with naked eye is a difficult task. Hence, there is a requirement of an application that can detect leaf diseases accurately. There are various automated applications to identify plant leaf diseases. Most of them used texture representations extracted from leaf images with conventional machine learning models [3–5].

Most of the recent works in literature used feature extraction techniques like Histogram of Oriented Gradients (HOG), Local Binary Patterns (LBP), Gray-Level Co-occurrence Matrix (GLCM) are used in literature to extract texture-based features from plant leaf images [6, 7]. These features were fed to popular classifiers like Support Vector Machine (SVM) to categorize different types of diseased leaves [8, 9]. However, using these features directly with ML models results in reduced performance. So, in this work we investigate to reduce the dimensions of texture features and blend them to get composite representation with pepper leaf dataset [10–12].

Initially, we performed necessary pre-processing to remove background noise obtained during image acquisition. Later, we extracted two types of texture-based features from pepper leaf images using HOG and LBP feature engineering methods and applied Principal Component Analysis (PCA) dimensionality reduction technique to get reduced representations of HOG and LBP features. In our experiments, we observed that HOG features are better than LBP. Using reduced representations lead to improve performance. When LBP features are fused with HOG, composite representations are obtained. These representations contain more discriminant information which help classification models to identify pepper leaf diseases accurately. Our proposed fused representation achieved a highest accuracy of 89.11% with HGB Classifier.

2 Related Work

This part of the chapter provides an overview of various methodologies employed for detecting plant leaf diseases in past. First part of this section describes various pre-processing techniques used in literature followed by feature engineering methods algorithms that are used for classification.

In recent past, several pre-processing techniques have been applied on plant leaf images to correctly identify the type of plant diseases. Most of the previous works used image processing techniques and applied smoothing, sharpening filters the enhance the image and used several filters to remove additive noise from the images [13, 14]. ROI segmentation is major task employed to detect and segment diseased portions from images to improve the performance of automated plant leaf disease diagnosis systems [15–17].

Texture-based features obtained from images play a vital role and effects the performance of image classification systems. Histogram of Oriented Gradients (HOG), GIST, Scale Invariant Feature Transform (SIFT), and Local Binary Patterns (LBP) are majorly employed feature engineering algorithms to obtain intensity and texture-based features [6, 18, 19]. These feature engineering methods are employed in various tasks like medical image classification, scene classification, object recognition, and leaf disease identification [20–23]. Most of machine learning algorithms like K-Nearest Neighbor Classifier (K-NN), Decision Tree Classifier, Random Forest, Support Vector Machine (SVM), and Naïve Bayes Classifier are trained on these texture-based features for classification purpose [8].

A K-Nearest Neighbor (K-NN) Classifier with Gray-Level Co-occurrence Matrix (GLCM) texture features of plant leaf images were used to identify plant leaf diseases [24, 25]. Another machine learning-based system was proposed for grapes plant leaf disease detection by Harshal Waghmare et al. First, background of all images was removed and segmentation is performed as a pre-processing step. A high-pass filter is applied on segmented images to analyze disease part of the leaf. Local Binary Patterns (LBP)-based texture features are extracted from pre-processed images and these features were used to identify different types of grape plant diseases using Support Vector Machine (SVM) Classifier [20, 26]. A cotton leaf disease detection and classification technique based on machine learning and image processing tools are proposed by Pooja et al. Initially, Region of Interest (ROI) is segmented from plant leaf images using image processing tools and features are extracted. These features were passed to SVM Classifier to identify the type of disease [27, 28].

In this work, we use HOG and LBP feature extracted from pepper leaf images and fuse them to create composite representation. These representations are then projected to lower dimension using PCA. Then we apply various popular classification algorithms like Logistic Regression, Naïve Bayes Classifier, Decision Tree Classifier, and Support Vector Machine (SVM) with linear and Radial Basis Function (RBF) kernel and HistGradientBoosting Classifier for classification purpose.

3 Proposed Methodology

This part of the chapter provides an illustration of various stages of proposed method for pepper leaf disease detection. Our proposed work consists of four phases, followed one after other. They are data pre-processing, feature extraction, dimensionality reduction and classification.

3.1 Data Pre-processing

Data acquired from real world consist of random noise in background. So, background subtraction is performed on pepper leaf images to remove random background noise. This is done by creating suitable mask for every image present in dataset and then background removal operation is performed by using corresponding masks. Figure 1a represents images from original dataset and Fig. 1b represents background removed images. These processed images are passed to feature extraction phase.

(a) Pepper leaf images with background noise

(b) Pepper leaf images after background removal

Fig. 1 Pepper leaf images before and after pre-processing

3.2 Feature Extraction

Feature extraction is an important phase in any machine learning task. In our work, we use two different feature extraction techniques, Histogram of Oriented Gradients (HOG) and Local Binary Patterns (LBP) which extracts texture-based features from pepper leaf images.

3.2.1 Feature Extraction from Pepper Leaf Images Using HOG

The HOG feature descriptor counts the occurrences of gradient orientation in localized portions of an image. Initially, all processed images of dimension (256×256) are reshaped to (64×128) dimensions. Next, changes in X and Y directions of images (gradients) are computed by dividing the entire image into (8×8) patches. Next, magnitude and orientations are computed by using gradients. Then, Histogram of Gradients are calculated for each (8×8) cells and these cells are combined to create (16×16) cells. The gradients of these cells are normalized to get a vector of (1×36) dimension for each cell. Finally, for every image of dimension (64×128) we get a feature vector of 3780 dimensions. This feature descriptor is normalized using min-max normalization method. Figure 2 represents Histogram of Oriented Gradients computer for a given pepper leaf image.

3.2.2 Feature Extraction from Pepper Leaf Images Using LBP

Local Binary Patterns (LBPs) compute texture features from local regions instead of computing global texture features as in the case of Gray-Level Co-occurrence Matrix (GLCM). Initially, all processed images of dimension (256×256) are reshaped to

Fig. 2 Histogram of gradients for a given input pepper leaf image

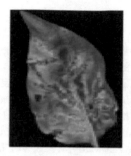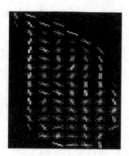

(128 × 128) dimensions. Next, all these images are converted to gray scale. LBP histogram is obtained from those images by appropriately selecting p and r values, where p represents the number of points in neighborhood of a pixel and r is the radios. Finally, for every image of dimension (128 × 128) we get a feature vector of 26 dimensions. This feature descriptor is normalized using min-max normalization method. We used OpenCV module of python to extract LBP features.

3.3 Dimensionality Reduction

In this phase, all the features of dimension 3780 obtained from HOG feature extraction technique and 26 dimensions obtained from LBP are projected into lower dimensional space with 512 and 13 dimensions for HOG and LBP, respectively. For this, we used Principal Component Analysis. In the case of limited data, high-dimensional features may lead to curse of dimensionality. To resolve this problem, we included this module in our work. Figure 3 represents the architecture of proposed method for pepper leaf disease detection.

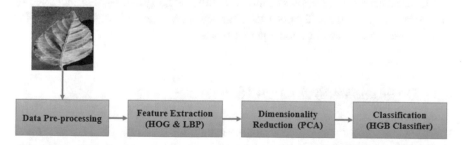

Fig. 3 Architecture of proposed system for pepper leaf disease detection

3.4 Classification

In this work, we used popular classification models like Logistic Regression, Naïve Bayes Classifier, Decision Tree Classifier, Support Vector Machine (SVM) with linear and Radial Basis Function (RBF) kernel and HistGradientBoosting Classifier. We test the performance extracted features before and after applying dimensionality reduction. We observed that SVM with RBF kernel and HistGradientBoosting kernels perform better than other classifiers for both HOG and LBP features in both the cases of dimensionality reduction. Finally, we fused HOG and LBP features to form a composite feature representation of dimension 3806. With these features HistGradientBoosting classifiers achieved a highest accuracy of 89.11% and outperformed all other models.

4 Proposed Methodology

This section provides clear picture of experiments conducted and results obtained using proposed method. First overview of dataset used for experiments is described followed by evaluation metrics used to measure the performance of proposed method. Finally, a summary of experiments and their results is provided.

4.1 Pepper Leaf Disease Dataset

We used Pepper Leaf Disease dataset, part of Plant Village dataset which contains 54,306 samples of 26 type of diseased leaf images belonging to 14 types of plant species. This dataset contains 2475 samples of pepper plant representing both healthy and diseased leaves. 997 samples belong to healthy category and 1478 samples belong to bacterial spot category. Totally 1980 samples are considered for training and 495 samples are used for testing model performance.

4.2 Performance Evaluation Measures

Different classification model performance evaluation measures like accuracy, precision, recall, $f1$ score are calculated to prove the efficiency of proposed method on test data. These measures can be computed using Confusion Matrix.

4.3 Result Analysis

We conducted our experiments in three different ways to check the performance of classification models with HOG, LBP, and fused features before and after applying PCA for pepper leaf disease detection.

4.3.1 Experiments Without Dimensionality Reduction

These experiments are conducted to check how well HOG and LBP features can detect pepper leaf diseases before dimensionality reduction.

From Table 1, it is clear that HistGradientBoosting Classifier outperformed other ML by achieving accuracy of 84.81%. Comparatively, Decision Tree Classifier could not perform well. SVM with RBF kernel also obtained an accuracy of 83.06% which is second highest measure. From Table 2, we can observe that HGB classifier achieved 83.87% accuracy with LBP features and Naïve Bayes classifier obtained lower accuracy. From both experiments we can conclude that HOG features perform better than LBP features for the task of pepper leaf disease detection and HGB classifier out performed all other models used, with both types of features.

Table 1 Performance of different ML algorithms with HOG features

Model	Accuracy	Precision	Recall	$F1$ Score
Logistic Regression	75.4	75	75	75
Naïve Bayes	70.97	78	71	66
Decision Tree	67.74	67	68	67
SVM—Linear	77.82	78	78	78
SVM—RBF	83.06	84	83	83
HGB classifier	84.81	85	84	85

Table 2 Performance of different ML algorithms with LBP features

Model	Accuracy	Precision	Recall	$F1$ score
Logistic Regression	80.24	80	80	80
Naïve Bayes	67.74	69	68	68
Decision Tree	72.58	73	73	73
SVM—linear	81.05	81	81	81
SVM—RBF	81.85	82	82	82
HGB Classifier	83.87	84	84	84

Table 3 Performance of different ML algorithms with HOG features

Model	Accuracy	Precision	Recall	F1 Score
Logistic Regression	73.39	73	73	73
Naïve Bayes	76.21	76	76	76
Decision Tree	62.5	62	62	62
SVM—linear	77.98	79	78	79
SVM—RBF	84.27	84	84	84
HGB Classifier	85.47	85	85	85

Table 4 Performance of different ML algorithms with LBP features

Model	Accuracy	Precision	Recall	F1 Score
Logistic Regression	79.44	79	79	79
Naïve Bayes	82.26	82	82	82
Decision Tree	77.02	77	77	77
SVM—linear	80.24	80	80	80
SVM—RBF	83.47	83	83	83
HGB Classifier	84.27	84	84	84

4.3.2   Experiments After Applying Dimensionality Reduction

These experiments are conducted to check how well HOG and LBP features can detect pepper leaf diseases after dimensionality reduction (Tables 3 and 4).

From previous experiments, it is clear that after applying PCA, there is significant improvement in the performance of classification models with both, HOG and LBP features. HGB classifier followed same trend and outperformed other classification models with both HOG and LBP features after applying PCA. Even after reduced dimension, there is significant improvement in all measures used to test efficiency of models.

4.3.3   Experiments with Fused Features of HOG and LBP

This experiment is conducted to check the performance of ML models with composite representation obtained after blending LBP features with HOG features.

From Table 5, it is clear that HGB classifier trained on fused feature descriptor obtained 89.11% accuracy which is highest when compared with the performance of same classifier trained on HOG and LBP features before and after applying PCA. So, we conclude that, fused texture representations of pepper leaf images help to identify diseases accurately rather than using conventional usage of LBP and HOG features.

Table 5 Performance of different ML algorithms with fused HOG & LBP features

Model	Accuracy	Precision	Recall	$F1$ Score
Logistic Regression	80.24	80	80	80
Naïve Bayes	75.81	76	76	76
Decision Tree	69.35	70	69	69
SVM—linear	81.05	81	81	81
SVM—RBF	88.71	89	89	89
HGB Classifier	89.11	89	89	89

5 Conclusion

Pepper is most commonly used ingredient in dishes. Identifying pepper leaf diseases is a challenge for formers. There is a high requirement to automate the process of detecting pepper leaf diseased for correct usage of pesticides and reduce loss of yield. In this paper, we investigated the performance of various classification models with two different types texture-based features. During our experiments, we observed that models are able to perform well with reduced representations of HOG and LBP features rather than using them directly. We also observed that fused representation of HOG and LBP features helped the models to perform well, and there is 4% improvement in accuracy with fused features. In our experiments, we also observed that HGB classifier outperforms other ML algorithm in every case.

References

1. Z.B. Husin, A.H.B.A. Aziz, A.Y.B.M. Shakaff, R.B.S.M. Farook, Feasibility study on plant chili disease detection using image processing techniques, in *IEEE 3rd International Conference on Intelligent System Modeling and Simulation ISMS, Kota Kinabalu*, pp 291–296 (2012)
2. G. Balakrishna, N.R. Moparthi, Study report on indian agriculture with IoT. Int. J. Electr. Comput. Eng. **10**(3), 2322 (2020)
3. J.G.A. Barbedo, A review on the main challenges in automatic plant disease identification based on visible range images. Biosyst. Eng. **144**, 52–60 (2016)
4. K. Mannepalli, P.N. Sastry, M. Suman, Accent recognition system using deep belief networks for Telugu speech signals. Int. J. Speech Technol. **19**(1), 87–93 (2017)
5. T. Rajesh Kumar, G.R. Suresh, S. Kanaga Suba Raja, Conversion of non audible murmur to normal speech based on full-rank Gaussian mixture model. J. Comput. Theor. NanoSci. 1546–1955 **15**(1), 185–190 (2018)
6. S. Mahapatra, S. Kannoth, R. Chiliveri, R. Dhannawat, Plant Leaf Classification and disease recognition using svm, a machine learning approach. Sustain. Humanosph. **16**(1), 1817–1825 (2020)
7. Ayushree, G.N. Balaji, Comparative analysis of coherent routing using machine learning approach in MANET. Smart Comput. Inf. 731–741 (2018)

8. M. Bhagat, D. Kumar, I. Haque, H.S. Munda, R. Bhagat, Plant leaf disease classification using grid search based SVM, in *2nd International Conference on Data, Engineering and Applications (IDEA)*. IEEE, Feb 2020, pp. 1–6

9. G.D. Puri, D. Haritha, Framework to avoid similarity attack in big streaming data. Int. J. Electr. Comput. Eng. **8**(5), 2920–2925 (2018)

10. S. Anjali Devi, S. SivaKumar, Comprehensive survey on sentiment analysis based on workflow foundation. J. Adv. Res. Dyn. Control Syst. **10**(9 Special Issue), 1189–120 (2018)

11. T. Rajesh Kumar, T. Vamsidhar, B. Harika, T. Madan Kumar, R. Nissy, Students performance prediction using data mining techniques, in *IEEE Explorer (ICISS-2019)*, 978-1-5386-7798-8 (2019)

12. V. Talasila, T. Rajesh Kumar, C.P. Sai, S. Satya Sai, Ayyappa, Predicting the risk of heart failure with EHR sequential data modelling. Int. J. Recent Technol. Eng. (IJRTE), 2277–3878, **6**(7), 458–461 (2019)

13. A. Asfarian, Y. Herdiyeni, A. Rauf, K.M. Mutaqin, Paddy diseases identification with texture analysis using fractal descriptors based on Fourier spectrum, in *IEEE International Conference on Computer, Control, Informatics and Its Applications IC3INA*, Jakarta, pp. 77–81 (2013)

14. H.S.A. Bommadevara, Y. Sowmya, G. Pradeepini, Heart disease prediction using machine learning algorithms. Int. J. Innov. Technol. Explor. Eng. **8**(5), 270–272 (2019)

15. S.D. Khirade, A.B. Patil, Plant disease detection using image processing, in *IEEE International Conference on Computing Communication Control and Automation (ICCUBEA)* (2015), pp. 768–771

16. T. Rajesh Kumar, G.R. Suresh, S. KanagaSubaraja, C. Karthikeyan, Taylor-AMS features and deep convolutional neural network for converting non-audible murmur to normal speech. Comput. Intell. 1–24 (2020)

17. B. Dudi, V. Rajesh, An efficient algorithm for medicinal plant recognition. Int. J. Pharm. Res. **10**(3), 87–93 (2018)

18. R. Patil, S. Kumar, A bibliometric survey on the diagnosis of plant leaf diseases using artificial intelligence (2020)

19. B. Dudi, V. Rajesh, Medicinal plant recognition based on cnn and machine learning. Int. J. Adv. Trends Comput. Sci. Eng. **8**(4), 628–631 (2019)

20. J.D. Bodapati, N. Veeranjaneyulu, S.N. Shareef, S. Hakak, M. Bilal, P.K.R. Maddikunta, O. Jo, Blended multi-modal deep convnet features for diabetic retinopathy severity prediction. Electronics **9**(6), 914 (2020)

21. V. Dondeti, J.D. Bodapati, S.N. Shareef, V. Naralasetti, Deep convolution features in non-linear embedding space for fundus image classification deep convolution features in non-linear embedding space for fundus image classification, June 2020, pp. 307–313. https://doi.org/10.18280/ria.340308

22. J.D. Bodapati, N.S. Shaik, V. Naralasetti, N.B. Mundukur, Joint training of two-channel deep neural network for brain tumor classification. Signal Image Video Process. 1–8 (2020)

23. J.D. Bodapati, N. Veeranjaneyulu, S. Shaik, Sentiment analysis from movie reviews using LSTMs. Ingénierie Des Systèmes D Inf. **24**(1), 125–129 (2019)

24. J. Trivedi, Y. Shamnani, R. Gajjar, Plant leaf disease detection using machine learning, in *International Conference on Emerging Technology Trends in Electronics Communication and Networking*. Springer, Singapore, Feb 2020, pp. 267–276

25. S. Inthiyaz, M.V.D. Prasad, R.U.S. Lakshmi, N.S. Sai, P.P. Kumar, S.H. Ahammad, Agriculture based plant leaf health assessment tool: A deep learning perspective. Int. J. Emerg. Trends Eng. Res. **7**(11), 690–694 (2019)

26. M. Anila, G. Pradeepini, Study of prediction algorithms for selecting appropriate classifier in machine learning. J. Adv. Res. Dyn. Control Syst. **9**(Special Issue 18), 257–268 (2017)

27. H. Waghmare, R. Kokare, Y. Dandawate, Detection and classification of diseases of grape plant using opposite colour local binary pattern feature and machine learning for automated decision support system, in *2016 3rd international conference on signal processing and integrated networks (SPIN)*. IEEE Feb 2016, pp. 513–518

28. M.N. Shariff, B. Saisambasivarao, T. Vishvak, T. Rajesh Kumar, Biometric user identity verification using speech recognition based on ANN/HMM. J. Adv. Res. Dyn. Control Syst. **9**(12 Special issue), 1739–1748 (2017)

Tracking Missing Objects in a Video Using YOLO3 in Cloudlet Network

M. Srilatha⊙, N. Srinivasu⊙, and B. Karthik

Abstract In real time, people are using CCTV for monitoring activities continuously but sometimes theft is taking place. In this scenario, people need to roll back the video and need to identify when it happened. But in practice, it is a difficult and time-consuming process to identify a missing object in the video within the local machine because of the lack of computing resources. To solve this problem, we are presenting an algorithm in this paper to notify missing objects in a video offloaded from a mobile or CCTV using YOLO3 object detection in a cloudlet network. In the area of cloud computing, a cloudlet is a data center in the local network with a rich set of computing resources available for mobile users.

Keywords YOLO · Object detection · Missing object · Tracking video · Cloudlet · Cloud computing

1 Introduction

The video surveillance systems should have security and its requirements are increasing day by day. The normal human-based surveillance systems are not enough to track suspicious behavior. In paper [1], the authors summarized various methods available for detecting objects, classification of objects, and tracking the movements of an object in the video. Each method listed in this paper at various phases of monitoring and tracking a video requires a lot of computing resources. This problem we are overcoming by offloading computation onto the cloudlet.

M. Srilatha · N. Srinivasu (✉)
Department of CSE, Koneru Lakshmaiah Education Foundation, Guntur, India
e-mail: srinivasu28@kluniversity.in

M. Srilatha
VR Siddhartha Engineering College, Vijayawada, India

B. Karthik
Department of Mechanical Engineering, VR Siddhartha Engineering College, Vijayawada, India

© The Author(s), under exclusive license to Springer Nature Singapore Pte Ltd. 2021   371
S. K. Saha et al. (eds.), *Smart Technologies in Data Science and Communication*,
Lecture Notes in Networks and Systems 210,
https://doi.org/10.1007/978-981-16-1773-7_30

In the area of video surveillance, very few number papers are available on the identification of missing or suspected objects. The authors in paper [2] presented an algorithm for detecting missing objects and unattended objects in a video. Here the authors proposed a missing object rule for identifying the missing object by calculating variations in the current edges with recorded trained edges of a static object in a video. Next, another rule is also specified in this paper to detect an unattended object by comparing current edges and with recorded edges. With this rule, if current object edges are not matched with any one of the training edges then it results in an unattended object. Both the rules are implemented by tracking the center mass point of an image with the recorded central pixel of an object. Concerning this paper, we are proposing an efficient algorithm for detecting objects using YOLO for better results.

Nowadays you only look once (YOLO) [3] is used by researchers for detecting real-time objects efficiently. According to the article [4], we have used the YOLO object detection algorithm for finding a missing object in this paper.

In the field of cloud computing and mobile application development, new technology was introduced by Satyanarayanan et al. [5] called cloudlet. It is a resource-intensive and interactive mobile application by providing powerful computing resources to mobile devices with lower latency. By using this cloudlet, we can offload the computation in the local network and the people can connect with this network using Wi-Fi or Bluetooth, etc. The cloudlet is an interactive device between the user and cloud computing.

This paper is organized as follows: In Sect. 2, we mentioned the research work carried out in the field of YOLO object detection and existed algorithms for offloading computation on to the cloudlet. In sect. 3 presented the methodology implemented to find the missing object, and in sect. 4 discussed the efficiency of results generated using our approach.

2   Related Work

2.1   Object Detection Using CNN, R-CNN, Faster R-CNN, and YOLO

In object detection, we need to generate bounding boxes around the required object for better accuracy in an image. Using traditional convolution neural networks (CNN) [6], it is difficult to classify different objects using bounding boxes because of different spatial locations and other aspect ratios. To overcome this problem, R-CNN and YOLO have been developed for efficient results.

In R-CNN [7], an image is feed into CNN for a convolution feature map. From this map, around 2000 regions are extracted for a single image, so it will take more time for generating bounding boxes for a single image. For an extension to this algorithm, faster R-CNN is proposed, in which convolution will be done only once. But still, it will not check the entire image. To solve this problem, YOLO algorithm [8] was designed for finding bounding boxes and to classify these boxes based on probability. In the YOLO version 1 and version 2, the performance of detecting small objects was poor. It has been improved in YOLO version 3 [9] for real-time video.

The authors in the paper [10] used the YOLO algorithm to find the number of vehicles moving in a single lane. The limitations mentioned in this paper were the processing time required for tracking vehicles is more. This can be solved by offloading video on to resource-rich machine-like cloudlet.

2.2 Offloading Computations onto the Cloudlet

The concept of offloading is to transfer the computational task to a separate external platform. This makes cloud computing more powerful. But sometimes, [11] it is not good to offload computation to the cloud because of the availability of poor hardware performance, availability of required computing resources, and other network-related issues. Therefore, these problems are solved in mobile cloud computing by creating a data center in the local network called cloudlet with a rich set of computing resources.

In paper [12], the authors experimented on offloading multimedia data in various dimensions of power consumption, network delay, and bandwidth utilization. The performance of this work can be extended by adding the number of cores or GPU to the cloudlet.

In the video decoding, the process bandwidth requirement is reduced tremendously over cloud computing using GPU at user premises [13] as shown in Fig. 1.

Using this system, more mobile users can be connected to cloudlet and they can offload different tasks. In this paper, we are using this system to upload video from mobile devices or CCTV for notifying missing objects.

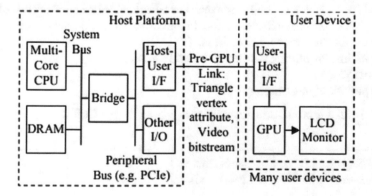

Fig. 1 Enhancing performance using GPU at user premises [13]

The authors in the paper [14] addressed the issues in the distributed surveillance application in a cloudlet network like bandwidth limitation, nature of heterogeneous data, high latency communication. Concerning this paper in our proposed system, we are detecting and notifying heterogeneous objects by calculating bounding boxes and probabilities of each box (confidence threshold) over the multiple objects.

2.3 Object Detection in Edge Computing

In the field of cloud computing, edge computing is a new technology for reducing computational cost, reducing latency, better response time, data security, and scalability [15].

The authors in paper [15] implemented an algorithm for object detection at an edge server (like a cloudlet, cloud clones, etc.). The authors in the paper were used the R-CNN algorithm for object detection. The performance of this approach can be enhanced using YOLO3 instead of R-CNN.

3   Methodology

Based on an earlier literature survey, we are presenting an algorithm for detecting objects, tracking a static missing object in a video using YOLO3 single object detection algorithm based on bounding box in cloudlet. The structure of the proposed system is shown in Fig. 2.

In this experiment, we make use of YOLO3 for feature extraction in a frame. In this experiment, we make use of YOLO3 because it supports the Darknet-53-layer convolution network for object detection. This layered architecture generates better results for small objects.

In this methodology, mobile users need to be connected with the nearest cloudlet by using Wi-Fi or Bluetooth connection. In the case of CCTV, the network connection is to be established using an IP address.

Algorithm for finding a missing object in a video
Input: Video
Output: Notifying Missing

1.   The video is taken from a mobile or is captured from CCTV.

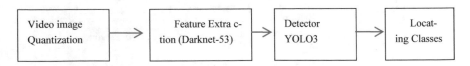

Fig. 2 Proposed systems

2. In the process of video quantization, the offloaded video is divided into several frames. In this experiment, we are generating frames based on the time interval. For every 5 s, a new frame is created.

3. In the feature extraction phase, each image is divided into 416 * 416 grids. For each cell setoff, bounding boxes are predicted and the confidence score is calculated for each grid. This score reflects how confident the cell contains an object. The confidence score fixed with 0.5

4. While in the phase of the detector the coordinate values f_x, f_y, f_w, f_h are calculated for each grid. Here, f_x, f_y are the center of the grid, and f_w, f_h height, and width.

5. For each grid, the confidence score and coordinate values are stored to classify the object.

6. Steps 3 and 4 are repeated till the end of the video.

7. In each cycle, new values are calculated identifying the class of object. These values are compared with already recorded values.

8. In this process, if the confidence value of an object is less than 0.5 or if there is variation in the coordinate values, immediately it notifies the message as the object is missing.

4 Results and Discussion

In our experiments, we used the mscoco trained dataset for good accuracy and experimented over a local cloudlet. The processor was an Intel Core i7-9900 K CPU and the NVIDIA GeForce 1050Ti graphic card was used to accelerate training for predicting results effectively. The process of objecting missing objects is computed over cloudlet (Fig. 3).

First, we experimented with our algorithm for detecting cell phone in a video. By using 53-layered architecture, mean average factor is moderate in the case of frame size 416×416. If the frame size is 32×312, average mean factor is less and it is a time-consuming process. So, we fixed with fame size as 416.

With a time, limit of 5 s, a total of 30 frames is created. From each frame, bounding boxes are calculated (Fig. 4).

After the t-time interval, the value of the confidence score is less than 0.5 and there is variation in the coordinates of the bounding box. With these variations, the system will generate the missing message (Fig. 5).

This result shows detecting object class of type laptop by using bounding boxes (Fig. 6).

These results show that after some time interval the object gets misplaced. This is notified by displaying a message.

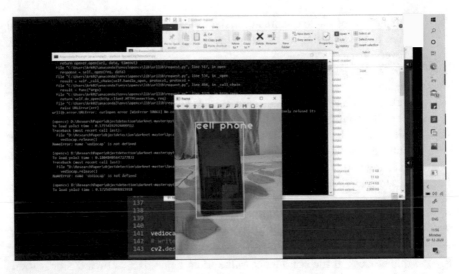

Fig. 3 Detecting object using YOLO3

Fig. 4 Detecting missing object coordinates f_x, f_y, f_w, f_h with a fixed frame and current frame

5 Conclusion

In this paper, we presented an approach for identifying missing static objects in real-time video surveillance. This work can be extended further for tracking multiple moving objects and identifying suspected objects in a streaming video for better security by enhancing more computing facilities.

Fig. 5 Detecting object using YOLO3

Fig. 6 Detecting missing object coordinates f_x, f_y, f_w, f_h with a fixed frame and current frame

References

1. R.C. Joshi, M. Joshi, A.G. Singh, S. Mathur, Object detection, classification and tracking methods for video surveillance: a review, in *4th International Conference on Computing Communication and Automation (ICCCA)*. IEEE (2018), pp. 1–7
2. T.Y. Lai, J.Y. Kuo, C.-H. Liu, Y.W. Wu, Y.-Y. Fanjiang, S.-P. Ma, Intelligent detection of missing and unattended objects in complex scene of surveillance videos, in *International Symposium on Computer, Consumer and Control*. IEEE (2012), pp. 662–665

3. J. Redmon et al., You look only once: unified real-time object detection. arXiv preprint arXiv: 1506.02640 (2015)
4. https://analyticsindiamag.com/Top8AlgorithmsForObjectDetection/
5. M. Satyanarayanan, P. Bahl, R. Caceres, N. Davies, The case for VM-based cloudlets in mobile computing. IEEE Perv. Comput. **8**(4), 14–23 (2009)
6. R.L. Galvez, A.A. Bandala, E.P. Dadios, R.R.P. Vicerra, J.M.Z. Maningo: Object detection using convolutional neural networks, in *IEEE Region 10 Conference, Jeju, Korea (South)* (2018), pp. 2023–2027
7. B. Liu, W. Zhao, Q. Sun, Study of object detection based on Faster R-CNN, in *2017 Chinese Automation Congress (CAC)*. IEEE (2017), pp. 6233–6236
8. W. Yang, Z. Jiachun, Real-time face detection based on YOLO, in *1st IEEE International Conference on Knowledge Innovation and Invention (ICKII)*, pp. 221–224. IEEE (2018)
9. Y.Q. Huang, J.C. Zheng, S.D. Sun, C.F. Yang, J. Liu, Optimized YOLOv3 algorithm and its application in traffic flow detections. Appl. Sci. **10**(9), 3079 (2020)
10. C.S. Asha, A.V. Narasimhadhan, Vehicle counting for traffic management system using YOLO and correlation filter, in *IEEE International Conference on Electronics, Computing and Communication Technologies (CONECCT)*. IEEE (2018), pp. 1–6
11. K. Akherfi, M. Gerndt, H. Harroud, Mobile cloud computing for computation offloading: issues and challenges. Appl. Comput. Inf. **14**(1), 1–16 (2018)
12. T. Huang, F. Ruan, S. Xue, L. Qi, Y. Duan, Computation offloading for multimedia workflows with deadline constraints in cloudlet-based mobile cloud. Wireless Netw. 1–15 (2019)
13. T. Lin, S. Wang, Cloudlet-screen computing: a multi-core-based, cloud-computing-oriented, traditional-computing-compatible parallel computing paradigm for the masses, in *IEEE International Conference on Multimedia and Expo*. IEEE (2009), pp. 1805–1808
14. A.M. Ali, N.M. Ahmad, A.H.M. Amin, Cloudlet-based cyber foraging framework for distributed video surveillance provisioning, in *4th World Congress on Information and Communication Technologies (WICT 2014)*. IEEE (2014), pp. 199–204
15. J. Work Ren, Y. Guo, D. Zhang, Q. Liu, Y. Zhang, Real-life use cases for edge computing—IEEE Innovation at Distributed and efficient object detection in edge computing: challenges and solutions. IEEE Network **32**(6), 137–143 (2018)

Face Hallucination Methods—A Review

Venkata Naresh Mandhalaⓘ**, A. V. S. Pavan Kumar**ⓘ**,**
Debnath Bhattacharyyaⓘ**, and Hye-jin Kim**

Abstract Face hallucination is the method specifically used for faces which inherit super-resolution technique. Face hallucination helps to get a high-resolution image from a low-resolution image. Applications of this technique are image enhancement and face recognition security. Face hallucination became a widely used application in the identification of facial images in all the fields. In this paper, numerous approaches and methods used for face hallucination were discussed. A contemporary analysis was made using various approaches for enhancing low-resolution images to high-resolution images. Super-resolution technique is a potential application in face recognition system which is an active research area nowadays.

Keywords Face hallucination · Super-resolution · Image enhancement · Face recognition

1 Introduction

Image processing is one of the widely utilized in the stream of engineering. The growth of image processing is increasing day by day due to the advancements in the areas of digital imaging. Few important areas where image processing is widely used are security monitoring, remote sensing, and also in the field of medicine. Generally,

V. N. Mandhala (✉) · D. Bhattacharyya
Department of Computer Science and Engineering, Koneru Lakshmaiah Education Foundation, Vaddeswaram, Guntur, AP, India

A. V. S. Pavan Kumar
Department of Computer Science and Engineering, GIET University, Gunupur, India
e-mail: avspavankumar@giet.edu

H. Kim
Kookmin University, 77 Jeongneung-RO, Seongbuk-gu, Seoul 02707, Republic of Korea
e-mail: hyejinaa@daum.net

© The Author(s), under exclusive license to Springer Nature Singapore Pte Ltd. 2021   379
S. K. Saha et al. (eds.), *Smart Technologies in Data Science and Communication*,
Lecture Notes in Networks and Systems 210,
https://doi.org/10.1007/978-981-16-1773-7_31

it was classified into two categories; they are biometric and non-biometric authentication systems, which includes fingerprint, palm print, iris, and also voice recognition comes under biometric identification and point of scale (POS) is considered as non-biometric identifications.

Face recognition is considered as one of the important areas in the fields of biometric authentication system. It refers to the processing of recognition of the face of a human in the system. The main input to the system is the image which may contain a face or non-face image and output is determined by the input and the type of the method we follow to recognize the face. In order to perform this, the features play a vital role in extracting the matching features and recognize according to the algorithm used to analyze the relative position, shape, and size as well as other important features related to the face. Face recognition is a difficult task because it requires a lot of information related to the image which contains the face and how it discriminates one image with another [1].

Face hallucination (FH) inherits the technique called super-resolution utilized for the purpose of facial recognition system which can be used for recognizing the faces more efficiently and effectively. The super-resolution is a widely used technique in the areas of image processing especially for the low-resolution images to upgrade them to high-resolution images. Early researchers [2, 3] have had provided an excellent information by their work done on super-resolution which raise the basic issues. Though it was widely used in the image processing, there is a scope to examine this concept beyond the boundaries. The concept of true resolution comes into the picture we need to concentrate on the way the input is specified whether it is a single or double line can be determined with the utmost precision than the width by calculating its centroid of the image.

When we consider an image which contains the face, the structure is explained in the form of components related to the input image that correlate to the facial contours and considerable smooth regions. Most of the algorithms proposed till now meant to locate these characteristics of the face and process the alignment in both frontal and other different poses related to the facial image. In the generic method, super-resolution algorithms aim in high-resolution and low-resolution patches taken from the input face image to recover the missing details [4].

Many face hallucination algorithms are proposed by several researchers [5–11]. Liu et al. introduced "a two-step method for face hallucination which included residual compensation and global image reconstruction" [12]. "The face hallucination method based on Eigen transformation" was proposed by Wang and Tang [9]. The high resolution of the image is constructed by a linear combination of training images and their combination coefficients are calculated by principal component analysis (PCA). Because of that, the loss of local details may occur and more likely weighing coefficients are pretended to be same for the entire image by many artifacts. The reconstruction of HR images is done by using several methods by considering the sequence input images either a low-resolution or a single-frame image [13].

Face recognition targets at verifying whether two facial images are from the same identity by designing discriminative features and similarities [14]. In the empirical

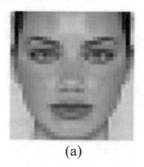 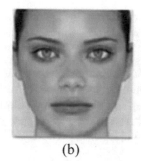 

(a)   (b)   (c)

Fig. 1 How face hallucination work

studies [15], the face recognition determines that at least a minimum of face resolution between 32 × 32 and 64 × 64 is mandatory for any standard face recognition algorithms. Torralba et al. [16] reported "a significant performance drop when the image resolution is decreased below 32 × 32 pixels". It is natural to expect that hallucinated face images can improve the recognition performance for LR facial images. Unfortunately, we find that this expectation does not hold in a lot of cases.

When we want to perform any facial recognition process, the main source of inputs are the images with faces. To do so, we capture the faces by various methods and sometimes while capturing the faces, due to various reasons captures images may contain small size, blurred images, illumination problem and most of the time low-resolution images which lead to unclear structure of the input image and it becomes a challenging issue to perform a task. Baker and Kanade [6] proposed a face super-resolution method which will serve the purpose of above-said problems.

Figure 1 contains three images that depict (a) an input image with low resolution, (b) applying face hallucination, and (c) is an exact original face obtained as an output after applying face hallucination.

In the last twenty years, several specific face hallucination algorithms are developed to improve the quality of the image. Though the current face hallucination strategies have succeeded a good edge, there is a need for the improvement in this area to be developed.

Most of the algorithms sometimes follow few common steps: The first step generally used to maintain the features of the face in the global images using maximum a posteriori (MAP). The other step is used on the remaining image to improve the results of the first step [5]. Moreover, most of the algorithms use square measure supported a collection of both the high-resolution and low-resolution training images, which is used for facial image synthesis that includes super-resolution approach.

The face hallucination algorithms are generally based on the following three constraints; they are:

Data constraint: The image that we obtain as an output should exactly be near to the original image which is processed with the algorithms as shown in Fig. 1.

Global constraint: The image that is obtained should consist of the all features which are common to a human face. Without the global constraint, the output that is achieved may be noisy.

Local constraint: The image that is obtained as an output should have features that are specific to the face and should coincide with the local features of the photorealistic image. Without the local constraint, the achieved output may be too smooth.

The following lists of methods are used to enhance the facial features to improve the image resolution through [11]:

a. Two-step approach [5]
b. Bayes' theorem-based face hallucination [6]
c. Learnt image models for super-resolution from multiple views [7]
d. Sparse coding-based face hallucination [8]
e. Eigen transformation for face hallucination [9]
f. Face hallucination based on MCA [10].

2 Literature Review

Qu et al. [17] "focused mainly on notations that are used for neighbor embedding based face hallucination, which in turn applied to present RIKNN method exploits both manifold of LR and HR to find optimal neighbors. As a compliment, it also discussed about the details of the RIKNN method in face hallucination, and their computation complexities are analyzed. Figure 2 depicts the observations and the mechanism which was followed in the first k nearest neighbor (k-NN) which in turn gives the same for the of LR and HR patch".

Farrugia et al. [18] "A new approach named LM-CSS which stands for Linear Model of Coupled Sparse Support method (as shown in Fig. 3). This method is mainly used to convert a facial image of the high-resolution from the test image of low-resolution. It first performs a smooth approximation used generally to the ground-truth in Euclidean space on the image of high-resolution. Thus by taking the assumption into the consideration the patches that are present on the HR present on a high-resolution means, to gain the optimal support for the representation of the first approximation is a good approach to rebuild the ground truth. It also allows us to show to be validated in other domains. This method which holds the coupled sparse allows us to support and then used the model to improve each patch using multivariate ridge regression". Farrugia et al. proposed a method which mainly differs from conventional methods.

Hu [19] "A novel deep image method called image regression approach which mainly focused on the unviewed sketch-photo hallucination problem solution. It consists of the two-branch network models which deal with the local features and the whole structure of image. The problems that occurs in the unviewed or forensic sketches are non-rigid misalignment is solved by using a generator which integrates the spatial transformer network [20], and synthesization on the clearer images by

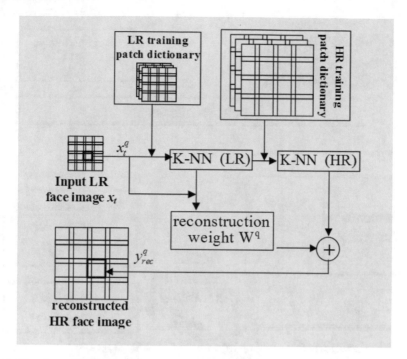

Fig. 2 Flowchart of RIKNN method

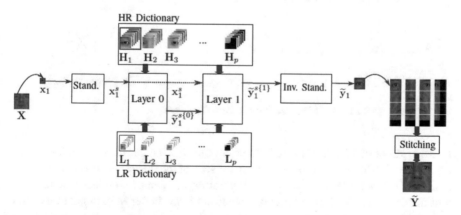

Fig. 3 Flowchart for (LM-CSS) method

implementing the content-based and adversarial training. At the outset the network generates the quality images with high intensity than the existing and conventional approached which are under usage".

Ko [21] stated that "To generate the HR patch from an LR patch a feed-forward neural network is trained by also considering its surrounding pixels (as shown in Fig. 4), to improve the weights for various clusters of patches a multitask deep

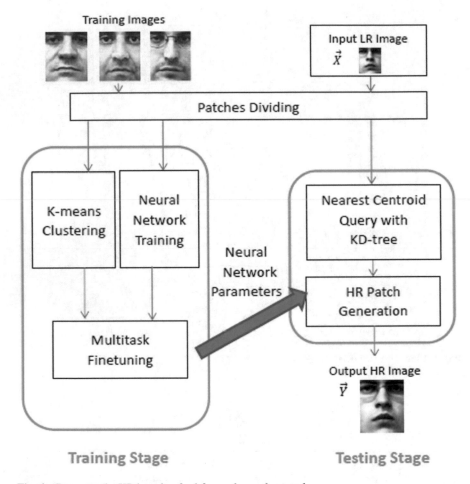

Fig. 4 Generates the HR by using feed-forward neural network

learning structure is also employed. To prove this, the images from FEI database are considered for the experimental results. It consists of the images from 200 individuals and the 200 frontal-face images with a regular expression of each faces are utilized, and finally concludes that the proposed method is far better in computational time and gives a better performance".

Patch-based multitask deep learning method is matched with the latest hallucination method which includes anchored patch based [7] and Chen's position patch-based method [6]. Not only by considering the well-known methods it also compares with general super-resolution methods with our method on the same database, which also includes the recent deep learning method SRCNN [22]".

Yang et al. [4] "To represent each face images in facial components and their contours and also the smooth regions by the local image structures for face hallucination, the matching gradients is used to maintain the structure of the image for the

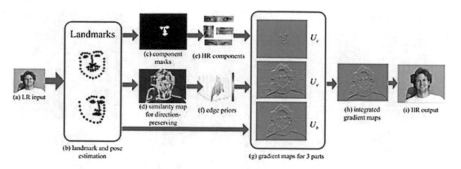

Fig. 5 Flowchart for face hallucination gradient method

reconstructed high-resolution output. The input images are aligned in such a way that to generate accurate images and the high-frequency information of the images for storing the structural consistency for every facial components, t. Statistical priors to produce the salient structures in the high-resolution images the contours are used. For the images that contains the smooth regions where the image gradients are stored for further usage the patch matching is well utilized method. The gradient algorithms used on the multi- PIE data set, the image variations with landmarks of every face in the dataset, and by considering the images manually that generates training images with the glasses labels. A group of images with 2184 of which each image with 320 × 240 at upright and also the frontal images of around 289 single images is utilized as the training dataset for face hallucination implementation that produces the hallucinated images of remarkable quality. (as shown in Fig. 5), The results in the local image structure by matching gradients give a remarkable results and it also produces the remarkable results by having component-level alignment and the glasses labels".

Hui et al. [23] "The novel correspondence-based FH is mainly focused on the problems of evaluating the high-resolution (HR) images by considering the single low resolution (LR) input. (as shown in Fig. 6), By using the shape and texture the difficulties that are leading to the problem in the face hallucination technique are identified and used to estimate in an appropriate manner and works well though we consider the small set of training data and it allows us for the implantation in real-time environment which is a good solution".

3 Conclusion

This paper mainly focused on the developments related to face hallucination. The method mainly focused to enhance the input image and to achieve a high-quality image as an output by applying super-resolution method on the low-resolution image contents. This paper presents various methods associated to face hallucination, using sparse representation, FH using eigen transformation and also low-resolution facial images, deep joint, K-NN search strategy, face hallucination for

LR Reference Sample   Flow Field   Warped Sample

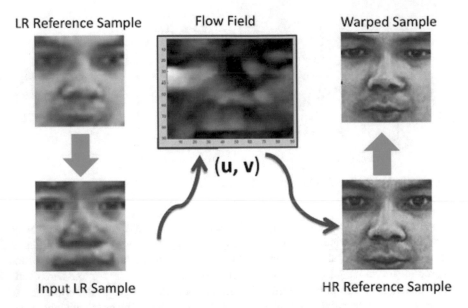

Input LR Sample   HR Reference Sample

Fig. 6 Correspondence-based FH

single frame, unviewed sketches, patch-based multitask deep NN and concluded with structured, linear models of coupled sparse support and correspondence-based face hallucination. From this, we can determine their pro's and con's.

References

1. V.N. Mandhala, D. Bhattacharyya, T.H. Kim, Hybrid face recognition using image feature extractions: a review. Int. J. Bio-Sci. Bio-Technol. **6**(4), 223–234 (2014)
2. M.P. Autee, M.S. Mehta, M.S. Desai, V. Sawant, A. Nagare, A review of various approaches to face hallucination. Procedia Comput. Sci. **45**, 361–369 (2015)
3. S.K. Vikram Mutneja, A general review of face hallucination methods. Int. J. Adv. Res. Comput. Eng. Technol. (IJARCET) **4**(4) (2015)
4. C.Y. Yang, S. Liu, M.H. Yang, Structured face hallucination, in *Proceedings of the IEEE Conference on Computer Vision and Pattern Recognition* (2013), pp. 1099–1106
5. C. Liu, H.Y. Shum, W.T. Freeman, Face hallucination: theory and practice. Int. J. Comput. Vis. **75**(1), 115–134 (2007)
6. T. Kanade, J.F. Cohn, Y. Tian, Comprehensive database for facial expression analysis, in *Proceedings Fourth IEEE International Conference on Automatic Face and Gesture Recognition (Cat. No. PR00580)*. IEEE (2000, March), pp. 46–53
7. D. Capel, A. Zisserman, Super-resolution from multiple views using learnt image models, in *Proceedings of the 2001 IEEE Computer Society Conference on Computer Vision and Pattern Recognition. CVPR 2001*. IEEE (2001), vol. 2, pp. II-II
8. J. Yang, H. Tang, Y. Ma, T. Huang, Face hallucination via sparse coding. 2008 15th IEEE International Conference on Image Processing. IEEE (2008, October), pp. 1264–1267
9. X. Wang, X. Tang, Hallucinating face by eigen transformation. IEEE Trans. Syst. Man Cybern. Part C (Appl. Rev.) **35**(3), 425–434 (2005)

10. Y. Liang, X. Xie, J.H. Lai, Face hallucination based on morphological component analysis. Signal Proces. **93**(2), 445–458 (2013)
11. Y. Hu, K.M. Lam, G. Qiu, T. Shen, From local pixel structure to global image super-resolution: a new face hallucination framework. IEEE Trans. Image Process. **20**(2), 433–445 (2010)
12. C. Liu, H.Y. Shum, C.S. Zhang, A two-step approach to hallucinating faces: global parametric model and local nonparametric model, in *Proceedings of the 2001 IEEE Computer Society Conference on Computer Vision and Pattern Recognition. CVPR 2001. IEEE* (2001, December), vol. 1, pp. I–I
13. S.C. Park, M.K. Park, M.G. Kang, Super-resolution image reconstruction: a technical overview. IEEE Signal Process. Mag. **20**(3), 21–36 (2003)
14. Y.M. Lui, D. Bolme, B.A. Draper, J.R. Beveridge, G. Givens, P.J. Phillips, A meta-analysis of face recognition covariates. 2009 IEEE 3rd International Conference on Biometrics: Theory, Applications, and Systems. IEEE (2009, September), pp. 1–8
15. M. Turk, A. Pentland, Eigenfaces for recognition. J. Cogn. Neurosci. **3**(1), 71–86 (1991)
16. A. Torralba, R. Fergus, W.T. Freeman, 80 million tiny images: A large data set for nonparametric object and scene recognition. IEEE Trans. Pattern Anal. Mach. Intell. **30**(11), 1958–1970 (2008)
17. S. Qu, R. Hu, S. Chen, J. Jiang, Z. Wang, M. Zhang, Super-resolution for face image with an improved k-NN search strategy. China Commun. **13**(4), 151–161 (2016)
18. R.A. Farrugia, C. Guillemot, Face hallucination using linear models of coupled sparse support. IEEE Trans. Image Process. **26**(9), 4562–4577 (2017)
19. C. Hu, D. Li, Y.Z. Song, T.M. Hospedales, Now you see me: deep face hallucination for unviewed sketches. BMVC 1–11 (2017)
20. J. Wu, S. Ding, W. Xu, H. Chao, *Deep Joint Face Hallucination and Recognition.* (2016), pp. 1–10. arXiv preprint arXiv:1611.08091
21. W.J. Ko, S.Y. Chien, Patch-based face hallucination with multitask deep neural network. 2016 IEEE International Conference on Multimedia and Expo (ICME). IEEE (2016, July), pp. 1–6
22. J.S. Park, S.W. Lee, An example-based face hallucination method for single-frame, low-resolution facial images. IEEE Trans. Image Process. **17**(10), 1806–1816 (2008)
23. Z. Hui, W. Liu, K.M. Lam, A novel correspondence-based face-hallucination method. Image Vis. Comput. **60**, 171–184 (2017)

Image Encryption for Secure Internet Transfer

G. Dheeraj Chaitanya, P. Febin Koshy, D. Dinesh Kumar, G. Himanshu, K. Amarendra, and Venkata Naresh Mandhala

Abstract Information security has always been an important aspect of consideration in daily life communication and with the advent of COVID-19 and communication, including sensitive data like corporate files, personal data, etc., increasingly moving online, it has become more important than ever to provide secure communication methods. One such method is using encryption and steganography. Encryption is the process of converting ordinary plaintext data into an alternative form that is unreadable by perpetrators without authorization. Furthermore, steganography is the practice of hiding a file, message, video, or image within another such file. In this paper, we introduce a web application that will enable the user to encrypt the data in a file of his choice and then hide it within a picture or GIF. This will provide the layman an application that is easy to use to communicate securely, even without extensive technical knowledge. The encryption algorithm used will be the AES-256 algorithm and the steganographic algorithm will be Syndrome-Trellis Codes (STCs) for their high embedding efficiency. The results display that the system designed offers a significant security level.

Keywords Steganography · STC · AES-256

1 Introduction

With the advent of COVID-19, it is very important now that communication be held online. Several companies are moving the communication infrastructure online. So, in some cases, it is desired that the communication being held between two parties be made secret and not accessible to the public [1–5]. In simpler words, digital communication is the proper transfer of data from one client to another. While there

G. Dheeraj Chaitanya · P. Febin Koshy · D. Dinesh Kumar · G. Himanshu · K. Amarendra (✉) ·
V. N. Mandhala
Department of Computer Science and Engineering, Koneru Lakshmaiah Education Foundation,
Guntur, Andhra Pradesh, India
e-mail: amarendra@kluniversity.in

are various methods that people have been using to achieve this, their results are usually poor or not up to the mark.

Digital communication comes with several benefits. It is particularly useful in the pressing times because the need for physical contact is drastically reduced. Adding on, digital communication requires very little resources. All it needs is an electronic device and Internet connectivity. Furthermore, sending data can be done in the matter of seconds, which aids in quicker completion of work [6–9].

The data can be of various kinds from phone number to hashes, all these are from a lower perspective are just ASCII codes. ASCII codes represent text in computers, telecommunications equipment, and other devices [10]. These codes even though simply provide a lot of information and play a vital role in communication. The current systems for transferring of data technically work though it offers little security or protection whatsoever.

The data being communicated can fall into the wrong hands through several ways. Some ways are: Man in the Middle attacks, if the data is being communicated through an unsafe, public channel.

This is where encryption and steganography come in. Encryption is the practice of converting ordinary plaintext data into unintelligible data so that an attacker cannot decipher it. Steganography is the means by which secret data is hidden in files like a picture, video, or audio.

So, in more simpler words steganography is the act of hiding data inside data, which to common eyes is just a single set of data [11–14]. The major benefit with steganography is that it does not attract attention to the hidden data whatsoever. The hidden data does not stand out unlike encryption where one can clearly see that some data is present which is not accessible but still their presence is felt. Steganography gets around this with a stealthy approach.

Most of the time encryption data can be identified without much attention due to how it stands out with regular data. Whereas data hidden in a file for instance, a image will not grab much attention as there are almost no visible changes to the image the data is hidden with. This makes steganography a great approach for hiding data and facilitates the transfer of data from one client to another. Along with being a more steal their approach, it also gives a good amount of security because we can implement various authentication mechanics in order to facilitate that only the sender and receiver can decrypt the image with hidden data [15–18]. This overall makes steganography a great tool for secure communication over the Internet.

Looking into current systems in place for the general public to use, there is almost nothing for the general community to use without learning any technicalities. The current system available for secure Internet transfer offers less or none protection to the data whatsoever [19–23]. The security part of the system in place is where we are gonna adapt and improvise. We propose to use steganography in order for secure data transmission. This will also be in a way that anyone can use a GUI application to send and receive data securely without knowing any technical aspects of the process. Therefore, we think that this project provides tamper-proof data transfer to end users all with just click in the developed web application that we will be forming along with a complete easy to use GUI while keeping security in mind.

Table 1 File details

File name	Dimensions	Size (KB)
Sample.gif	512 * 512	13.3
Sample2.gif	225 * 225	2.92

2 Materials and Experimental Procedures

2.1 Materials

The Graphics Interchange Format (GIF) is a bitmap image format that has come into widespread usage on the World Wide Web due to its wide support and portability between applications and operating systems [24–26]. This format supports up to 8 bits per pixel for each image, allowing a single image to reference its own palette of up to 256 different colors chosen from the 24-bit RGB color space. It also supports animations and allows a separate palette of up to 256 colors for each frame.

Portable Network Graphics (PNG) is a raster-graphics file format that supports lossless data compression. PNG supports palette-based images (with palettes of 24-bit RGB or 32-bit RGBA colors), grayscale images (with or without alpha channel for transparency), and full-color non-palette-based RGB or RGBA images.

We made use of two Portable Network Graphics (PNG) images which were extracted from two Graphics Interchange Formats (GIFs). The details of the files used are represented in Table 1.

2.2 Methods

Existing Methods. At present, there exists no software or application that is easy to use for the average layman with easy functionality [27–30]. They must use codes and run them in their specific environment to produce desired results. This is a cumbersome process.

Proposed Method. What we propose to do is create an application that is easy to use, whose methodology is shown in Fig. 1, wherein the user just has to input the secret data he wants to communicate and then the data will get encrypted using AES-256

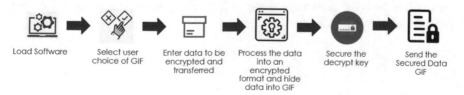

Load Software   Select user choice of GIF   Enter data to be encrypted and transferred   Process the data into an encrypted format and hide data into GIF   Secure the decrypt key   Send the Secured Data GIF

Fig. 1 The architecture of the web application

and a secret key of user's choice. Then, embedded into the GIF, of the user's choice, This GIF can be sent to the receiver through mail or other means.

The text of data and the choice of GIF the user selects from local computer are the input. The text is then encrypted using AES-256 algorithm and embedded into GIF using STCs. The decrypt key is noted down and the GIF is sent securely to the receiver.

System Implementation. The proposed work is implemented in Python. The frontend uses HTML, CSS and is run using Jinja. In the back end, the web framework to build the API is Flask. SQLite is used for the database implementation.

Testing Procedure. We need to take two dynamic GIF images, which are all color dynamic images with 8-bit index values. Then process the GIF and extracted images out of the GIF. Post-extraction, we take the data to be encrypted which, using AES-256 bit encryption, is stored in the extracted image. We then encrypt the image with a key, which would be needed to decrypt the image later down the experiment. Then, we compare the original image and the processed image with our data hidden in it and note down the difference.

2.3   Testing and Analysis

Breakdown of the GIF image into PNG. To start with, we ask the user to input their choice of GIF (Graphics Interchange Formats). We then send the specific GIF into the processing engine which defames the GIF into individual images which when viewed in a defined motion acts like a GIF. We number all the breakdown images from the GIF to facilitate easy processing later into the engine.

Encryption and Decryption of sensitive data through the file. We then ask the user for the data to be encrypted and stored inside the specified GIF. We run the data through the encryption engine which encrypts the data using AES-256 bit encryption. This data is passed on down the engine to be embedded onto the images derived from the GIF in order to facilitate hiding the data in the specified GIF. After the data is hidden, we combine all the images back in the same order to form a GIF, which to the world seems like any other GIF image. After this, we encrypt the GIF with the hidden data with a secure key that would be needed to decrypt the GIF and hence the data hidden inside it. This means that for the recipient of the secured data image; they should have the GIF image along with the decrypt key to successfully extract the data out of the GIF file.

3 Results and Discussion

We compared the original image and the processed image with our data hidden in it and found no visible difference. Using MSE comparison which is represented in Table 2, we found the error to be minimal. This means there is no visible distortion in the stego images (Fig. 2).

Table 2 MSE values for stego images

Cover image	MSE
Image 1	0.0255
Image 2	0.0236

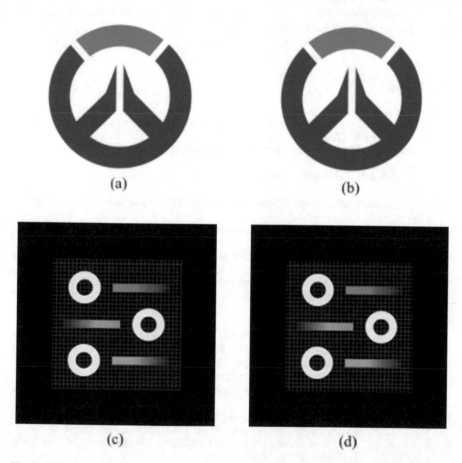

(a)　　　　　　　　　　　　　(b)

(c)　　　　　　　　　　　　　(d)

Fig. 2 Comparison of stego images. **a** Original image 1. **b** Stego 1 generated by app. **c** Original image 2. **d** Stego 2 generated by app

4   Conclusions

This paper presents an application that uses STC framework and AES-256 algorithm to perform steganography on GIF images. We encrypt the secret data that is to be transmitted and hide it in an image that will be securely transmitted to a receiver. STC framework has been used because it provides better security than other steganographic methods like LSB or diamond encoding. Even if the secret image is attacked and hidden data is acquired, the data cannot be decrypted due to the AES-256 algorithm. This ensures security. The software reduces the tediousness of using codes and runtime environments and also achieves the task of being easy to use while ensuring complete security. It not only keeps high stego-image quality but also considers large amounts of data into cover images for secret communication. The experimental results, using MSE value calculation, show that the proposed method has a better performance than previous and existing works.

References

1. J.B. Arun, R. Choudhary, *Image Encryption for Secure Data Transfer and Image Based Cryptography* (IJERT, 2015)
2. P.P. Dang, P.M. Chau, *Image Encryption for Secure Internet Multimedia Applications* (IEEE, 2017)
3. M. Kundalakesi, M. Harinee, Secure data transfer with image encryption. J. Sci. Res. Dev. (2018)
4. Y. Yiğit, M. Karabatak, *A Stenography Application for Concealing Student Information* (ISDFS, 2019)
5. R. Amirtharajan, R. Akila, P. Deepikachowdavarapu, *A Comparative Analysis of Image Steganography* (IJCA, 2010)
6. S. Thenmozhli, M. Chandra Sekaran, *Efficient Technique for Image Stenography Based on coordinates of pixels*. IOSR-JCE (2013)
7. N.V. Kalyankar, S.D. Khamitkar, P.U. Bhalchandra, S.N. Lokhande, N.K. Deshmukh, *Stenography Using Palette Images* (IEEE, 2014)
8. T. Pandikumar, T. Gebreslassie, *Information Security Using Image Based Steganography* (IRJET, 2018)
9. A. Patino-Vanegas, S.H. Contreras-Ortiz, J.C. Martinez-Santos, A low noise stenography method for medical images with QR encoding of patient information. SPIE Med. Imag. (2017)
10. J. Lin, Z. Qian, Z. Wang, X. Zhang, G. Feng, A new steganography method for dynamic GIF images based on palette sort. Wirel. Commun. Mob. Comput. (2020)
11. A.K. Sahu, G. Swain, *Reversible Image Steganography Using Dual-Layer LSB Matching* (Springer, Berlin, 2020)
12. B. Kusuma Priya, L.P. Maguluri, T. Srinivasarao, T.E. Rao, A systematic approach for data hiding using cryptography and steganography. World Acad. Res. Sci. Eng. (2020)
13. K. Amarendra, V.N. Mandhala, B.C. Gupta, G.G. Sudheshna, Anusha, Image steganography using lsb. Int. J. Sci. Technol. Res. (2019)
14. B. Mandal, A. Pradhan, G. Swain, Adaptive LSB substitution steganography technique based on PVD, in *Proceedings of the International Conference on Trends in Electronics and Informatics* (2019)
15. S. Sadiya Shireen, B. Murali Krishna, N. Lakshmi Prasanna, FPGA based RSA authenticated data hiding in image through steganography. Int. J. Innov. Technol. Explor. Eng. (2019)

16. A.K. Sahu, Swain, A novel n-rightmost bit replacement image steganography technique. 3D Res. (2019)
17. M. Chandra Sekhar, S.K. Chandini, V. Sai Rohith, V. Jhansi Lakshmi, P. Kumar, Data hiding using bit plane complexity segmentation steganography. Int. J. Eng. Technol. (UAE) (2018)
18. R. Dixit, Ravindranath, Encryption techniques & access control models for data security: a survey. Int. J. Eng. Technol. (UAE) (2018)
19. J. Ram Kumar, V. Avutu, Anurag, MSEC scheme for providing secure data transformation using coding technique. Int. J. Eng. Technol. (UAE) (2018)
20. A.K. Sahu, Swain, An improved data hiding technique using bit differencing and LSB matching. Internet Work. Indonesia J. (2018)
21. G. Swain, A data hiding technique by mixing MFPVD and LSB substitution in a pixel. Inf. Technol. Control (2018)
22. G. Swain, High-capacity image steganography using modified LSB substitution and PVD against pixel difference histogram analysis. Secur. Commun. Netw. (2018)
23. G. Swain, Digital image steganography using eight-directional PVD against RS analysis and PDH analysis. Adv. Multimed. (2018)
24. G. Swain, Adaptive and non-adaptive PVD steganography using overlapped pixel blocks. Arab. J. Sci. Eng. (2018)
25. K. Likitha, P.S.G. Aruna Sri, R. Phanitha, A secure data sharing and revocation in cloud using IDE. Int. J. Appl. Eng. Res. (2017)
26. K.T. Rao, Saidhbi, Impact of steganography in secure data transaction under private cloud- an analytical survey. J. Adv. Res. Dyn. Control Syst. (2017)
27. A. Roshini, K. Sri Sai Manish, R. Vedavyas, V. Lakshmidhar, V. Prathyusha, M. Kumar, Reversible steganography: data hiding with image enhancement. J. Adv. Res. Dyn. Control Syst. (2017)
28. A. Pradhan, K.R. Sekhar, Swain, Adaptive PVD steganography using horizontal, vertical, and diagonal edges in six-pixel blocks. Secur. Commun. Netw. (2017)
29. WSNs, T.V. Krishna Chowdary, Satyanarayana, A novel secured data transmission and authentication technique against malicious attacks. J. Adv. Res. Dyn. Control Syst. (2017)
30. K. Varalakshmi, P.M. Ashok Kumar, A. Rami Reddy, Kavitha, A novel data hiding technique based on image processing techniques. J. Adv. Res. Dyn. Control Syst. (2017)

Correction to: Smart Cyclones: Creating Artificial Cyclones with Specific Intensity in the Dearth Situations Using IoT

G. Subbarao, S. Hrushikesava Raju, Lakshmi Ramani Burra,
Venkata Naresh Mandhala, and P. Seetha Rama Krishna

Correction to:
Chapter "Smart Cyclones: Creating Artificial Cyclones with Specific Intensity in the Dearth Situations Using IoT"
in: S. K. Saha et al. (eds.),
Smart Technologies in Data Science and Communication,
Lecture Notes in Networks and Systems 210,
https://doi.org/10.1007/978-981-16-1773-7_28

In the original version of the chapter, the following belated corrections have been incorporated:

The author name "S. Hrushikesava Rao" has been changed to "S. Hrushikesava Raju" in the Frontmatter, Backmatter and in Chapter "Smart Cyclones: Creating Artificial Cyclones with Specific Intensity in the Dearth Situations Using IoT".

The correction/erratum chapter and the book have been updated with the change.

The updated version of this chapter can be found at
https://doi.org/10.1007/978-981-16-1773-7_28

Author Index

Printed in the United States
by Baker & Taylor Publisher Services